OpenIntro Statistics

First Edition

David M Diez
Postdoctoral Fellow
Department of Biostatistics
Harvard School of Public Health
david.m.diez@gmail.com

Christopher D Barr
Assistant Professor
Department of Biostatistics
Harvard School of Public Health
cdbarr@gmail.com

Mine Çetinkaya-Rundel
Assistant Professor of the Practice
Department of Statistics
Duke University
cetinkaya.mine@gmail.com

ISBN: 9-781461-062615

Contents

Preface

OpenIntro is an organization focused on developing free and affordable education materials. *OpenIntro Statistics*, our first project, is intended for introductory statistics courses at the high school through university levels. The chapters are as follows:

1. **Introduction to data.** Data structures, variables, summaries, graphics, and basic data collection techniques.
2. **Probability (special topic).** The basic principles of probability. An understanding of this chapter is not required for the main content in Chapters 3-8.
3. **Distributions of random variables.** Introduction to the normal model and other key distributions.
4. **Foundations for inference.** General ideas for statistical inference in the context of estimating the population mean.
5. **Large sample inference.** Inferential methods for one or two sample means and proportions using the normal model, and also contingency tables via chi-square.
6. **Small sample inference.** Inference for means using the t distribution, as well as simulation and randomization techniques for proportions.
7. **Introduction to regression.** An introduction to linear regression. Most of this chapter could be covered after Chapter 1.
8. **Multiple regression and ANOVA.** An introduction to multiple regression and one-way ANOVA for an accelerated course.

This textbook was written to allow flexibility in choosing and ordering course topics. The material is divided into two pieces: main text and special topics. The goal of the main text is to allow people to move towards statistical inference and modeling sooner rather than later. Special topics, labeled in the table of contents and in section titles, may be added to a course as they arise naturally in the curriculum.

OpenIntro Statistics would also serve as a helpful supplement in a course preparing students for the Advanced Placement Statistics exam, either though the textbook or use of the online resources outlined below.

Examples, exercises, and appendices

Examples and within-chapter exercises have been clearly labeled throughout the textbook and may be identified by their distinctive bullets:

● **Example 0.1** Large filled bullets signal the start of an example.

Full solutions to examples are provided and often include an accompanying table or figure.

⊙ **Exercise 0.2** Large empty bullets signal readers that an exercise has been inserted into the text for additional practice and guidance. Students may find it useful to fill in the bullet after understanding or successfully completing the exercise. Solutions are provided for the vast majority of within-chapter exercises in footnotes[1].

There are a large selection of exercises at the end of each chapter useful for practice or homework assignments. Many of these questions have multiple parts, and odd-numbered questions include brief solutions in Appendix B. These end-of-chapter exercises are also available online in a public question bank at **openintro.org**, and the available selection is constantly growing based on teacher contributions. Numbered citations in end-of-chapter exercises may be found in Appendix B.

Probability tables for the normal, t, and chi-square distributions are in Appendix C, and PDF copies of these tables are also available from **openintro.org** for anyone to use, share, or modify.

Online resources and getting involved

We encourage anyone learning or teaching statistics to visit **openintro.org** and get involved by using the many online resources, which are all free and rapidly expanding, or by creating new material. Students can test their knowledge with practice quizzes for each chapter, or try an application of concepts learned using real data. A companion *R* package has also been released, which includes most data sets introduced in this book[2]. Teachers can download the source files for labs, slides, data sets, textbook figures, or create their own custom quizzes and problem sets for students to take at **openintro.org**. Anyone can download a PDF or the source files of this textbook for modifying and sharing. All of these products are free, and we want to be clear that anyone is welcome to use these online tools and resources with or without this textbook as a companion.

We value your feedback. If there is a particular component of the project you especially like or think needs improvement, we want to hear from you. You may find our contact information in the About section of the website.

Acknowledgements

This project would not be possible without the dedication and volunteer hours of all those involved. No one has received any monetary compensation from this project, and we hope you will join us in extending a *thank you* to all those volunteers below.

The authors would like to thank Filipp Brunshteyn, Rob Gould, and Chris Pope for their involvement and contributions to OpenIntro. A special thank you to Andrew Bray, who has developed the R labs and diligently worked to translate end-of-chapter exercises to the project website. We also deeply appreciate the contribution of Meenal Patel, who has helped raise the professional profile of OpenIntro by designing a business system and website for the project. We are also very grateful to Dalene Stangl, Dave Harrington, Jan de Leeuw, Kevin Rader, and Philippe Rigollet for providing us with valuable feedback on this textbook.

---

[1] Solutions are usually located down here!

[2] Diez DM, Barr CD, Çetinkaya-Rundel M. 2011. `openintro`: OpenIntro data sets and supplement functions. http://cran.r-project.org/web/packages/openintro

Chapter 1

Introduction to data

Scientists seek to answer questions using rigorous methods and careful observations. These observations – collected from the likes of field notes, surveys, and experiments – form the backbone of a statistical investigation and are called **data**. Statistics is the study of how best to collect, analyze, and draw conclusions from data. It is helpful to put statistics in the context of a general process of investigation:

1. Identify a question or problem.
2. Collect relevant data on the topic.
3. Analyze the data.
4. Form a conclusion.

Statistics as a subject focuses on making stages (2)-(4) objective, rigorous, and efficient. That is, statistics has three primary components: How best can we collect data? How should it be analyzed? And what can we infer from the analysis?

The topics scientists investigate are as diverse as the questions they ask. However, many of these investigations can be addressed with a small number of data collection techniques, analytic tools, and fundamental concepts in statistical inference. This chapter provides a glimpse into these and other themes we will encounter throughout the rest of the book. We introduce the basic principles of each branch and learn some tools along the way. We will encounter applications from other fields, some of which are not typically associated with science but nonetheless can benefit from statistical study.

1.1 Case study: treating heart attack patients with a new drug

Section 1.1 introduces a classic challenge in statistics: evaluating the efficacy of a pharmaceutical treatment. Terms in this section, and indeed much of this chapter, will all be revisited later in the text, and in greater detail. The plan for now is simply to begin getting a sense of the role statistics can play in practice.

In this first section, we consider an experiment used to evaluate whether a drug, sulphinpyrazone, reduces the risk of death in heart attack patients[1]. In this study, we might start by writing the principle question we hope to answer:

[1] Anturane Reinfarction Trial Research Group. 1980. Sulfinpyrazone in the prevention of sudden death after myocardial infarction. New England Journal of Medicine 302(5):250-256.

> Does administering sulphinpyrazone to a heart attack patient reduce the risk of death?

The researchers who asked this question collected data on 1,475 heart attack patients. Each volunteer patient was randomly assigned to one of two groups:

Treatment group. Patients in the treatment group received the experimental drug, sulphinpyrazone.

Control group. Patients in the control group did not receive the drug but instead were given a **placebo**, which is a fake treatment with no chemical agents that is made to look real.

In the end, there were 733 patients in the treatment group and 742 patients in the control group. The patients were not told which group they were in, and the reason for this secrecy is that patients who know they are being treated often times show improvement (or slower degeneration) regardless of whether the treatment actually works. This improvement is called a **placebo effect**. If patients in the control group were not given a placebo, we would be unable to sort out whether any observed improvement was due to the placebo effect or the treatment's effectiveness.

After 24 months in the study, each patient was either still alive or had died; this information describes the patient's **outcome**. So far, there are two relevant characteristics about each patient: patient `group` and patient `outcome`. We could organize these data into a table. One common organization method is shown in Table 1.1, in which each patient is represented by a row, and the columns relate to the information known about the patients.

Patient	group	outcome
1	treatment	lived
2	treatment	lived
⋮	⋮	⋮
1475	control	lived

Table 1.1: Three patients from the sulphinpyrazone study.

Considering data from each patient individually would be a long, cumbersome path towards answering the original research question. Instead, it is often more useful to perform a data analysis, considering all of the data at once. We first might summarize the raw data in a more helpful way, like that shown in Table 1.2. In this table, we can quickly see what happened over the entire study. For instance, to identify the number of patients in the treatment group who died, we could look at the intersection of the `treatment` row and the `died` column: 41.

		outcome		
		lived	died	Total
group	treatment	692	41	733
	control	682	60	742
	Total	1374	101	1475

Table 1.2: Descriptive statistics for the sulphinpyrazone study.

⊙ **Exercise 1.1** Of the 733 patients in the treatment group, 41 died. Using these two numbers, compute the proportion of patients who died in the treatment group. Answer in the footnote[2].

We can compute summary statistics from the summary table. A **summary statistic** is a single number summarizing a large amount of data[3]. For instance, the primary results of the study could be placed in two summary statistics: the proportion of people who died in each group.

Proportion who died in the treatment group: $41/733 = 0.056$.

Proportion who died in the control group: $60/742 = 0.081$.

These two summary statistics are useful in evaluating whether the drug worked. There is an observed difference in the outcomes: the death rate was 2.5% lower in the treatment group. We will encounter many more summary statistics throughout this first chapter.

Here we now move into the fourth stage of the investigative process: drawing a conclusion. We might ask, does this 2.5% difference in death rates provide convincing evidence that the drug worked? Even if the drug didn't work, we wouldn't necessarily observe the exact same death rate in the two groups. Maybe the difference of 2.5% was just due to chance. Regrettably, our analysis does not indicate whether what we are seeing is real or a random fluctuation. We will have to wait until a later section before we can make a more formal assessment.

We conduct a more formal analysis in Section 1.8 (optional) for this drug study so that we can draw a more careful conclusion from the data. However, this analysis will not make much sense before we discuss additional principles, ideas, and tools of statistics in Sections 1.2-1.7.

1.2 Data basics

Effective presentation and description of data is a first step in most analyses. This section introduces one structure for organizing data as well as data terminology that will be used throughout this book.

1.2.1 Observations, variables, and cars

Table 1.3 displays rows 1, 2, and 54 of a data set concerning cars from 1993. These observations (measurements) of 54 cars will be referred to as the `cars` data set[4].

Each row in the table represents a single car or **case**[5] and contains six characteristics or measurements for that car. For example, car 54 is a midsize car that can hold 5 people.

Each column of Table 1.3 represents an attribute known about each case, and these attributes are called **variables**. For instance, the `mpgCity` variable holds the city miles per gallon rating of every car in the data set.

In practice, it is especially important to ask clarifying questions to ensure important aspects of the data are understood. For instance, it is always important to be sure we know what each variable means and the units of measurement. Descriptions of all six car variables are given in Table 1.4.

[2] The proportion of the 733 patients who died is $41/733 = 0.056$.

[3] Formally, a summary statistic is a value computed from the data. Some summary statistics are more useful than others.

[4] Lock RH. 1993. 1993 New Car Data. Journal of Statistics Education 1(1).

[5] A case may also be called a **unit of observation** or an **observational unit**.

	type	price	mpgCity	drivetrain	passengers	weight
1	small	15.9	25	front	5	2705
2	midsize	33.9	18	front	5	3560
⋮	⋮	⋮	⋮	⋮	⋮	⋮
54	midsize	26.7	20	front	5	3245

Table 1.3: Three rows from the `cars` data matrix.

variable	**description**
`type`	car type (`small`, `midsize`, or `large`)
`price`	the average purchase price of the vehicles in $1000's (positive number)
`mpgCity`	rated city mileage in miles per gallon (positive number)
`drivetrain`	the drivetrain, also called the powertrain (`front`, `rear`, `4WD`)
`passengers`	passenger capacity (positive whole number, taking values `4`, `5`, or `6`)
`weight`	car weight in pounds (positive number)

Table 1.4: Variables and their descriptions for the `cars` data set.

1.2.2 Data Matrices

The data in Table 1.3 represent a **data matrix**, which is a common way to organize data. Each row of a data matrix corresponds to a separate case, and each column corresponds to a variable. A data matrix for the drug study introduced in Section 1.1 is shown in Table 1.1 on page 2, where patients were the cases and there were two recorded variables.

Data matrices are convenient for recording data as well as analyzing data using a computer. In data collection, if another individual or case is added to the data set, an additional row can be easily added. Similarly, additional columns can be added for new variables.

⊙ **Exercise 1.2** Researchers collected body measurements for bushtail possums in Eastern Australia[6]. They trapped 104 possums and recorded age, gender, head length, and four other pieces of information for each possum. How might these data be organized in a data matrix? Answer in the footnote[7].

1.2.3 Types of variables

Examine the `type`, `price`, `drivetrain`, and `passengers` variables in the `cars` data set. Each of these variables is inherently different from the other three yet many of them share certain characteristics.

First consider `price`, which is said to be a **numerical** variable since it can take a wide range of numerical values, and it is sensible to add, subtract, or take averages with those values. On the other hand, we do not classify a variable reporting telephone area codes as numerical since their average, sum, and difference have no clear meaning.

[6] Lindenmayer DB, Viggers KL, Cunningham RB, and Donnelly CF. 1995. Morphological variation among columns of the mountain brushtail possum, Trichosurus caninus Ogilby (Phalangeridae: Marsupiala). Australian Journal of Zoology 43:449-458.

[7] Here each possum represents a case, and there are seven pieces of information recorded for each case. A table with 104 rows and seven columns could hold these data, where each row represents a possum and each column represents a particular type of measurement or recording.

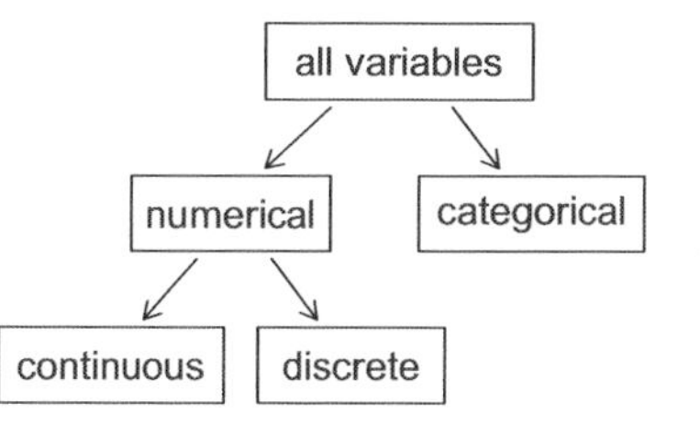

Figure 1.5: Breakdown of variables into their respective types.

The `passengers` variable is also numerical, although it seems to be a little different than `price`. The variable `passengers` can only take whole positive numbers (`1`, `2`, ...) since it is not possible to have 4.5 passengers. The variable `passengers` is said to be **discrete** since it only can take numerical values with jumps (e.g. 3 or 4, but not any number in between). On the other hand, `price` is said to be **continuous**.

The variable `drivetrain` can only take a few different values: `front`, `rear`, and `4WD`. Because the responses themselves are categories, `drivetrain` is called a **categorical** variable[8]. The three possible values (`front`, `rear`, `4WD`) are called the variable's **levels**.

Finally, consider the `type` variable describing whether a car is `small`, `medium`, or `large`. This variable seems to be a hybrid: it is a categorical variable but the levels have some inherent ordering. A variable with these properties is called an **ordinal** variable. To simplify analyses, any ordinal variables in this book will be treated as categorical variables.

● **Example 1.3** Data were collected about students in a statistics course. Three variables were recorded for each student: number of siblings, student height, and whether the student had previously taken a statistics course. Classify each of the variables as continuous numerical, discrete numerical, or categorical.

The number of siblings and student height represent numerical variables. Because the number of siblings is a count, it is discrete. Height varies continuously, so it is a continuous numerical variable. The last variable classifies students into two categories – those who have and those who have not taken a statistics course – which makes this variable categorical.

⊙ **Exercise 1.4** In the sulphinpyrazone study from Section 1.1, there were two variables: `group` and `outcome`. Are these numerical or categorical variables?

1.2.4 Relationships among variables

Many analyses are motivated by a researcher looking for a relationship between two or more variables. A biologist studying possums in Eastern Australia may want to know the answers to some of the following questions.

(1) If a possum has a shorter-than-average head, will its skull width usually be smaller or larger than the average skull width?

(2) Will males or females, on average, be longer?

(3) Which population of possums will be larger on average: those living in Victoria or in other locations?

---

[8] Sometimes also called a **nominal** variable.

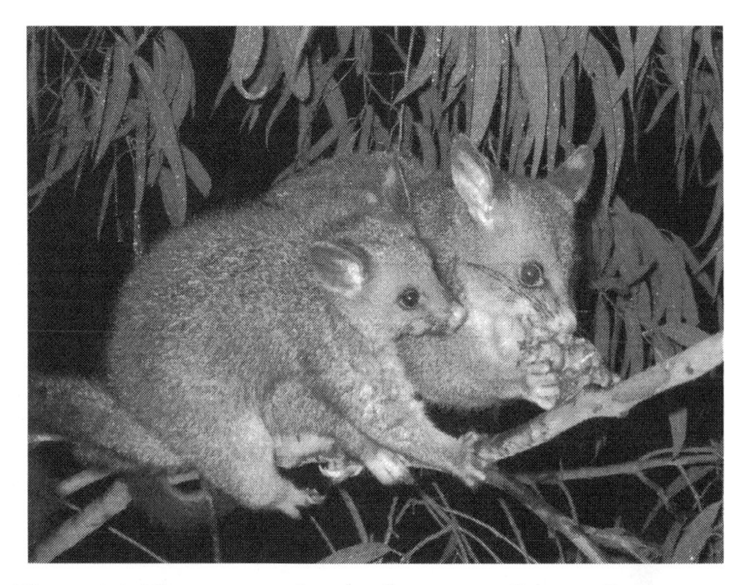

Figure 1.6: The common brushtail possum of Australia.

Photo by wollombi on Flickr: http://flickr.com/photos/wollombi/58499575/

	pop	sex	age	headL	skullW	totalL	tailL
1	Vic	m	8	94.1	60.4	89.0	36.0
2	Vic	f	6	92.5	57.6	91.5	36.5
3	Vic	f	6	94.0	60.0	95.5	39.0
⋮	⋮	⋮	⋮	⋮	⋮	⋮	⋮
104	other	f	3	93.6	59.9	89.0	40.0

Table 1.7: Four rows from the `possum` data set.

(4) Does the proportion of males differ based on location, i.e. from Victoria to the other locations?

To answer these questions, data must be collected. Four observations from such a data set is shown in Table 1.7, and descriptions of each variable are presented in Table 1.8. Examining summary statistics could provide insights to each of the four questions about possums. Additionally, graphs can be used to visually summarize data and are useful for answering such questions as well.

Scatterplots are one type of graph used to study the relationship between two numerical variables. Figure 1.9 compares the variables `headL` and `skullW`. Each point on the plot represents a single possum. For instance, the highlighted dot corresponds to Possum 1 from Table 1.7, which has a head length of 94.1mm and a skull width of 60.4mm. The scatterplot suggests that if a possum has a short head, then its skull width also tends to be smaller than the average possum.

⊙ **Exercise 1.5** Examine the variables in the `cars` data set, which are described in Table 1.4 on page 4. Create two questions about the relationships between these variables that are of interest to you.

variable	description
pop	location where possum was trapped (Vic or other)
sex	possum's gender (m or f)
age	age, in years (whole number, data range: 1 to 9)
headL	head length, in mm (data range: 82.5 to 103.1)
skullW	skull width, in mm (data range: 50.0 to 68.6)
totalL	total length, in cm (data range: 75.0 to 96.5)
tailL	tail length, in cm (data range: 32.0 to 43.0)

Table 1.8: Variables and their descriptions for the possum data set.

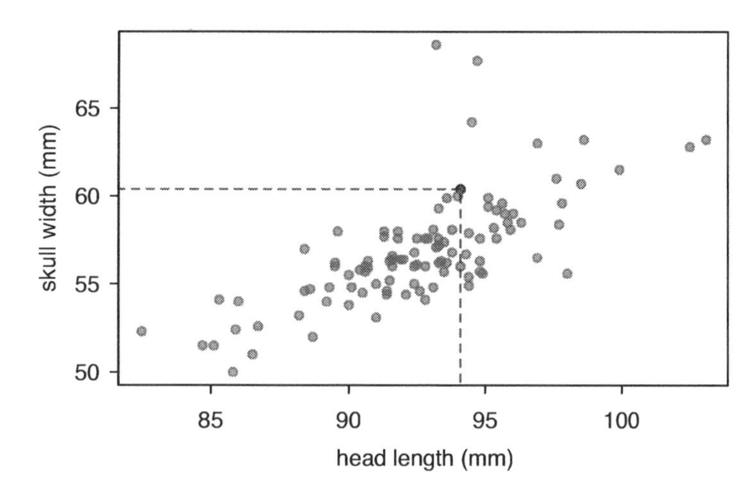

Figure 1.9: A scatterplot showing skullW against headL. The possum with a head length of 94.1mm and a skull width of 60.4mm is highlighted.

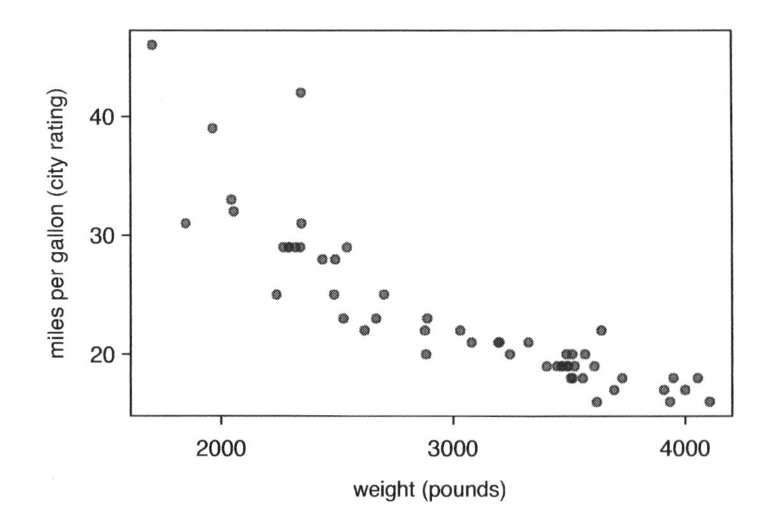

Figure 1.10: A scatterplot of `mpgCity` versus `weight` for the `cars` data set.

1.2.5 Associated and independent variables

The variables `headL` and `skullW` are said to be associated because the plot shows a discernible pattern. When two variables show some connection with one another, they are called **associated** variables. Associated variables can also be called **dependent** variables and vice-versa.

● **Example 1.6** Examine the scatterplot of `weight` and `mpgCity` in Figure 1.10. Are these variables associated?

It appears that the heavier a car is, the worse mileage it gets. Since there is some relationship between the variables, they are associated.

Because there is a downward trend in Figure 1.10 – larger weights are associated with lower mileage – these variables are said to be **negatively associated**. A **positive association** is shown in the possum data represented in Figure 1.9, where longer heads are associated with wider skulls.

If two variables are not associated, then they are said to be **independent**. That is, two variables are independent if there is no evident connection between the two. It is also possible for cases – such as a pair of possums or a pair of people – to be independent. For instance, if possums 1 and 2 are not siblings, do not compete for resources in the same territory, and show no other natural connections, then they can be called independent.

> **Associated or independent, not both**
> A pair of variables are either related in some way (associated) or not (independent). No pair of variables is both associated and independent. These same definitions hold true for a pair of cases as well.

1.3 Examining numerical data

The `cars` data set represents a *sample* from a larger set of cases. This larger set of cases is called the **population**. Ideally data would be collected from every case in the population. However, this is rarely possible due to high costs of data collection. As a substitute, statisticians collect subsets of the data called **samples** to gain insights into the population. The `cars` data set represents a sample of all cars from 1993, and the `possum` data set represents a sample from all possums in the Australian states of Victoria, New South Wales, and Queensland. In this section we introduce summary statistics and graphics as a first step in analyzing numerical data from a sample to help us understand certain features of the population as a whole.

1.3.1 Scatterplots for paired data

A **scatterplot** provides a case-by-case view of data for two numerical variables. In Section 1.2.4, a scatterplot was used to examine how head length and skull width were related in the `possum` data set. Another scatterplot is shown in Figure 1.11, comparing `price` and `weight` for the `cars` data set.

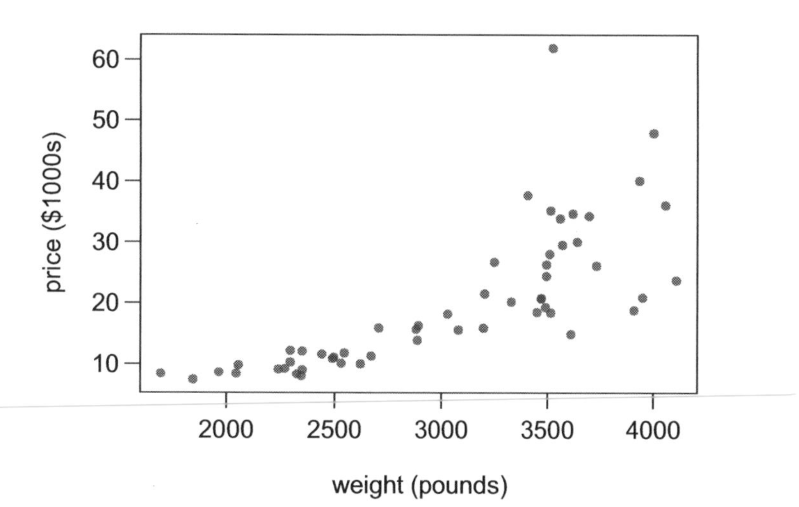

Figure 1.11: A scatterplot of `price` versus `weight` for the `cars` data set.

In any scatterplot, each point represents a single case. Since there are 54 cases in `cars`, there are 54 points in Figure 1.11.

⊙ **Exercise 1.7** What do scatterplots reveal about the data, and how might they be useful?

Some associations are more linear, like the relationship between `skullW` and `headL`, shown in Figure 1.9 on page 7. Others, like the one seen in Figure 1.10 can be curved.

⊙ **Exercise 1.8** Describe two variables that would have a horseshoe shaped association in a scatterplot. One example is given in the footnote[9].

[9]Consider the case where your vertical axis represents something "good" and your horizontal axis represents something that is only good in moderation. Health and water consumption fit this description since water becomes toxic when consumed in excessive quantities.

1.3.2 Dot plots and the mean

Sometimes two variables is one too many: only one variable may be of interest. In these cases, a dot plot provides the most basic of displays. A dot plot is a one-variable scatterplot, and a sample dot plot is shown in Figure 1.12.

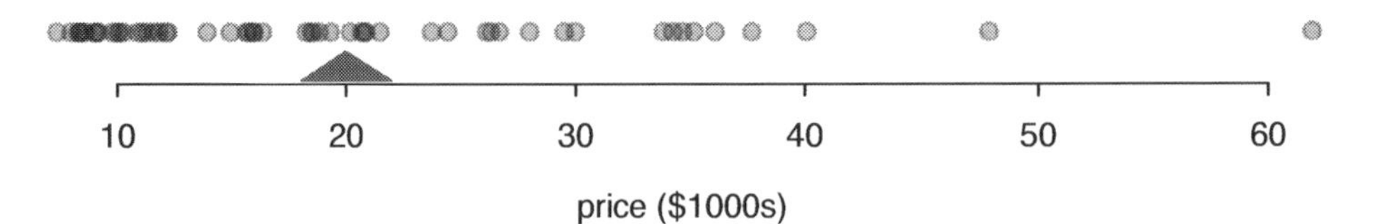

Figure 1.12: A dot plot of `price` for the `cars` data set. The triangle marks the sample's mean price.

The **mean**, sometimes called the average, is a way to measure the center of a **distribution** of data. To find the mean price of the cars in the sample, add up all the prices and divide by the number of cases.

$$\bar{x} = \frac{15.9 + 33.9 + \cdots + 26.7}{54} = 19.99259 \tag{1.9}$$

$\bar{x}$
sample mean

The sample mean is often labeled $\bar{x}$. The letter x is being used as an abbreviation for `price`, and the bar says it is the sample average of price. It is useful to think of the mean as the balancing point of the distribution. The sample mean is shown as a blue triangle in Figure 1.12.

> **Mean**
> The sample mean of a numerical variable is computed as the sum of all of the observations divided by the number of observations:
>
> $$\bar{x} = \frac{x_1 + x_2 + \cdots + x_n}{n} \tag{1.10}$$
>
> where $x_1, x_2, \ldots, x_n$ represent the n observed values.

n
sample size

⊙ **Exercise 1.11** Examine equations (1.9) and (1.10) above. What does x_1 correspond to? And x_2? Can you infer a general meaning to what x_i might represent? Answers in the footnote[10].

⊙ **Exercise 1.12** What was n in the `cars` data set? Answer in the footnote[11].

μ
population mean

The *population* mean is also computed in the same way, however, it has a special label: μ. The symbol μ is the Greek letter *mu* and represents the average of all observations in the population. Sometimes a subscript, such as $_x$, is used to represent which variable the population mean refers to, i.e. μ_x.

[10] x_1 corresponds to the price of the first car (15.9), x_2 to the price of the second car (33.9), and x_i corresponds to the price of the ith car in the data set.
[11] The sample size is $n = 54$.

● **Example 1.13** The average price of all cars from 1993 can be estimated using the sample data. Based on the `cars` sample, what would be a reasonable estimate of μ_x, the mean price of cars from 1993?

The sample mean may provide a good estimate of μ_x. While this estimate will not be perfect, it provides a *point estimate* of the population mean.

1.3.3 Histograms and shape

Dot plots show the exact value for each observation. This is useful for small data sets, but can become problematic for larger samples. Rather than showing the value of each observation, we might prefer to think of the value as belonging to a *bin*. For example, in the `cars` data set, we could create a table of counts for the number of cases with prices between \$5,000 and \$10,000, then the number of cases between \$10,000 to \$15,000, and so on. Observations that fall on the boundary of a bin (e.g. \$10,000) are allocated to the lower bin. This tabulation is shown in Table 1.13. To make the data easier to see visually, these binned counts are plotted as bars in Figure 1.14. This binned version of the dot plot is called a **histogram**.

Price	5-10	10-15	15-20	20-25	25-30	30-35	···	55-60	60-65
Count	11	11	10	7	6	3	···	0	1

Table 1.13: The counts for the binned `price` data, where `price` is in thousands of dollars.

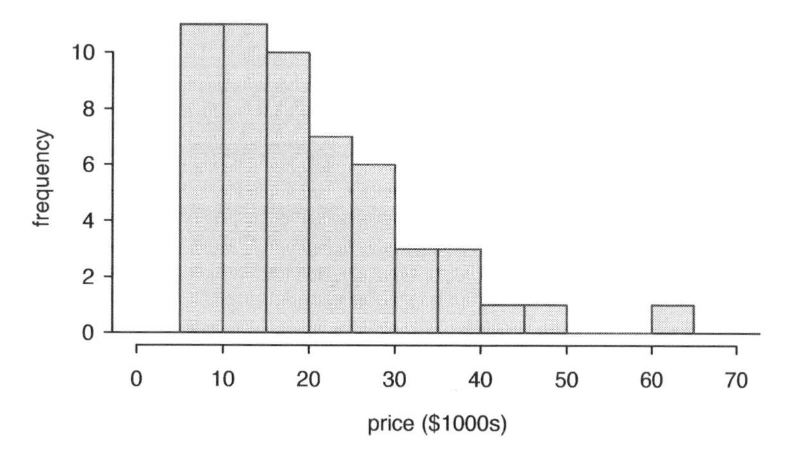

Figure 1.14: Histogram of `price`.

Histograms provide a view of the **data density**. Higher bars represent where the data are relatively more common. For instance, there are many more cars that cost less than \$15,000 than cars that cost at least \$50,000 in the data set. The bars make it especially easy to see how the density of the data changes from one price to another: in general, the higher the price, the fewer the cars.

Histograms are especially convenient for describing the shape of the data distribution. Figure 1.14 shows that most cars have a lower price, while fewer cars have higher prices.

When data trail off to the right in this way and have a longer right tail, the shape is said to be **skewed to the right**[12].

Data sets with the reverse characteristic – a long, thin tail to the left – are said to be **left skewed**. It might also be said that such a distribution has a long left tail. Data sets that show roughly equal trailing off in both directions are called **symmetric**.

> **Long tails to identify skew**
> When data trail off in one direction, it is called a **long tail**. If a distribution has a long left tail, it is left skewed. If a distribution has a long right tail, it is right skewed.

⊙ **Exercise 1.14** Take a look at Figure 1.12 on page 10. Can you see the skew in the data? Is it easier to see the skew in Figure 1.12 or Figure 1.14?

⊙ **Exercise 1.15** Besides the mean (since it was labeled), what can you see in Figure 1.12 that you cannot see in 1.14? Answer in the footnote[13].

In addition to looking at whether a distribution is skewed or symmetric, histograms can be used to identify modes. A **mode** is represented by a prominent peak in the distribution[14]. There is only one prominent peak in the histogram of `price`.

Figure 1.15 shows histograms that have one, two, and three prominent peaks. Such distributions are called **unimodal**, **bimodal**, and **multimodal**, respectively. Any distribution with more than 2 prominent peaks is called multimodal. Notice that there was one prominent peak in the unimodal distribution with a second less prominent peak that was not counted since it only differs from its neighboring bins by a few observations.

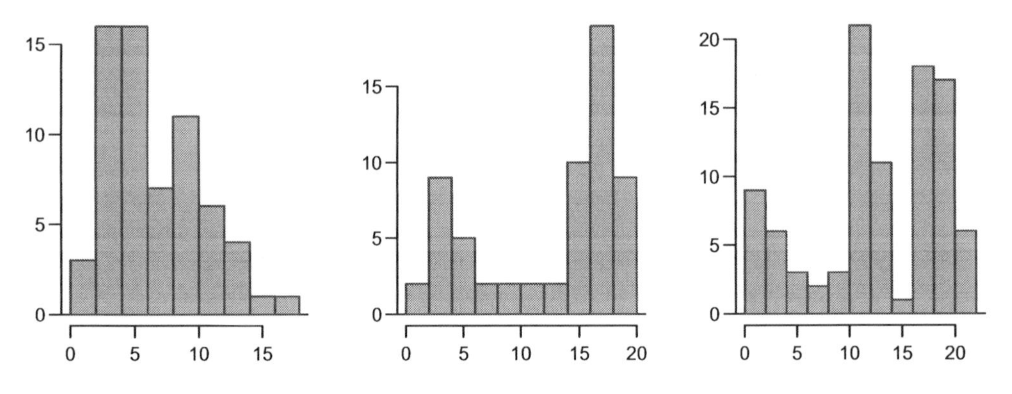

Figure 1.15: Counting only prominent peaks, the distributions are (left to right) unimodal, bimodal, and multimodal.

[12] Other ways to describe data skewed to the right: **right skewed**, **skewed to the high end**, or **skewed to the positive end**.

[13] The individual prices.

[14] Another definition of mode, which is not typically used in statistics, is the value with the most occurrences. It is common to have *no* observations with the same values in a data set, which makes this other definition useless for many real data sets.

⊙ **Exercise 1.16** Figure 1.14 reveals only one prominent mode in `price`. Is the distribution unimodal, bimodal, or multimodal?

⊙ **Exercise 1.17** Height measurements of young students and adult teachers at a K-3 elementary school were taken. How many modes would you anticipate in this height data set? Answer in the footnote[15].

> **TIP: Looking for modes**
> Looking for modes isn't about finding a clear and correct answer about the number of modes in a distribution, which is why *prominent* is not defined mathematically here. The importance of this examination is to better understand your data and how it might be structured.

1.3.4 Variance and standard deviation

The mean was introduced as a method to describe the center of a data set but the data's variability is also important. Here, we introduce two measures of variability: the variance and the standard deviation. Both of these are very useful in data analysis, even though their formulas are a bit tedious to compute by hand.

We call the distance of an observation from its mean its **deviation**. Below are the 1st, 2nd, and 54th deviations for the `price` variable:

$$\begin{aligned} x_1 - \bar{x} &= 15.9 - 20 = -4.1 \\ x_2 - \bar{x} &= 33.9 - 20 = 13.9 \\ &\vdots \\ x_{54} - \bar{x} &= 26.7 - 20 = 6.7 \end{aligned}$$

If we square these deviations and then take an average, the result is about equal to the sample **variance**, denoted by s^2:

s^2
sample variance

$$s^2 = \frac{(-4.1)^2 + (13.9)^2 + \cdots + (6.7)^2}{54 - 1} = \frac{16.8 + 193.2 + \cdots + 44.9}{53} = 132.4$$

(We divide by $n - 1$ (rather than dividing by n) when computing the variance.) Notice that squaring the deviations does two things: (i) it makes large values much larger, seen by comparing $(-4.1)^2$, 13.9^2, and 6.7^2, and (ii) it gets rid of any negative signs.

The **standard deviation** is defined as the square root of the variance:

$$s = \sqrt{132.4} = 11.5$$

s
sample standard deviation

A subscript of $_x$ may be added to the the variance and standard deviation – that is, s_x^2 and s_x – as a reminder that these are the variance and standard deviation of the observations represented by x_1, x_2, ..., x_n. These may be omitted when it is clear which data the standard deviation is referencing.

[15]There might be two height groups visible in the data set: one of the students and one of the adults. That is, the data might be bimodal.

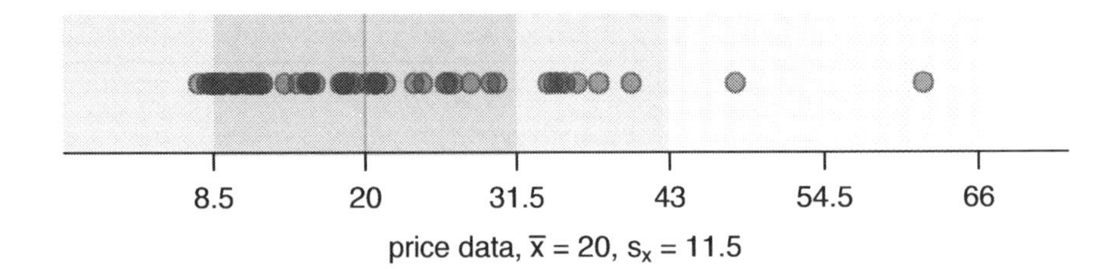

Figure 1.16: In the `price` data, 40 of 54 cars (74%) are within 1 standard deviation of the mean, $20,000. Additionally, 52 of the 54 cars (96%) and 53 of the 54 prices (98%) are within 2 and 3 standard deviations, respectively.

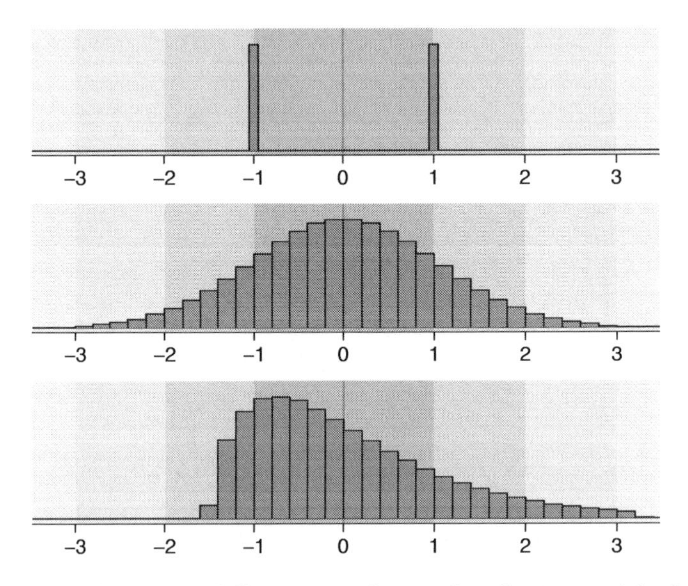

Figure 1.17: Three very different population distributions with the same mean $\mu = 0$ and standard deviation $\sigma = 1$.

> **Variance and standard deviation**
> The variance is roughly the average squared distance from the mean. The standard deviation is the square root of the variance.

Formulas and methods used to compute the variance and standard deviation for a population are similar to those used for a sample[16]. However, like the mean, the population values have special symbols: σ^2 for the variance and σ for the standard deviation. The symbol σ is the Greek letter *sigma*. As with the sample variance and standard deviation, subscripts such as $_x$ can be added to specify which data sets the population variance and standard deviation reference.

σ^2
population variance

σ
population standard deviation

The standard deviation is useful in considering how close the data are to the mean. Usually about 70% of the data are within one standard deviation of the mean and 95% within two standard deviations. However, these percentages can and do vary from one distribution to another. Figure 1.17 shows several different distributions that have the same center and variability but very different shapes.

[16]The only difference is that the population variance has a division by n instead of $n - 1$.

⊙ **Exercise 1.18** On page 11, the concept of shape of a distribution was introduced. A good description of the shape of a distribution should include modality and whether the distribution is symmetric or skewed to one side. Using Figure 1.17 as an example, explain why such a description is important.

● **Example 1.19** Describe the distribution of the `price` variable, shown in Figure 1.14 on page 11. The description should incorporate the center, variability, and shape of the distribution, and it should also be placed in context of the problem: the price of cars. Also note any especially unusual cases.

The distribution of car prices is unimodal and skewed to the high end. Many of the prices fall near the mean at $20,000, and most fall within one standard deviation ($11,500) of the mean. There is one very expensive car that costs more than $60,000.

In practice, the variance and standard deviation are sometimes used as a means to an end, where the "end" is being able to accurately estimate the uncertainty associated with a sample statistic. For example, in Chapter 4 we use the variance and standard deviation to assess how close the sample mean is to the population mean.

TIP: standard deviation describes variability
Standard deviation is complex mathematically. However, it is not conceptually difficult. It is useful to remember that usually about 70% of the data are within one standard deviation of the mean and about 95% are within two standard deviations.

1.3.5 Box plots, quartiles, and the median

A box plot summarizes a data set using five statistics while also plotting unusual observations. Figure 1.18 provides a vertical dot plot alongside a box plot of the `price` variable from the `cars` data set.

The first step in building a box plot is drawing a rectangle to represent the middle 50% of the data. The total length of the box, shown vertically in Figure 1.18, is called the **interquartile range** (IQR, for short). It, like the standard deviation, is a measure of variability in data. The more variable the data, the larger the standard deviation and IQR. The two boundaries of the box are called the **first quartile** (the 25th percentile, i.e. 25% of the data fall below this value) and the **third quartile** (the 75th percentile), and these are often labeled Q_1 and Q_3, respectively.

Interquartile range (IQR)
The **interquartile range (IQR)** is the length of the box in a box plot. It is computed as

$$IQR = Q_3 - Q_1$$

where Q_1 and Q_3 are the 25th and 75th percentiles.

The line splitting the box denotes the **median**, or the value that splits the data in half. Figure 1.18 shows 50% of the data falling below the median (dark hollow circles) and other 50% falling above the median (light-colored filled circles). There are 54 car prices in the data set (an even number) so the data are perfectly split into two groups of 27.

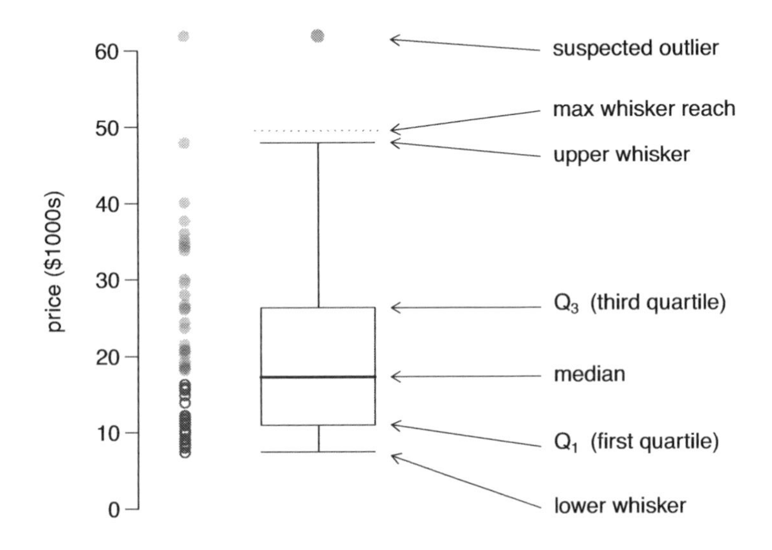

Figure 1.18: A vertical dot plot next to a labeled box plot of the `price` data. The median ($17,250), splits the data into the bottom 50% and the top 50%, marked in the dot plot by dark-colored hollow circles and light filled circles, respectively.

We take the median in this case to be the average of the two observations closest to the 50th percentile: $\frac{\$16{,}300+\$18{,}200}{2} = \$17,250$. When there are an odd number of observations, there will be exactly one observation that splits the data into two halves, and in this case that observation is the median (no average needed).

Median: the number in the middle

If the data are ordered from smallest to largest, the **median** is the observation right in the middle. If there are an even number of observations, there will be two values in the middle, and the median is taken as their average.

⊙ **Exercise 1.20** What percent of the data fall between Q_1 and the median? How much between the median and Q_3? Answers in the footnote[17].

Extending out from the box, the **whiskers** attempt to capture the data outside of the box, however, their reach is never allowed to be more than $1.5 * IQR$.[18] They grab everything within this reach. In Figure 1.18, the upper whisker cannot extend to the last point, which is beyond $Q_3 + 1.5 * IQR$, and so it extends only to the last point below this limit. The lower whisker stops at the lowest price, $7,400, since there is no additional data to reach; the lower whisker limit is not shown in the figure because the plot does not extend down to $Q_1 - 1.5 * IQR$. In a sense, the box is like the body of the box plot and the whiskers are like its arms trying to reach the rest of the data.

Any observation that lies beyond the whiskers is labeled with a dot. The purpose of

[17] Since Q_1 and Q_3 capture the middle 50% of the data and the median splits the data in the middle, 25% of the data fall between Q_1 and the median, and another 25% falls between the median and Q_3.

[18] While the choice of exactly 1.5 is arbitrary, it is the most commonly used value for box plots.

labeling these points – instead of just extending the whiskers to the minimum and maximum observed values – is to help identify any observations that appear to be unusually distant from the rest of the data. Unusually distant observations are called **outliers**. In the case of the `price` variable, the car with price \$61,900 is a potential outlier.

Outliers are extreme
An **outlier** is an observation that appears extreme relative to the rest of the data.

TIP: Why it is important to look for outliers
Examination of data for possible outliers serves many useful purposes, including

1. Identifying extreme skew in the distribution.
2. Identifying data collection or entry errors. If there was a car price listed as \$140,000, it would be worth reviewing the observation to see whether it was really \$14,000.
3. Providing insights into interesting phenomena with the data.

⊙ **Exercise 1.21** The observation \$61,900, a suspected outlier, was found to be an accurate observation. What would such an observation suggest about the nature of vehicle prices?

⊙ **Exercise 1.22** Using Figure 1.18 on the facing page, estimate the following values for `price` in the `cars` data set: (a) Q_1, (b) Q_3, and (c) IQR.

1.3.6 Robust statistics

How would sample statistics of the `cars` data set be affected if \$200,000 was observed instead of \$61,900 for the most expensive car? Or what if \$200,000 had been in the sample instead of the cheapest car at \$7,400? These two scenarios are plotted alongside the original data in Figure 1.19, and sample statistics are computed under each of these scenarios in Table 1.20.

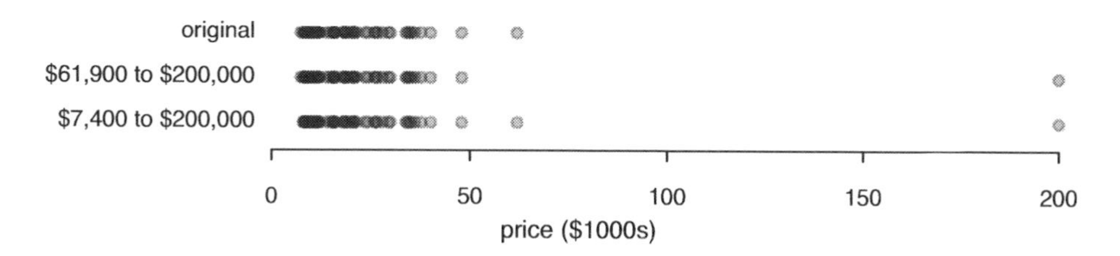

Figure 1.19: Dot plots of the original price data and two modified price data sets.

⊙ **Exercise 1.23** (a) Which is more affected by extreme observations, the mean or median? Table 1.20 may be helpful. (b) Is the standard deviation or IQR more affected by extreme observations?

	robust		not robust	
scenario	median	IQR	$\bar{x}$	s
original `price` data	17.25	15.30	19.99	11.51
move \$61,900 to \$200,000	17.25	15.30	22.55	26.53
move \$7,400 to \$200,000	18.30	15.45	26.12	35.79

Table 1.20: A comparison of how the median, IQR, mean ($\bar{x}$), and standard deviation (s) change when extreme observations are in play.

The median and IQR are called **robust estimates** because extreme observations have little effect on their values. The mean and standard deviation are much more affected by changes in extreme observations.

⊙ **Exercise 1.24** Why doesn't the median or IQR change from the original `price` data to the second scenario of Table 1.20?

⊙ **Exercise 1.25** Why are robust statistics useful? If you were searching for a new car and cared about price, would you be more interested in the mean vehicle price or the median vehicle price when considering the price for a regular car?

1.3.7 Transforming data (special topic)

When data are extremely skewed, we sometimes transform them so they are easier to model. Consider the histogram of salaries for Major League Baseball players' salaries from 2010, which is shown in Figure 1.21(a).

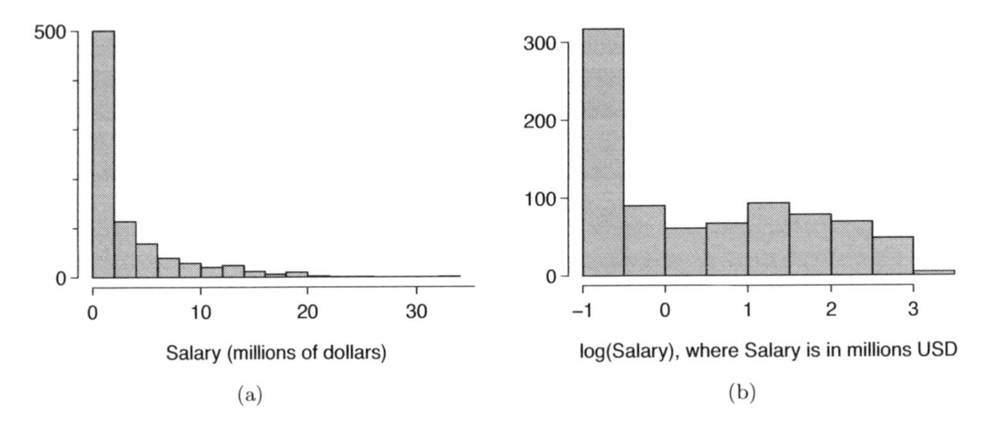

Figure 1.21: (a) Histogram of MLB player salaries for 2010, in millions of dollars. (b) Histogram of the log-transformed MLB player salaries for 2010.

● **Example 1.26** The histogram of MLB player salaries is useful in that we can see the data are highly skewed and centered (as gauged by the median) at about \$1 million. What isn't useful about this plot?

Most of the data are collected into one bin in the histogram and the data are so strongly skewed that some details in the data are obscured.

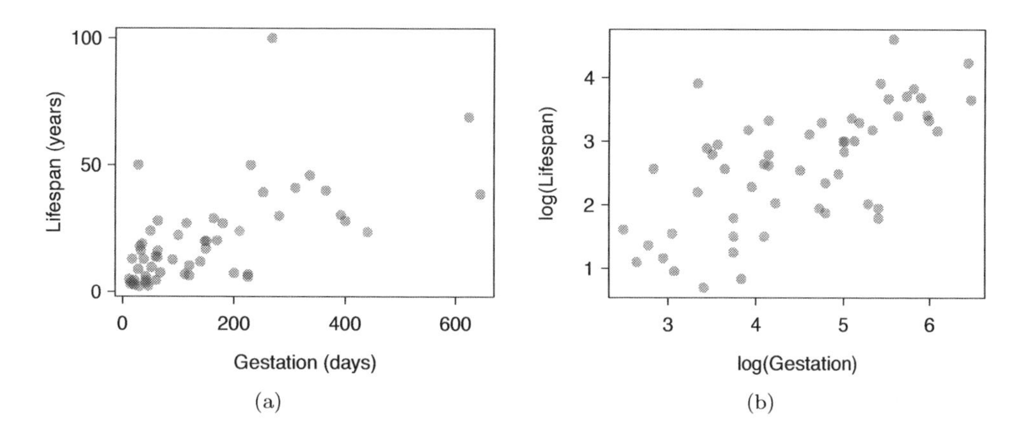

Figure 1.22: (a) Scatterplot of `Lifespan` against `Gestation` for 55 mammals. (b) A scatterplot of the same data but where each variable has been log-transformed.

There are some standard transformations that are often applied when much of the data cluster near zero (relative to the larger values in the data set) and all observations are positive. A **transformation** is a rescaling of the data using a function. For instance, a plot of the natural logarithm[19] of player salaries results in a new histogram in Figure 1.21(b). Transformed data are sometimes easier to work with when applying statistical models because the transformed data are much less skewed and outliers are usually less extreme.

Transformations can also be applied to one or both variables in a scatterplot. A scatterplot is shown in Figure 1.22(a) representing mammal lifespans against the time of gestation (time spent in the womb). There seems to be a positive association between these two variables. In Chapter 7, we might want to use a straight line to model the data. However, we'll find that the data in their current state cannot be modeled very well. Figure 1.22(b) shows a scatterplot where both the `gestation` and `lifespan` variables have been transformed using a log (base e) transformation. While there is a positive association in each plot, the transformed data show a steadier trend, which is easier to model than the untransformed data.

Other transformations other than the logarithm can be useful, too. For instance, the square root ($\sqrt{\text{original observation}}$) and inverse ($\frac{1}{\text{original observation}}$) are regularly used by statisticians. Common goals in transforming data are to see the data structure differently, reduce skew, assist in modeling, or straighten a nonlinear relationship in a scatterplot.

1.4 Considering categorical data

Like numerical data, categorical data can also be organized and analyzed. In this section, tables and other basic tools for categorical data analysis are introduced that will be used throughout this book.

[19] Statisticians often write the natural logarithm as log. You might be more familiar with it being written as ln.

1.4.1 Contingency tables

Table 1.23 summarizes two variables from the `cars` data set: `type` and `drivetrain`. A table that summarizes data for two categorical variables in this way is called a **contingency table**. Each number in the table represents the number of times a particular combination of variable outcomes occurred. For example, the number 19 corresponds to the number of cars in the data set that are small *and* have front wheel drive. Row and column totals are also included. The **row totals** equal the total counts across each row (e.g. $19+0+2 = 21$), and **column totals** are total counts down each column.

A table for a single variable is called a **frequency table**. Table 1.24 is a frequency table for the `type` variable. If we replaced the counts with percentages or proportions, the table would be called a **relative frequency table**.

	front	rear	4WD	total
small	19	0	2	21
midsize	17	5	0	22
large	7	4	0	11
total	43	9	2	54

Table 1.23: A contingency table for `type` and `drivetrain`.

small	midsize	large
21	22	11

Table 1.24: A frequency table for the `type` variable.

⊙ **Exercise 1.27** Why is Table 1.24 redundant if Table 1.23 is provided?

1.4.2 Bar plots and proportions

A bar plot is a common way to display a single categorical variable. The left panel of Figure 1.25 shows a **bar plot** for the vehicle type. In the right panel, the counts were converted into proportions (e.g. $21/54 = 0.389$ for `small`), showing the fraction of the whole that are in each category.

⊙ **Exercise 1.28** Which of the following statements would be more useful to an auto executive? (1) 21 cars in our sample were small vehicles. (2) 38.9% of the cars in our sample were small vehicles. Comment in the footnote[20].

Table 1.26 shows the row proportions for Table 1.23. The **row proportions** are computed as the counts divided by their row totals. The count 17 at the intersection of `midsize` and `front` is replaced by $17/22 = 0.773$, i.e. 17 divided by its row total, 22. So what does 0.773 represent? It corresponds to the proportion of midsize vehicles in the sample that have front wheel drive.

A contingency table of the column proportions is computed in a similar way, where each **column proportion** is computed as the count divided by the corresponding column total. Table 1.27 shows such a table, and here the value 0.442 represents the proportion of front wheel drive cars in the sample that are small cars.

[20] Even if the sample size (54) was provided in the first statement, the auto exec would probably just be trying to figure out the proportion in her head.

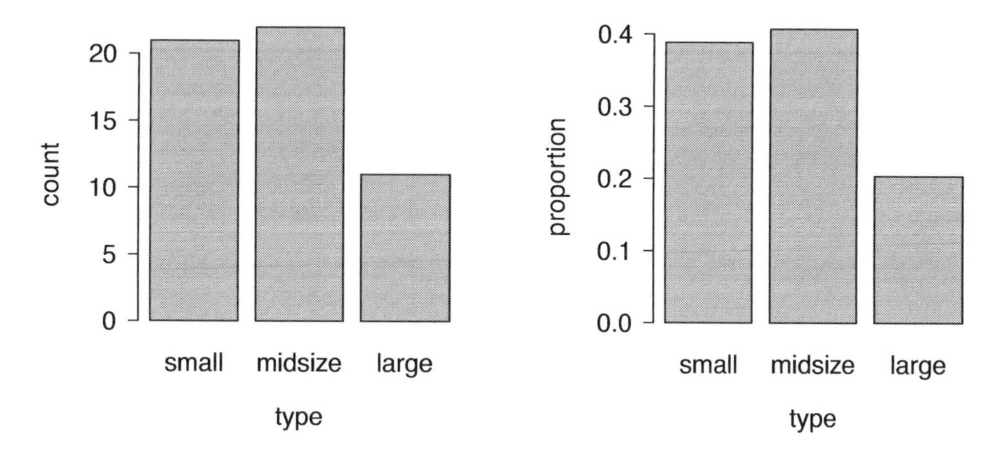

Figure 1.25: Two bar plots of `type`. The left panel shows the counts and the right panel the proportions in each group.

	front	rear	4WD	total
`small`	$19/21 = 0.905$	$0/21 = 0.000$	$2/21 = 0.095$	1.000
`midsize`	$17/22 = 0.773$	$5/22 = 0.227$	$0/22 = 0.000$	1.000
`large`	$7/11 = 0.636$	$4/11 = 0.364$	$0/11 = 0.000$	1.000
total	$43/54 = 0.796$	$9/54 = 0.167$	$2/54 = 0.037$	1.000

Table 1.26: A contingency table with row proportions for the `type` and `drivetrain` variables.

⊙ **Exercise 1.29** What does 0.364 represent in Table 1.26? Answer in the footnote[21]. What does 0.444 represent in Table 1.27?

⊙ **Exercise 1.30** What does 0.796 represent in Table 1.26? Answer in the footnote[22]. What does 0.407 represent in the Table 1.27?

● **Example 1.31** Researchers suspect the proportion of male possums might change by location. A contingency table for the `pop` (living location) and `sex` variables from the `possum` data set are shown in Table 1.28. Based on these researchers' interests, which would be more appropriate: row or column proportions?

The interest lies in how the `sex` changes based on `pop`. This corresponds to the row proportions: the proportion of males/females in each location.

Example 1.31 points out that row and column proportions are not created equal. It is important to consider each before settling on one to ensure that the most useful table is constructed.

---

[21] 0.364 represents the proportion of large cars in the sample that have rear wheel drive.

[22] 0.796 represents the proportion of cars in the sample that are front wheel drive vehicles.

	front	rear	4WD	total
small	19/43 = 0.442	0/9 = 0.000	2/2 = 1.000	21/54 = 0.389
midsize	17/43 = 0.395	5/9 = 0.556	0/2 = 0.000	22/54 = 0.407
large	7/43 = 0.163	4/9 = 0.444	0/2 = 0.000	11/54 = 0.204
total	1.000	1.000	1.000	1.000

Table 1.27: A contingency table with column proportions for the `type` and `drivetrain` variables.

	f	m	total
Vic	24	22	46
other	19	39	58
total	43	61	104

Table 1.28: A contingency table for `pop` and `sex` from the `possum` data set.

1.4.3 Segmented bar and mosaic plots

Contingency tables using row or column proportions are especially useful for examining how two categorical variables are related. Segmented bar and mosaic plots provide a way to put these tables into a graphical form. To reduce complexity, this section we will only consider vehicles with front and rear wheel drive, as shown in Table 1.29.

A **segmented bar plot** is a graphical display of contingency table information. For example, a segmented bar plot representing Table 1.29(a) is shown in Figure 1.30(a), where we have first created a bar plot using the `type` variable and then broken down each group by the levels of `drivetrain`. The row proportions of Table 1.29(b) have been translated into a standardized segmented bar plot in Figure 1.30(b).

The choice to describe each level of `drivetrain` within the levels of the `type` variable was arbitrary; we could have first created a bar plot of `type` and then broken it up using each level of `drivetrain`. Just like considering row and column proportions, it is good to evaluate both ways one might construct a segmented bar plot before choosing a final form.

⊙ **Exercise 1.32** Why is only one level of `drivetrain` shown for small vehicles in Figure 1.30?

A **mosaic plot** is a graphical display of contingency table information that is similar to a bar plot for one variable or a segmented bar plot when using two variables. Figure 1.31(a)

	front	rear	total
small	19	0	19
midsize	17	5	22
large	7	4	11
total	43	9	52

(a)

	front	rear
small	1.00	0.00
midsize	0.77	0.23
large	0.64	0.36
total	0.83	0.17

(b)

Table 1.29: (a) Contingency table for `type` and `drivetrain` where the two vehicles with `drivetrain` = `4WD` have been removed. (b) Row proportions for Table (a).

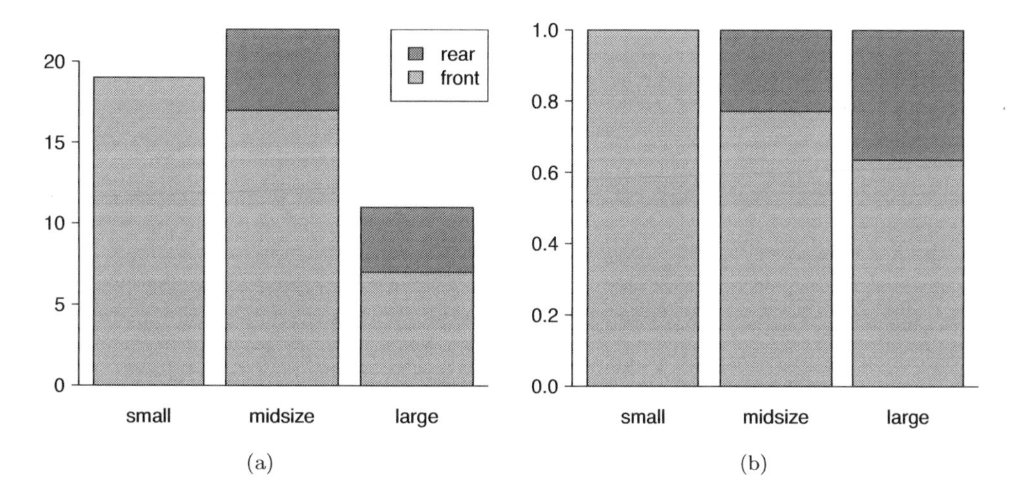

Figure 1.30: (a) Segmented bar plot for vehicle type, where the counts have been broken down by `type` then `drivetrain`. (b) Standardized version of Figure (a).

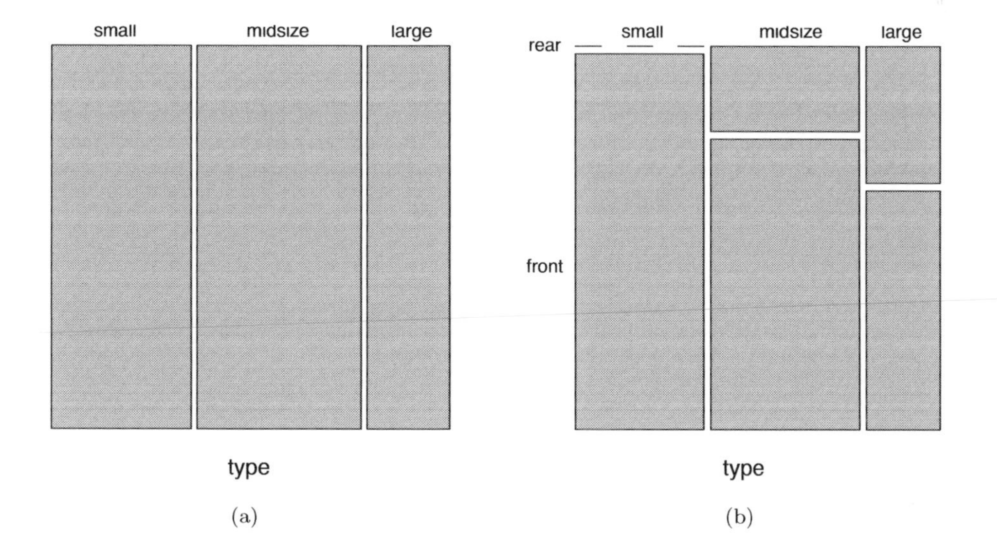

Figure 1.31: The one-variable mosaic plot for `type` and the two-variable mosaic plot for both `type` and `drivetrain`.

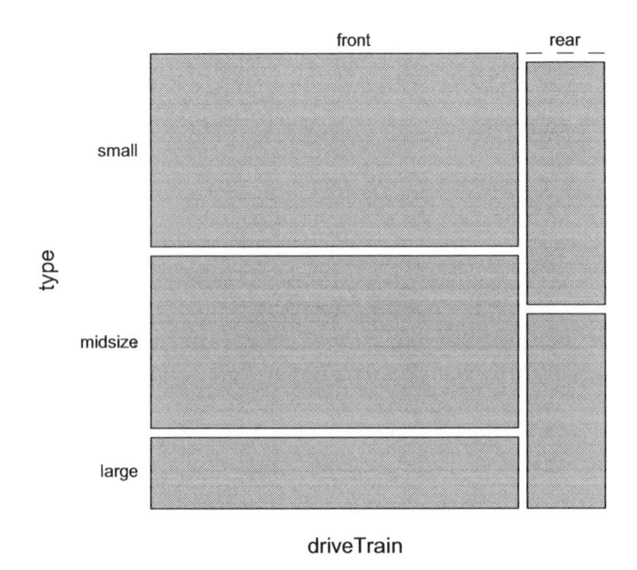

Figure 1.32: Mosaic plot where type is broken up within the drivetrain.

shows a mosaic plot for the `type` variable from Table 1.29(a). Each column represents a level of `type`, and the column widths correspond to the proportion of cars of each type. For instance, there are fewer small cars than midsize cars, so the small car column is slimmer.

This one-variable mosaic plot is further broken into pieces in Figure 1.31(b) using the `drivetrain` variable. Each column is split proportionally according to the drivetrain for vehicles of that particular type. For example, the second column, representing only midsize cars, was broken into midsize cars with front and rear drivetrains. Notice that a dashed line is used to represent small cars with rear wheel drive because there were none. As another example, the top of the third column represents large cars with rear wheel drive, and the lower part of the third column represents large cars with front wheel drive. Because each column is broken apart in very different places, this suggests the proportion of vehicles with front wheel drive differs with vehicle `type`. That is, `drivetrain` and `type` show some connection and are therefore associated.

In a similar way, a mosaic plot representing column proportions of Table 1.29(a) can be constructed as shown in Figure 1.32.

⊙ **Exercise 1.33** Describe how the mosaic plot shown in Figure 1.32 was constructed. Answer in the footnote[23].

1.4.4 The only pie chart you will see in this book

While pie charts are well known, they are not typically as useful as other charts in a data analysis. A **pie chart** is shown in Figure 1.33 on the next page alongside a bar plot. It is more difficult to compare group sizes in a pie chart than in a bar plot.

⊙ **Exercise 1.34** Using the pie chart, is it easy to tell which level, `midsize` or `small`, has a larger proportion in the sample? What about when using the bar plot?

[23]First, the cars were split up by `drivetrain` into two groups represented by the columns. Then the `type` variable splits each of these columns into the levels of `type`.

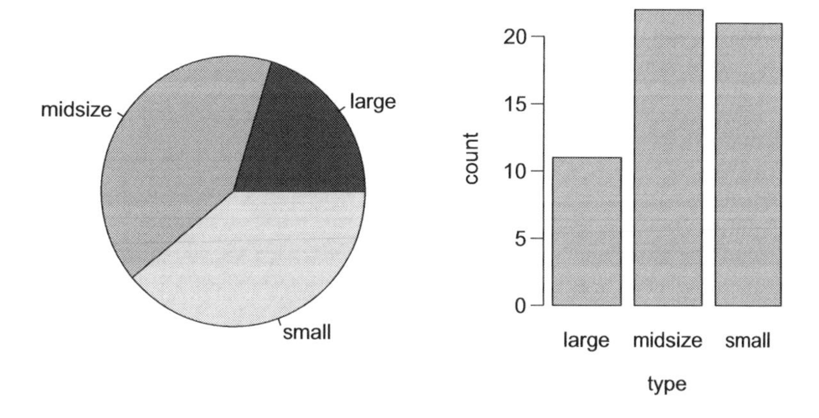

Figure 1.33: A pie chart and bar plot of `type` for the data set `cars`.

small		midsize		large
15900	11600	33900	28000	20800
9200	10300	37700	35200	23700
11300	11800	30000	34300	34700
12200	9000	15700	61900	18800
7400	11100	26300	14900	18400
10100	8400	40100	26100	29500
8400	10900	15900	21500	19300
12100	8600	15600	16300	20900
8000	9800	20200	18500	36100
10000	9100	13900	18200	20700
8300		47900	26700	24400

Table 1.34: The data from the `price` variable split up by `type`.

1.4.5 Comparing numerical data across groups

Some of the more interesting investigations can be considered by examining numerical data across groups. The methods required aren't really new. All that is required is to make a numerical plot for each group. Here two convenient methods are introduced: side-by-side box plots and hollow histograms.

From the data set `cars`, we will compare vehicle price according to vehicle type. There are three levels of `type` (`small`, `midsize`, and `large`), and the vehicle prices can be split into each of these groups, as shown in Table 1.34.

The **side-by-side box plot** is a traditional tool for comparing across groups. An example is shown in the left panel of Figure 1.35, where there are just three box plots – one for each `type` – placed into one plotting window and drawn on the same scale.

Hollow histograms are another useful plotting method for numerical data. These are just the outlines of histograms of each group put on the same plot, as shown in the right panel of Figure 1.35.

⊙ **Exercise 1.35** Use each plot in Figure 1.35 to compare the vehicle prices across groups. What do you notice about the approximate center of each group? What do you notice about the variability between groups? Is the shape relatively consistent between groups? How many *prominent* modes are there for each group?

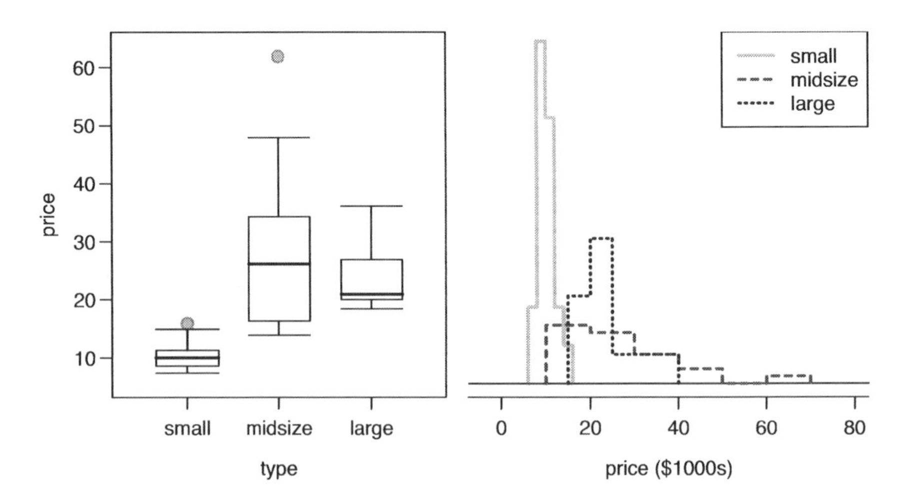

Figure 1.35: Side-by-side box plot (left panel) and hollow histograms (right panel) for `price` where the groups represent each level of `type`.

⊙ **Exercise 1.36** What components of each plot in Figure 1.35 do you find most useful?

1.5 Overview of data collection principles

The first step in conducting research is to identify topics or questions that are to be investigated. A clearly laid out research question is helpful in identifying what subjects or cases are to be studied and what variables are important. This information provides a foundation for *what* data will be helpful. It is also important that we consider *how* data are collected so that they are trustworthy and help achieve the research goals.

1.5.1 Populations and samples

Consider the following three research questions:

1. What is the average mercury content in swordfish in the Atlantic Ocean?
2. Over the last 5 years, what is the average time to degree for UCLA undergraduate students?
3. Does the drug sulphinpyrazone reduce the number of deaths in heart attack patients?

In each research question, some population of cases is considered. In the first question, all swordfish in the Atlantic ocean are relevant to answering the question. Each fish represents a case, and all of these fish represent the **population** of cases. Often times, it is too expensive to collect data for every case in a population. Instead a sample is taken. A **sample** represents a subset of the cases and is often a small fraction of the population. For instance, 60 swordfish (or some other number) in the population might be selected, and this sample data may be used to provide an estimate of the population average, i.e. an answer to the research question.

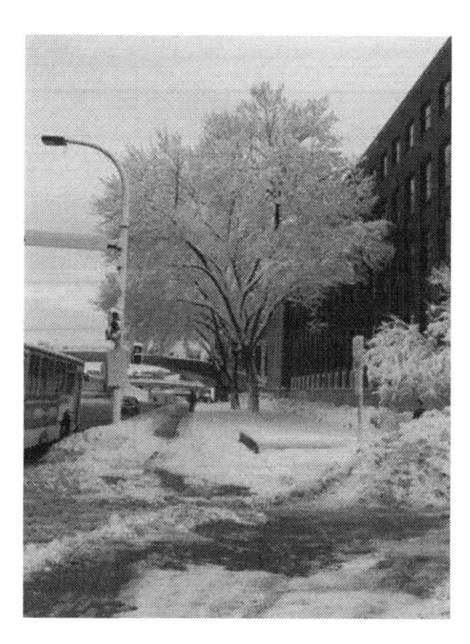

Figure 1.36: In February 2010, some media pundits cited one large snow storm as valid evidence against global warming. As comedian Jon Stewart pointed out, "It's one storm, in one region, of one country."

February 10th, 2010.

⊙ **Exercise 1.37** For the second and third questions above, identify what is an individual case and also the population under consideration. Answers in the footnote[24].

1.5.2 Anecdotal evidence

Consider the following possible responses to our three research questions:

1. A man on the news got mercury poisoning from eating swordfish, so the average mercury concentration in swordfish must be dangerously high.
2. I met two students who took more than 10 years to graduate from UCLA, so it must take longer to graduate at UCLA than at many other colleges.
3. My friend's dad had a heart attack and died after they gave him sulphinpyrazone. The drug must not work.

Each of the conclusions are based on some data. However, there are two problems. First, the data only represent one or two cases. Second and more importantly, it is unclear whether these cases are actually representative of the population. Data collected in this haphazard fashion are called **anecdotal evidence**.

> **Anecdotal evidence**
> Data collected in a haphazard fashion. Such evidence may be true and verifiable but often times represents extraordinary cases.

Anecdotal evidence typically is composed of unusual cases that we recall based on their striking characteristics. For instance, we are more likely to remember the two folks we met who took 10 years to graduate than the six others who graduated in four years.

[24](2) First, notice that this question is only relevant to students who complete their degree; the average cannot be computed using a student who never finished her degree. Thus, only UCLA undergraduate students who have graduated in the last five years represent cases in the population under consideration. (3) A heart attack patient represents a case. The population represents all heart attack patients.

Instead of looking at the most unusual cases, we should examine a sample of many cases that represent the population.

1.5.3 Sampling from a population

The `cars` data set represents a sample of cars from 1993. All cars from 1993 represent the population, and the cars in the sample were *randomly* selected from the population. Random selection in this context is equivalent to how raffles are run. The name of each car from the population was written on a raffle ticket, and 54 tickets were drawn.

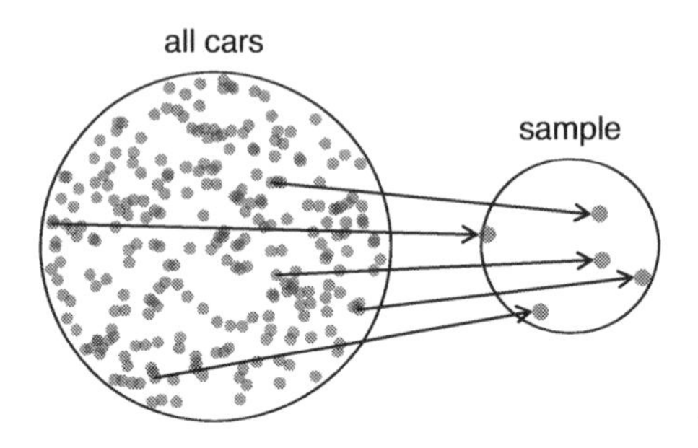

Figure 1.37: Cars from the population are randomly selected to be included in the sample.

Why pick a sample randomly? Why not just pick a sample by hand? Consider the following scenario.

⊙ **Exercise 1.38** Suppose a muscle car enthusiast is asked to select several cars for a study. What kind of cars do you think she might collect? Do you think her sample would be representative of all cars?

If someone was permitted to pick and choose exactly which cars were included in the sample, it is entirely possible that the sample could be skewed to that person's interests, which may be entirely unintentional. This introduces **bias** into a sample. Sampling randomly helps resolve this problem. The most basic random sample is called a **simple random sample**, and is the equivalent of using a raffle to select cases. This means that each case in the population has an equal chance of being included and there is no implied

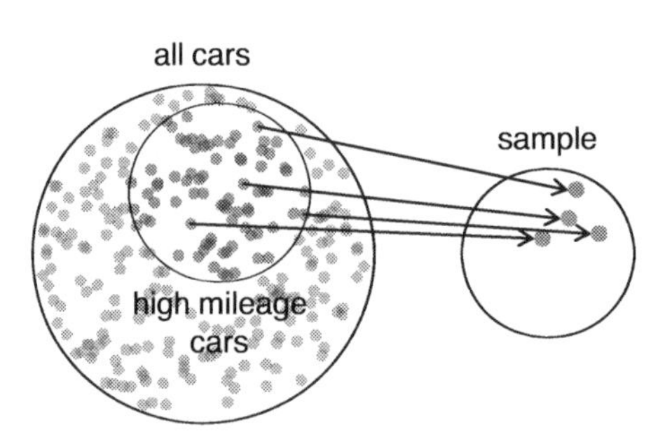

Figure 1.38: Instead of sampling from all cars from 1993, an environmentalist might inadvertently pick cars with high mileage disproportionally often.

connection between the cases in the sample. The act of taking a simple random sample helps eliminate bias, however, bias can still crop up in other ways.

Even when people are seemingly picked at random (for surveys, etc.), caution must be exercised if the **non-response** is high. For instance, if only 15% of the people randomly sampled for a survey actually respond, then it is unclear whether the results are **representative** of the entire population. This **non-response bias** can skew results.

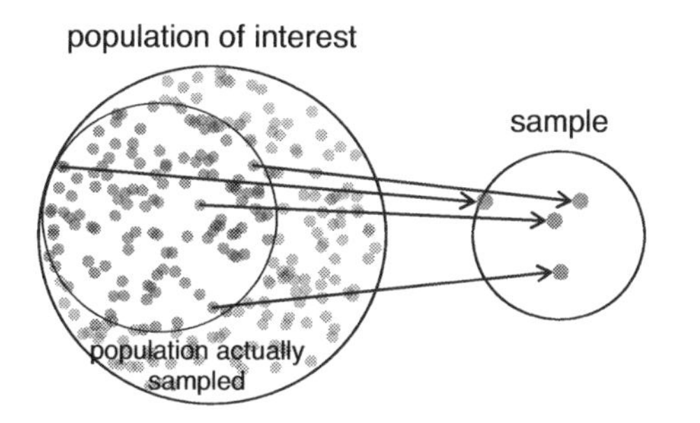

Figure 1.39: Surveys may result in only reaching a certain group within the population, and it is not obvious how to fix this problem.

Another common downfall is a **convenience sample**, where individuals who are easily accessible are more likely to be included in the sample. For instance, if a political survey is done by stopping people walking in the Bronx, this probably will not fairly represent all of New York City. It is often difficult to discern what sub-population a convenience sample represents.

⊙ **Exercise 1.39** We can easily access ratings for products, sellers, and companies through websites. These ratings are based only on those people who go out of their way to provide a rating. If a seller has a rating of 95% on Amazon, do you think this number might be artificially low or high? Why?

1.5.4 Explanatory and response variables

Consider the second question from page 5 for the `possum` data set:

(2) Will males or females, on the average, be longer?

This question might stem from the belief that a possum's gender might in some way affect its size. If we suspect a possum's sex might affect its total length, then `sex` is the **explanatory** variable and `totalL` is the **response** variable in the relationship[25]. If there are many variables, it may be possible to label a number of them as explanatory and the others as response variables.

TIP: Explanatory and response variables
To identify the explanatory variable in a pair of variables, identify which of the two is suspected of affecting the other.

explanatory variable —might affect→ response variable

[25]Sometimes the explanatory variable is called the **independent** variable and the response variable is called the **dependent** variable. However, this becomes confusing since a *pair* of variables might be independent or dependent, so we avoid this language.

Caution: association does not imply causation
Labeling variables as *explanatory* and *response* does not guarantee the relationship between the two is actually causal, even if there is an association identified between the two variables. We use these labels only to keep track of which variable we suspect affects the other.

In some cases, there is no explanatory or response variable. Consider the first question from page 5:

(1) If a possum has a shorter-than-average head, do you think its skull width will be smaller or larger than the average skull width?

This question does not have an explanatory variable since it is unclear whether `headL` would affect `skullW` or vice-versa, i.e. the direction is ambiguous.

1.5.5 Introducing observational studies and experiments

There are two primary types of data collection: observational studies and experiments.

Researchers perform an **observational study** when they collect data in a way that does not directly interfere with how the data arise. For instance, researchers may collect information using surveys, reviewing medical or company records, or follow a **cohort** of many similar individuals to consider why certain diseases might develop. In each of these cases, the researchers try not to interfere with the natural order of how the data arise. In general, observational studies can provide evidence of a naturally occurring association between variables, but they cannot show a causal connection.

When researchers want to establish a causal connection, they conduct an **experiment**. Usually there will be both an explanatory and a response variable. For instance, we may suspect administering a drug will reduce mortality in heart attack patients over the following year. To check if there really is a causal connection between the explanatory variable and the response, researchers will collect a sample of individuals and split the cases into groups. The cases in each group are *assigned* a treatment. When the groups are created using a randomization technique, the experiment is called a **randomized experiment**. For example, each heart attack patient in the drug trial could be randomly assigned (e.g. by flipping a coin) into one of two groups: the first group receives a placebo (fake treatment) and the second group receives the drug.

TIP: association $\neq$ causation
In general, association does not imply causation, and causation can only be inferred from a randomized experiment.

1.6 Observational studies and sampling strategies

1.6.1 Observational studies

The `possum` data set was from an observational study. While researchers captured the possums, their measurements and other variables were naturally occurring. Generally, data in observational studies are collected only by monitoring what occurs, while experiments require the primary explanatory variable in a study be assigned for each subject by the researchers.

Inferring causal conclusions from experiments is often reasonable. However, making the same causal conclusions from observational data can be treacherous and is not recommended. Thus, we can generally only infer associations from observational data.

⊙ **Exercise 1.40** Suppose an observational study tracked sunscreen use and skin cancer, and it was found that the more sunscreen someone used, the more likely the person was to have skin cancer. Does this mean sunscreen *causes* skin cancer?

Previous research tells us that using sunscreen actually reduces skin cancer risk, so maybe there is another variable that can explain this hypothetical association between sunscreen usage and skin cancer. One important piece of information absent is sun exposure. If someone is out in the sun all day, she is more likely to use sunscreen *and* more likely to get skin cancer.

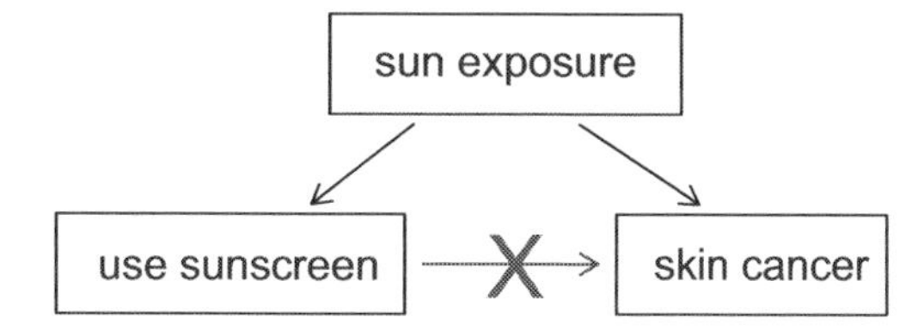

It just so happens that if someone is exposed to the sun they also usually use sunscreen. Exposure to the sun is unaccounted for in the investigation, giving the incorrect impression that sunscreen causes skin cancer.

Sun exposure is what is called a **lurking variable**[26], which is a variable that is correlated with both the explanatory and response variables. While one method to justify making causal conclusions from observational studies is to exhaust the search for lurking variables, there is no guarantee that all lurking variables can be examined or measured.

In the same way, the `possum` data set is an observational study with possible lurking variables of its own, and its data cannot easily be used to make causal conclusions.

⊙ **Exercise 1.41** There appears to be a real difference in the proportion of possums that are male based on location. However, it is unreasonable to conclude that this is a causal relationship because the data are observational. Suggest at least one lurking variable that might be the true cause for the differences in `sex`. One possibility is listed in the footnote[27].

Observational studies come in two forms: prospective and retrospective studies. A **prospective study** identifies individuals and collects information as events unfold. For instance, medical researchers may identify and follow a group of similar individuals over many years to assess the possible influences of behavior on cancer risk. **Retrospective studies** collect data after events have taken place, e.g. researchers may review past events in medical records. The `possum` data are an example of a retrospective study.

1.6.2 Three sampling methods (special topic)

Almost all statistical methods are based on the notion of implied randomness. If observational data are not collected in a random framework from a population, these statistical

[26]Also called a **confounding variable**, **confounding factor**, or a **confounder**.

[27]Some genes can affect one gender more than the other. If the `other` population has a gene that affects males more positively than females and this gene is less common in the `Vic` population, this might explain the difference in gender ratio for each level of `pop`.

methods – the estimates and computed errors – are not reliable. Here we consider three random sampling techniques: simple, stratified, and cluster sampling. We introduce these techniques in the context of Major League Baseball (MLB) salary data, where each player is a member of one of the league's 30 teams, and we consider taking a sample of 120 players.

Simple random sampling is probably the most intuitive form of random sampling. To take a simple random sample of the baseball players and their salaries, we could write each player's name on a ping pong ball, drop the ping pong balls into a bucket, shake the bucket around until we are sure it's all mixed up, then draw out balls from the bucket until we have the sample of 120 players. In general, a sample is referred to as "simple random" if each case in the population has an equal chance of being included in the final sample *and* knowing that a case is included in a sample does not provide useful information about which other cases/outcomes are included.

Stratified sampling is a divide-and-conquer sampling strategy. The population is divided up into groups called **strata** where the strata are chosen so that similar cases are grouped, then a second sampling method (such as simple random sampling) is employed within each stratum. In the baseball salary example, the teams could represent the strata; some teams have much more money (we're looking at you, Yankees). Then we might randomly sample four players from each team for a total of 120 players.

Stratified sampling is especially useful when the cases in each stratum are very similar in the measured outcome. For instance, the baseball teams might be useful strata since some teams have (much) more money and therefore tend to pay players more. The downside is that analyzing data from a stratified sample is more complex than analyzing data from a simple random sample, and the analysis methods introduced in this book would need to be extended.

● **Example 1.42** Why would it be good for cases within each stratum to be very similar?

We might get a more stable statistical estimate for the subpopulation in a stratum if the cases are very similar. Then when we summarize the entire population, these individual stable estimates for each subpopulation will help us build a reliable estimate for the full population.

A **cluster sample** is much like a two-stage simple random sample. We break up the population into many groups, called **clusters**. Then we sample a fixed number of clusters and collect a simple random sample in each cluster. This technique is very similar to stratified sampling in its process, except that there is no requirement to sample from every cluster while stratified sampling requires observations from every stratum.

Sometimes cluster sampling can be a more economical random sampling technique than alternatives. Also, unlike stratified sampling, cluster sampling is most helpful when there is a lot of case-to-case variability within a cluster but the clusters themselves don't look very different from one another (e.g. if neighborhoods represented clusters, then it is best if each neighborhood is very diverse). A downside of cluster sampling is that more advanced analysis techniques are typically required, though the methods in this book can be extended to handle such data.

● **Example 1.43** Describe a scenario where cluster sampling would be much easier than simple random or stratified sampling.

Suppose we are interested in estimating the malaria rate in a densely tropical portion of rural Indonesia. We believe there are 30 villages in that part of the Indonesian

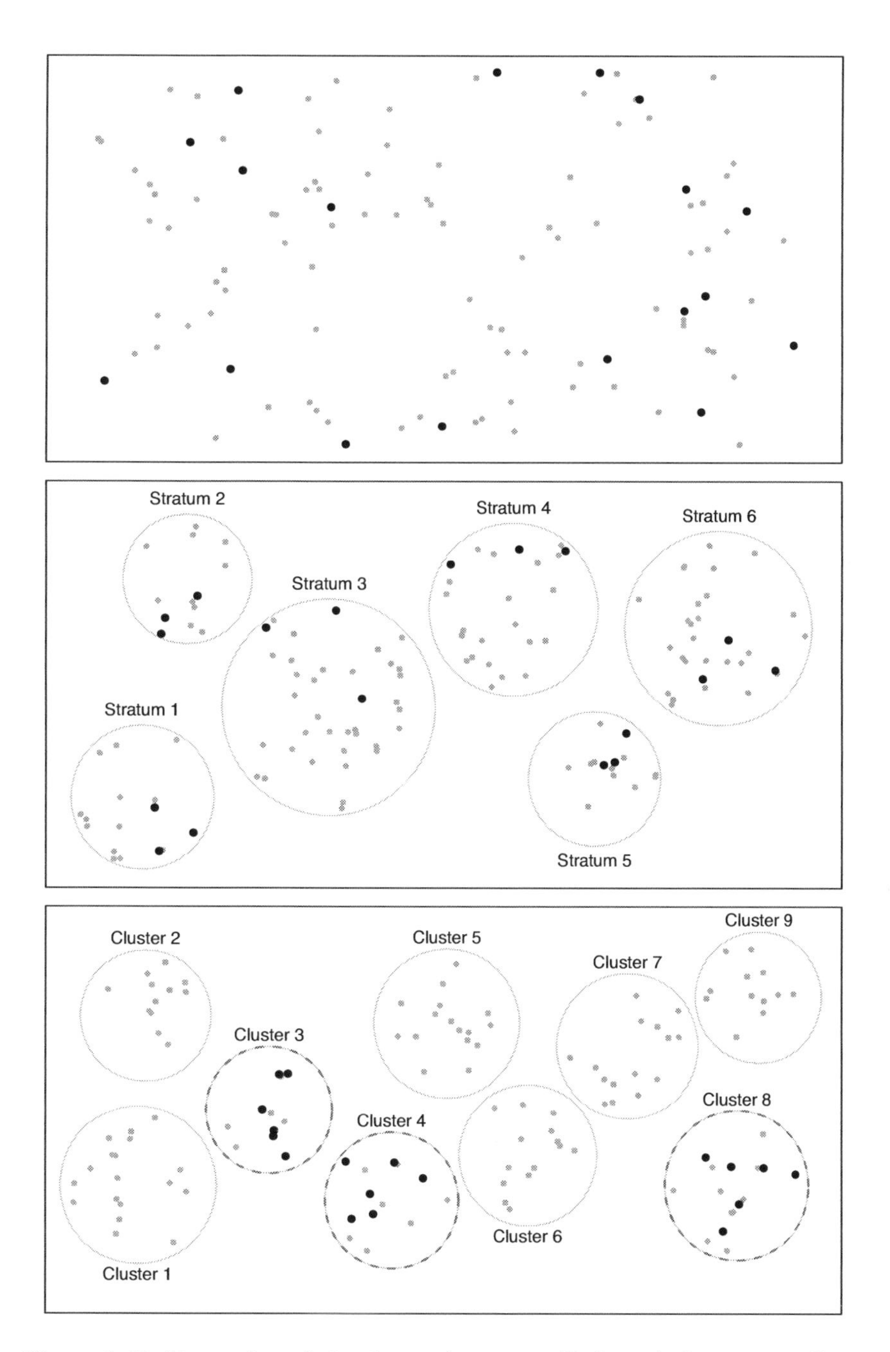

Figure 1.40: Examples of simple random, stratified, and cluster sampling. In the top panel, simple random sampling was used to randomly select the 18 cases. In the middle panel, stratified sampling was used: cases were grouped into strata, and then simple random sampling was employed within each stratum. In the bottom panel, cluster sampling was used, where data were binned into nine clusters, three of the clusters were randomly selected, and six cases were randomly sampled in each of these clusters.

jungle, each more or less similar to the next. Our goal is to test 150 individuals for malaria. In this case, a simple random sample would likely draw individuals from all 30 villages, which could make data collection extremely expensive. Stratified sampling would be a challenge since it is unclear how we would build stratum of similar individuals. However, cluster sampling seems like a very good idea. First, we might randomly select half the villages, then randomly select 10 people from each. This would probably reduce our data collection costs substantially in comparison to a simple random sample and would still give us reliable information.

⊙ **Exercise 1.44** Describe a second situation where cluster sampling would be more convenient than simple random or stratified sampling.

1.7 Experiments

Studies where the researchers assign treatments to cases are called **experiments**. When cases are randomly assigned to the treatment groups, it is called a **randomized experiment**. Randomized experiments are fundamentally important when trying to show a causal connection between two variables.

1.7.1 Principles of experimental design

Randomized experiments are generally built on four principles.

Controlling. Researchers assign treatments to cases, and they do their best to **control** any other differences in the groups. For instance, if a drug treatment in the form of a pill may be affected by how much water a patient drinks with the drug, the doctor may ask all patients to drink a 12 ounce glass of water with the pill to reduce one source of unnecessary variability between the cases.

Randomization. Researchers randomize patients into the groups to account for variables that cannot be controlled. In a clinical trial, some patients may be more susceptible to a disease than others due to their genetic make-up. Randomizing patients into the two treatment groups is one way to help even out the genetic differences in each group, and it also prevents accidental bias from entering the study.

Replication. The more cases researchers observe, the more accurately they can estimate the effect of the explanatory variable on the response. In a single study, we **replicate** by collecting a sufficiently large sample. Additionally, a group of scientists may replicate an entire study to verify an earlier finding.

Blocking. Researchers sometimes know or suspect variables other than the treatment that might influence the response. Under this circumstance, they may first group individuals based on this variable into **blocks** and then randomize cases within each block to the treatment groups. This strategy is often referred to as **blocking**. For instance, if we were looking at the effect of a drug on heart attacks, we might first split patients in the study into low-risk and high-risk blocks, then randomly assigning half the patients from each block to the control group and the other half to the treatment group, as shown in Figure 1.41. This strategy ensures each treatment group has an equal number of low-risk and high-risk patients.

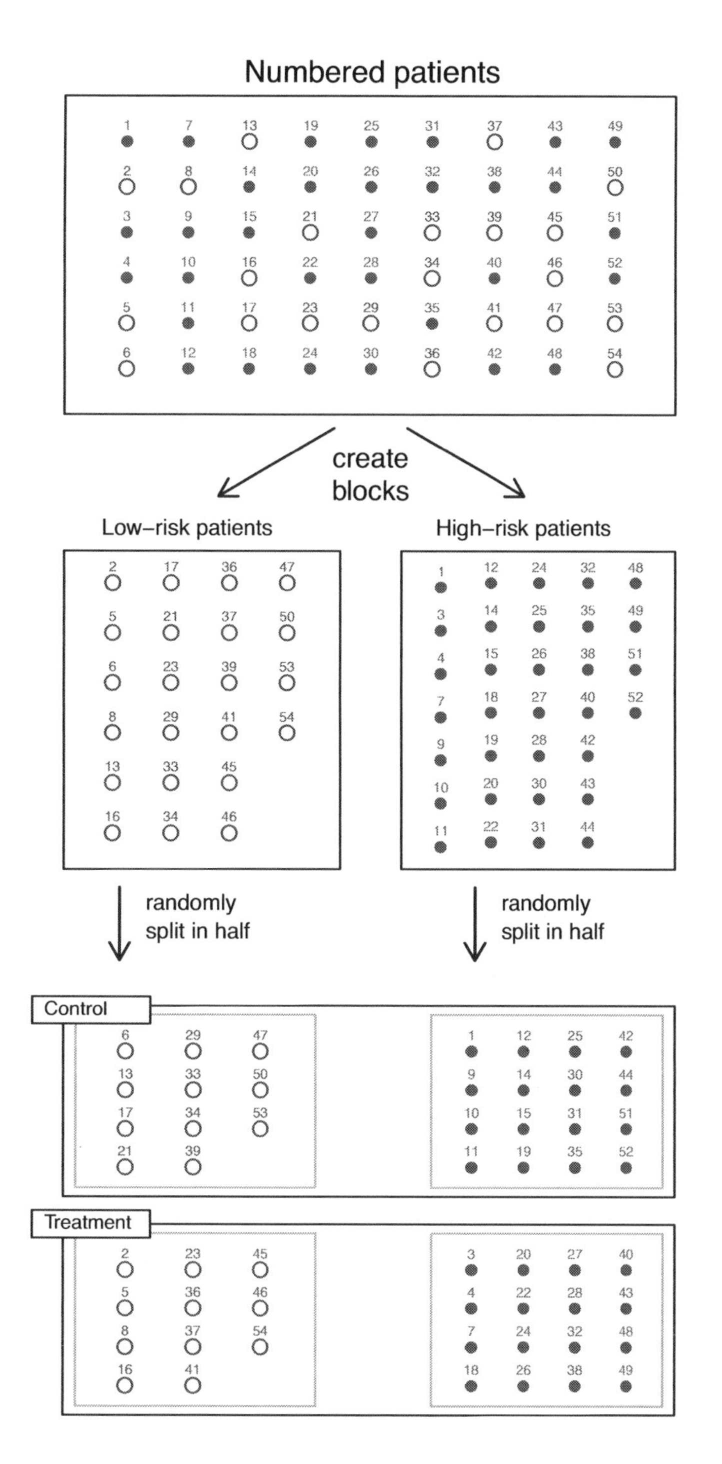

Figure 1.41: Blocking using a variable depicting patient risk. Patients are first divided into low-risk and high-risk blocks, then each block is evenly divided into the treatment groups using randomization. This strategy ensures an equal number of patients in each treatment group from both the low-risk and high-risk categories.

It is important to incorporate the first three experimental design principles into any study, and this book describes applicable methods for analyzing data from such experiments. Blocking is a slightly more advanced technique, and statistical methods in this book may be extended to analyze data collected using blocking.

1.7.2 Reducing bias in human experiments

Randomized experiments are the gold standard for data collection, but they do not ensure an unbiased perspective into the cause and effect relationships in all cases. Human studies are perfect examples where bias can unintentionally arise. Here we reconsider the sulphinpyrazone study described in Section 1.1.

Researchers wanted to examine whether a drug called sulphinpyrazone would reduce the number of deaths after heart attacks. They designed a randomized experiment because they wanted to draw causal conclusions about the drug's effect. Study volunteers[28] were randomly placed into two study groups. One group, the **treatment group**, received the drug. The other group, called the **control group**, did not receive any drug treatment.

Put yourself in the place of a person in the study. If you are in the treatment group, you are given a fancy new drug that you anticipate will help you. On the other hand, a person in the other group doesn't receive the drug and sits idly, hoping her participation doesn't increase her risk of death. These perspectives suggest there are actually two effects: the one of interest is the effectiveness of the drug and the second is an emotional effect that is difficult to quantify.

Researchers aren't interested in this emotional effect, which might bias the study. To circumvent this problem, researchers do not want patients to know which group they are in. When researchers keep the patients uninformed about their treatment, the study is said to be **blind**. But there is one problem: if a patient doesn't receive a treatment, she will know she is in the control group. The solution to this problem is to give fake treatments to patients in the control group. A fake treatment is called a **placebo**, and an effective placebo is the key to making a study truly blind. A classic example of a placebo is a sugar pill that is made to look like the actual treatment pill. Often times, a placebo results in a slight but real improvement in patients. This often positive effect has been dubbed the **placebo effect**.

The patients are not the only ones who should be blinded: doctors and researchers can accidentally bias a study. When a doctor knows a patient has been given the real treatment, she might inadvertently give that patient more attention or care than a patient that she knows is on the placebo. To guard against this bias (which again has been found to have a measurable effect in some instances), most modern studies employ a **double-blind** setup where doctors or researchers who interact with patients are, just like the patients, unaware of who is or is not receiving the treatment[29].

1.8 Case study: efficacy of sulphinpyrazone (special topic)

● **Example 1.45** Suppose your professor splits the students in class into two groups: students on the left and students on the right. If $\hat{p}_L$ and $\hat{p}_R$ represent the proportion

[28] Human subjects are more often called **patients**, **volunteers**, or **study participants**.

[29] There are always some researchers involved in the study who do know which patients are receiving which treatment. However, they do not have interactions with the patients and do not tell the blinded doctors who is receiving which treatment.

of students who own an Apple product on the left and right, respectively, would you be surprised if $\hat{p}_L$ did not exactly equal $\hat{p}_R$?

While the proportions would probably be close to each other, it would be unusual for them to be exactly the same. We would probably observe a small difference due to chance.

Exercise 1.46 If we don't think the side of the room a person sits on in class is related to whether the person owns an Apple product, what assumption are we making about the relationship between the `sideOfRoom` and `ownsAppleProduct` variables? Answer in the footnote[30].

1.8.1 Variability within data

The study examining the effect of sulphinpyrazone introduced in Section 1.1 was double-blinded, and the results are summarized in Table 1.42. The variables have been called `group` and `outcome`. Do these results mean the drug was effective at reducing deaths? In the observed groups, a smaller proportion of individuals died in the treatment group than the control group (0.056 versus 0.081), however, it is unclear whether that difference represents *convincing evidence* that the drug is effective.

		outcome		
		lived	died	Total
group	treatment	692	41	733
	control	682	60	742
	Total	1374	101	1475

Table 1.42: Summary results for the sulphinpyrazone study.

Example 1.47 Statisticians are sometimes called upon to evaluate the strength of evidence. When looking at the death rates in the study, what comes to mind as we try to determine whether the data show convincing evidence of a real difference?

The observed death rates (0.056 versus 0.081) suggest the drug may be effective since the treatment group has a lower proportion of deaths. However, the sample proportions are very close. Generally there is a little bit of fluctuation in sample data, and we wouldn't expect the sample proportions to be *exactly* equal even if the truth was that they were equal across the two treatments. As we look at the data, we want to ask, is the difference between the death rates so large that it probably isn't due to chance?

Example 1.47 is a reminder that the sample will not perfectly reflect the population. It is possible to see a small difference by chance. Small differences in large samples can be important and meaningful but it is unclear when we should say that a difference is so large it was probably not due to chance. Table 1.42 shows there were 19 fewer deaths in the treatment group than in the control group for the sulphinpyrazone study, a difference in death rates of 2.5% $\left(\frac{60}{742} - \frac{41}{733} = 0.025\right)$. Might this difference just be due to chance? Or is this convincing evidence that sulphinpyrazone works? We label these two competing claims, H_0 and H_A:

[30] We would be assuming the variables `sideOfRoom` and `ownsAppleProduct` are independent.

H_0: **Independence model.** The variables `group` and `outcome` are independent. They have no relationship, and the difference in death rates, 2.5%, was due to chance.

H_A: **Alternative model.** The `group` and `outcome` variables are *not* independent. The difference in death rates of 2.5% was not due to chance and the treatment did reduce the death rate.

Consider what it would mean to the study if the independence model, which says that the variables `group` and `outcome` are unrelated, is true. Each person was either going to live or die, and the drug had no effect on the outcome. The researchers were just randomly splitting up these individuals into two groups, very much like we split the class in half in Example 1.45. The researchers observed a difference of 2.5% by chance.

Consider the alternative: the treatment affected the outcome. We would expect to see some difference in the groups, with a lower percentage of deaths in the group of patients who received the drug.

If the data conflict so much with H_0 that the independence model cannot be deemed reasonable, we will reject it in favor the alternative model, H_A. In other words, we will not reject the position that H_0 is true unless the evidence from the study in favor of H_A is extremely convincing.

1.8.2 Simulating the study

Suppose H_0 is true. Under this model, the 2.5% difference would have been due to chance, and the group assignments would not have impacted the patient outcomes. We will use this thought experiment to simulate differences that are due to chance using a **randomization technique**.

We are going to recreate (simulate) the study under the scenario that patient outcome has nothing to do with the treatment. To do this, we are going to randomly reassign the patients into a fake treatment or fake control group. Then, since we know this fake group assignment has nothing to do with the patients' actual outcomes – the fake group assignment and the outcome are independent – any observed difference in death rates between the fake groups must be due to chance.

We run this **simulation** by taking 733 `treatmentFake` and 742 `controlFake` labels and randomly assign them to the patients. These label counts correspond to the number of `treatment` and `control` assignments in the actual study. We use a computer program to randomly assign these labels to the patients, and we organize these results into Table 1.43.

		outcome		
		lived	died	Total
groupFake	treatmentFake	686	47	733
	controlFake	688	54	742

Table 1.43: Simulation results, where any difference in death rates between `treatmentFake` and `controlFake` is purely due to chance.

⊙ **Exercise 1.48** What is the difference in death rates between the two fake groups in Table 1.43? How does this compare to the observed 2.5% in the real groups? Answer in the footnote[31].

[31] $54/742 - 47/733 = 0.0087$ or about 0.9%. This difference due to chance is smaller.

1.8.3 Checking for independence

We computed one possible difference under the independence model in Exercise 1.48, which represents one difference due to chance. We could repeat the simulation to get another difference from chance: -0.005. And another: -0.010. And another: 0.003. And so on until we repeat the simulation enough times that we have a good idea of what represents the *distribution of differences from chance alone*. Figure 1.44 shows a histogram of the differences found from 100 simulations.

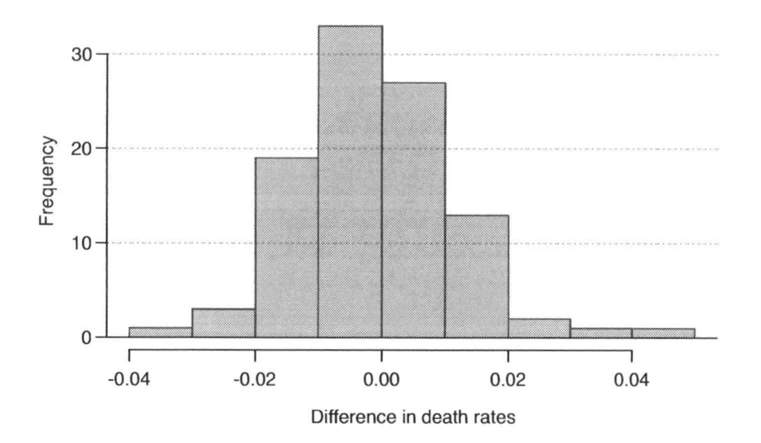

Figure 1.44: A histogram of differences from 100 simulations produced under the independence model, H_0, where `groupFake` and `outcome` are independent. Four of the one-hundred simulations had a difference of at least 2.5%.

● **Example 1.49** How often would you observe a difference of 2.5% (0.025) according to Figure 1.44? Often, sometimes, rarely, or never?

It appears that a difference of at least 2.5% due to chance alone would only happen about 4% of the time according to Figure 1.44. We might describe that as being a rare event.

The difference of 2.5% is a rare event, and this suggests two possible interpretations of the results of the study:

H_0 **Independence model.** The drug doesn't work, and we observed a difference that would only happen rarely.

H_A **Alternative model.** The drug does work, and what we observed was that the drug was actually working, which explains the large difference of 2.5%.

We should be skeptical that what was observed just happened to be a rare event, and we usually reject the null hypothesis in a situation like this and conclude the drug works.

One field of statistics, statistical inference, is built on evaluating whether such differences are due to chance. In statistical inference, statisticians evaluate which model is most reasonable given the data. Errors do occur – we might choose the wrong model. While we do not always choose correctly, statistical inference gives us tools to control and evaluate how often these errors occur. In Chapter 4, we give a formal introduction to the problem of model selection. However, we spend the next two chapters building a foundation of probability and theory necessary to make that discussion rigorous.

1.9 Exercises

1.9.1 Case study

1.1 A migraine is a common type of headache, which patients sometimes wish to treat with acupuncture. To determine whether acupuncture relieves migraine pain, researchers conducted a randomized controlled study where 100 adults suffering from migraine pain were randomly assigned to treatment and control groups (50 in each). Patients in the treatment group received twice-weekly treatment for 12 weeks with an acupuncture program that was specifically designed to treat migraines. Patients in the control group received placebo acupuncture (needle insertion at nonacupoint locations). Patients were asked at the end of the study whether or not they experienced a reduction in the severity of their migraines. The results are summarized in the contingency table below.

(a) What percent of patients in the control group experienced a reduction in the severity of their migraines? What percent in the treatment group?

(b) At first glance, does acupuncture appear to be effective treatment for migraines?

(c) Do the data provide convincing evidence that there is a real pain reduction for those patients in the treatment group? Or do you think that the observed difference might just be due to chance?

		Improvement		
		No	Yes	Total
Group	Control	22	28	50
	Treatment	15	35	50
	Total	37	63	100

1.2 A survey was conducted to study the smoking habits of UK residents. The table below shows how many of the males and females are smokers. [1].

		Gender	
		Female	Male
Smoke	No	731	539
	Yes	234	187

(a) What percent of males smoke? What percent of females?

(b) At first glance, do gender and smoking appear to be related?

(c) Do the data provide convincing evidence that there is a real relationship between gender and smoking? Or do you think that the observed difference might just be due to chance?

1.9.2 Data basics

In Exercises 1.3 to 1.6 identify

(a) the cases studied,

(b) the variables studied and their types, and

(c) the main research question of the study.

1.3 Researchers collected data to examine the relationship between pollutants and preterm births in Southern California. Ambient levels of of carbon monoxide (CO), nitrogen dioxide, ozone, and coarse particulate matter (PM_{10}) were measured during the study by air quality monitoring stations. Length of gestation data were collected on 143,196 births between the years 1989 and 1993. The analysis suggested that increased ambient PM_{10} and, to a lesser degree, CO concentrations may contribute to the occurrence of preterm births. [2]

1.4 The Buteyko method is a shallow breathing technique developed by Konstantin Buteyko, a Russian doctor, in 1952. Anecdotal evidence suggests that the Buteyko method can reduce asthma symptoms and improve quality of life. In a study aimed to determine the effectiveness of this method, 600 adult patients aged 18-69, who had been diagnosed with asthma and were currently being treated, were divided into two groups. One group practiced the Buteyko method and the other did not. Patients were scored on quality of life, activity, asthma symptoms, and medication reduction on a scale from 0 to 10. On average, the participants in the Buteyko group experienced a significant reduction in asthma symptoms, and a improvement in quality of life. [3]

1.5 While obesity is measured based on body fat percentage (more than 35% body fat for women and more than 25% for men), measuring body fat percentage accurately is difficult. Therefore body mass index (BMI), ratio $\frac{weight}{height^2}$, is often used as an alternative indicator for obesity. A common criticism of BMI is that it assumes the same relative body fat percentage regardless of age, sex, or ethnicity. In order to determine how useful BMI is for predicting body fat percentage across age, sex and ethnic groups, researchers studied 202 black and 504 white adults who resided in or near New York City, were ages 20-94 years, and had BMIs of 18-35 kg/m^2. Participants reported their age, sex and ethnicity and were measured for weight, height. Body fat percentage was measured by submerging the participants in water. [4]

1.6 A social scientist interested in the relationship between certain characteristics of voters and support for Proposition 23 conducted a survey before the 2010 midterm election on 1000 registered California voters. If enacted by voters, this proposition would freeze the provisions of California's clean air legislation. Along with whether or not they support Prop 23, respondents in this survey were asked their age, race, gender, highest degree earned, occupation, income, party affiliation, how many cars they own, and whether or not they own an alternative fuel vehicle (hybrid, biodiesel, etc.).

1.7 Exercise 1.2 introduced a study about the smoking habits of UK residents. Below is a data matrix displaying a portion of the data collected in this survey. Note that "£" stands for British Pounds Sterling and "cig" stands for cigarettes.

	gender	age	marital	grossIncome	smoke	amtWeekends	amtWeekdays
1	Female	42	Single	Under £2,600	Yes	12 cig/day	12 cig/day
2	Male	44	Single	£10,400 to £15,600	No	N/A	N/A
3	Male	53	Married	Above £36,400	Yes	6 cig/day	6 cig/day
⋮	⋮	⋮	⋮	⋮	⋮	⋮	⋮
1691	Male	40	Single	£2,600 to £5,200	Yes	8 cig/day	8 cig/day

(a) What does each row of the data matrix represent?

(b) How many participants were included in the survey?

(c) Identify the variables in the data set and give a short description of what each variable represents.

1.8 Sir Ronald Aylmer Fisher was an English statistician, evolutionary biologist and geneticist who worked on a data set that contained sepal length and width, and petal length and width from three species of iris flowers (setosa, versicolor and virginica). There were 50 flowers from each species in the data set. [5]

(a) How many cases were included in the data?

(b) How many numerical variables are included in the data? Indicate what they are, and if they are continuous or discrete.

(c) How many categorical variables are included in the data, and what are they? How many levels does each have, and what are they?

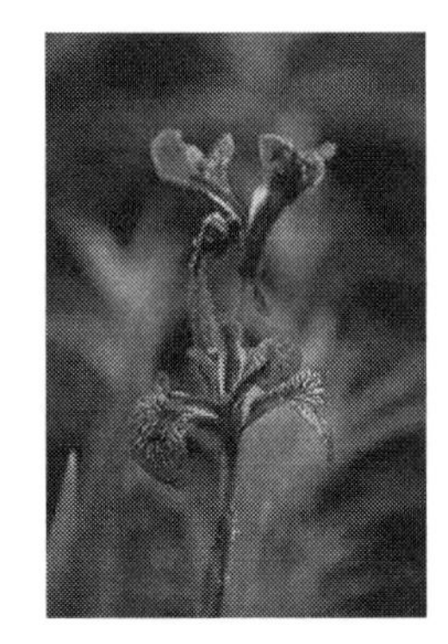

Photo by rtclauss on Flickr.

http://www.flickr.com/photos/rtclauss/3834965043/

1.9 Exercise 1.2 introduced a study about the smoking habits of UK residents, and a summary table for this study is shown in Exercise 1.7. Indicate if the variables in the study are numerical or categorical. If numerical, identify as continuous or discrete.

1.9.3 Examining numerical data

1.10 Data were collected on life spans (in years) and gestation lengths (in days) for 62 mammals. A scatterplot of life span vs. length of gestation is shown below. [6]

(a) What type of an association is apparent between life span and length of gestation?

(b) What type of an association would you expect to see if the axes of the plot were reversed, i.e. if we plotted length of gestation vs. life span?

(c) Are life span and length of gestation independent? Explain your reasoning.

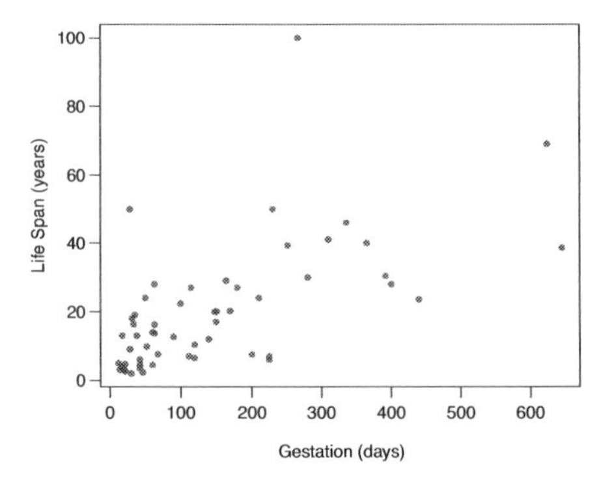

1.11 Office productivity is relatively low when the employees feel no stress about their work or job security. However high levels of stress can also lead to reduced employee productivity. Sketch a plot to represent the relationship between stress and productivity and explain your reasoning.

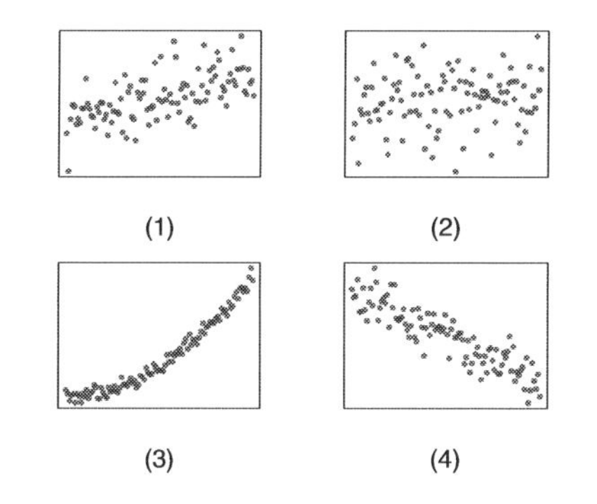

1.12 Indicate which of the plots to the right show

(a) linear association

(b) non-linear association

(c) positive association

(d) negative association

(e) no association

Each part may refer to more than one plot.

1.13 Identify which value represents the sample mean and which value represents the claimed population mean.

(a) A recent article in a college newspaper stated that college students get an average of 5.5 hrs of sleep each night. A student who was skeptical about this value decided to conduct a survey by randomly sampling 25 students. On average, the sampled students slept 6.25 hours per night.

(b) American households spent an average of about $52 in 2007 on Halloween merchandise such as costumes, decorations and candy. To see if this number has changed, researchers conducted a new survey in 2008 on 1,500 households and found that in the average amount a household spent on Halloween was $58.

1.14 A random sample from the smokers in the data set introduced in Exercise 1.2 discussed in Exercise 1.7 is provided below. Make a dot plot for the distribution of each of the variables listed. Then find the mean for each variable and mark it on the corresponding dot plot. If you cannot calculate the mean for a variable, indicate why.

gender	age	maritalStatus	grossIncome	smoke	amtWeekends	amtWeekdays
Female	51	Married	£2,600 to £5,200	Yes	20 cig/day	20 cig/day
Male	24	Single	£10,400 to £15,600	Yes	20 cig/day	15 cig/day
Female	33	Married	£10,400 to £15,600	Yes	20 cig/day	10 cig/day
Female	17	Single	£5,200 to £10,400	Yes	20 cig/day	15 cig/day
Female	76	Widowed	£5,200 to £10,400	Yes	20 cig/day	20 cig/day

1.15 In a class of 25 students, 24 of them take an exam in class and 1 student takes a make-up exam the next day. The professor grades the first batch of 24 exams and finds that the average score is 74 with a standard deviation of 8.9. The student who takes the make-up the next day scores an 64 on the exam.

(a) Will the average score increase or decrease?

(b) What will the revised average score be?

(c) Will the standard deviation of the data increase or decrease?

1.16 Workers at a particular mining site receive an average of 35 days paid vacation, which is lower than the national average. The manager of this plant is under pressure from a union to increase the amount of paid time off. However, he does not want to give more days off to the workers because that would be costly. Instead he decides he should fire 10 employees in such a way as to raise the average number of days off that are reported by his employees. In order to achieve this goal, should he fire employees who have the most number of days off, least number of days off, or those who have about the average number of days off?

1.17 Exercise 1.2 introduces a data set on the smoking habits of UK residents. Below are histograms displaying the distributions of the number of cigarettes smoked on weekdays and weekends. Describe the two distributions and compare them. *Hint: There are 421 smokers in this data set.*

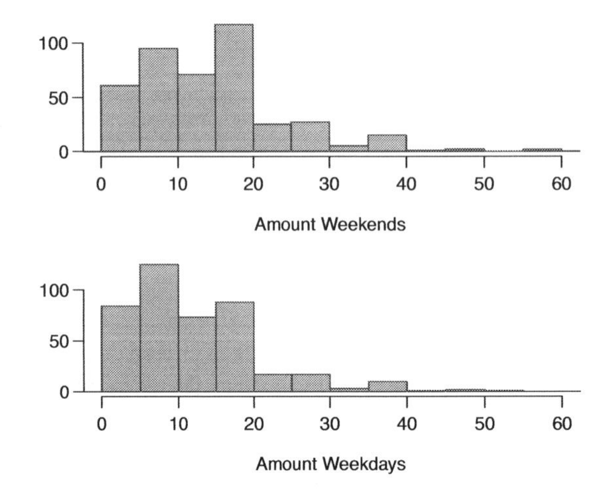

1.18 Below are the final scores of 20 introductory statistics students.

79, 83, 57, 82, 94, 83, 72, 74, 73, 71,
66, 89, 78, 81, 78, 81, 88, 69, 77, 79

Draw a histogram of these data and describe the distribution.

1.19 Find the standard deviation of the amount of cigarettes smoked on weekdays and on weekends by the 5 randomly sampled UK residents given in Exercise 1.14. Is the variability higher on weekends or on weekdays?

1.20 A factory quality control manager decides to investigate the percentage of defective items produced each day. Within a given work week (Monday through Friday) the percentage of defective items produced was 2%, 1.4%, 4%, 3%, 2.2%.

(a) Calculate the mean for these data.

(b) Calculate the standard deviation for these data, showing each step in detail.

1.21 Find the median in following data sets.

(a) 3, 5, 6, 7, 9

(b) 3, 5, 6, 7, 9, 10

1.22 Find the median in following data sets.

(a) -1, 6, 8, -2, -9, -3, 4, -3, 3

(b) 6, 6, -5, -5, 7, 1, 4, 0, 8, -8

1.23 Below is the five number summary for the data given in Exercise 1.18. Create a box plot of the data based on these values.

Min	Q1	Q2 (Median)	Q3	Max
57	72.5	78.5	82.5	94

1.24 A Harvard Business Review study on Twitter usage shows that the median number of tweets per day is 0.01. 25% of users have 0 tweets per day while the top 25% of users average 0.37 tweets per day. Justin Bieber tweets an average of 14.3 times per day. Would you consider him an outlier? Explain. [7, 8]

1.25 Describe the distribution in the histograms on the right and match them to the box plots.

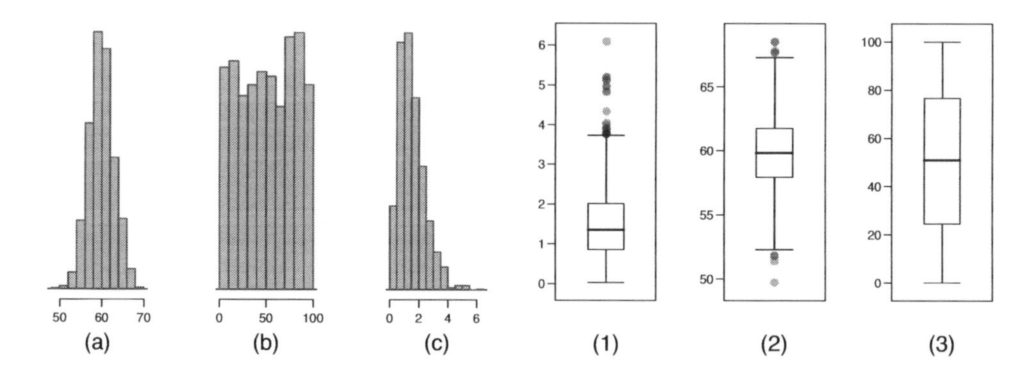

1.26 Compare the two plots below. What characteristics of the distribution are apparent in the histogram and not in the box plot? What characteristics are apparent in the box plot but not in the histogram?

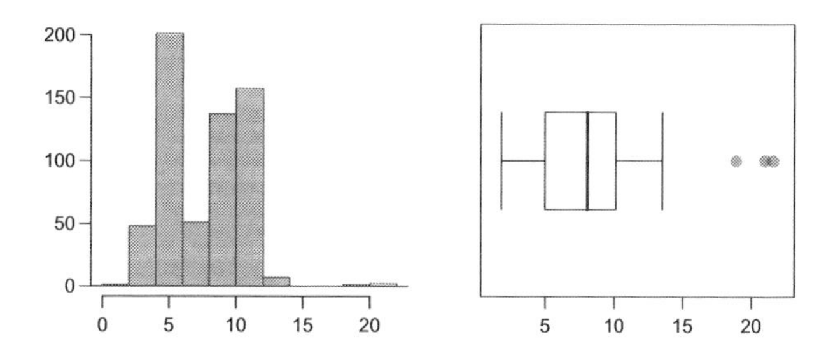

1.27 The infant mortality rate is defined as the number of deaths of infants under one year old in a given year per 1,000 live births in the same year. This rate is often used as an indicator of the level of health in a country. The histogram below shows the distribution of the infant death rate in 2010 for 224 countries as provided by the CIA factbook. [9]

(a) Estimate Q1, the median, and Q3 from the histogram.

(b) Would you expect the mean of this data set to be smaller or larger than the median? Explain your reasoning.

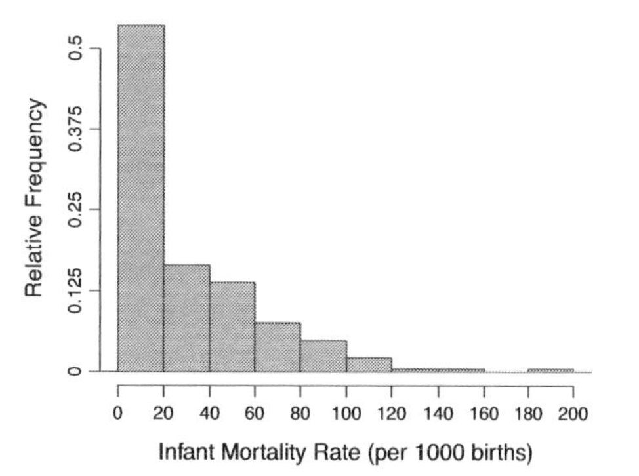

1.28 The histogram and the boxplot below (left) shows the distribution of finishing times for all runners of the New York Marathon between 1980 and 1999. The figure on the right provides boxplots of these data stratified by gender.

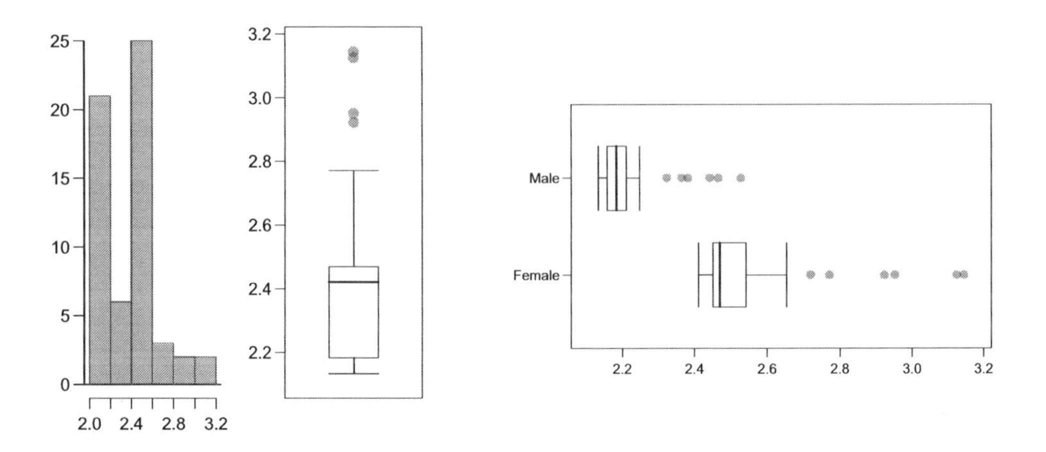

(a) What features of the distribution are apparent in the histogram and not the box plot on the left? What features are apparent in this box plot but not in the histogram?

(b) What may be the reason for the bimodal distribution? Explain.

(c) Compare the distribution of marathon times for men and women based on the box plot shown on the right.

1.29 The time series plot to the right is another way to look at the data from Exercise 1.28. Describe the trends apparent in the plot and comment on whether or not these trends were apparent in the histogram and/or the box plot.

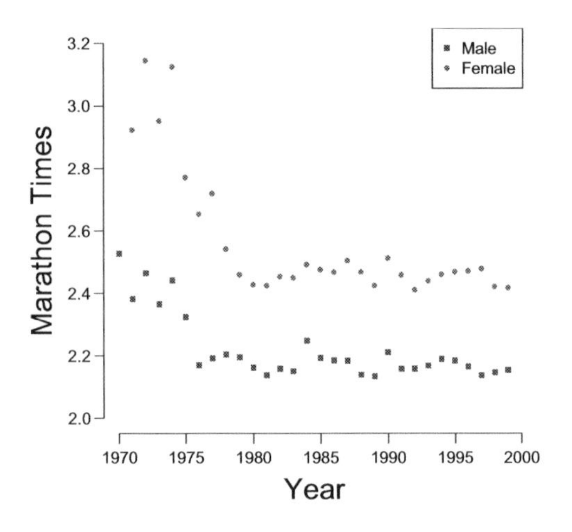

1.30 The first histogram below shows the distribution of the yearly incomes of 40 patrons at a college coffee shop. Suppose two new people walk into the coffee shop: one making \$225,000 and the other \$250,000. The second histogram shows the new income distribution. Summary statistics are also provided.

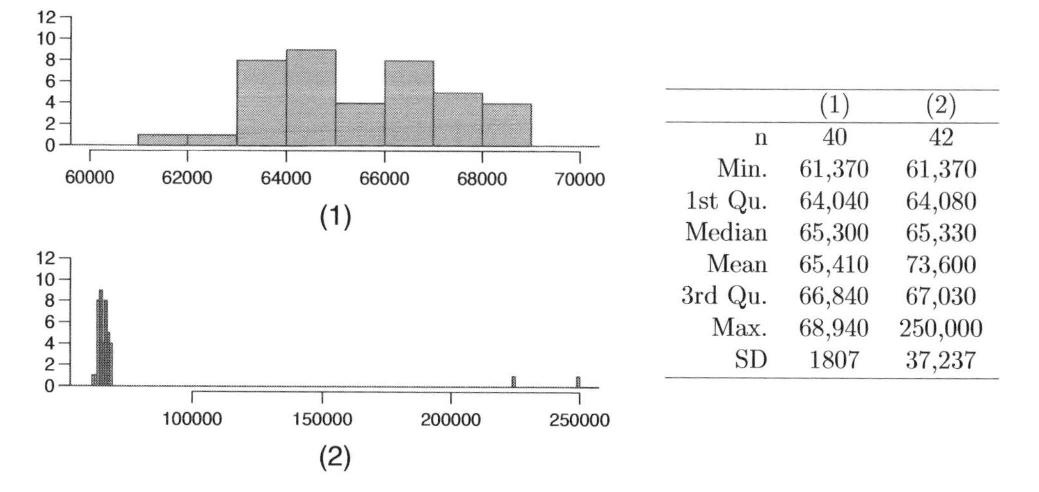

	(1)	(2)
n	40	42
Min.	61,370	61,370
1st Qu.	64,040	64,080
Median	65,300	65,330
Mean	65,410	73,600
3rd Qu.	66,840	67,030
Max.	68,940	250,000
SD	1807	37,237

(a) Would the mean or the median best represent what we might think of as a typical income for the 42 patrons at this coffee shop? What does this say about the robustness of the two measures?

(b) Would the standard deviation or the IQR best represent the amount of variability in the incomes of the 42 patrons at this coffee shop? What does this say about the robustness of the two measures?

1.31 For each of the following, describe whether you expect the distribution to be symmetric, right skewed or left skewed. Also specify whether you would use the mean or median to describe the center, and whether you would prefer to use the standard deviation or IQR to measure the spread.

(a) Housing prices in a country where 25% of the houses cost below \$350,000, 50% of the houses cost below \$450,000, 75% of the houses cost below \$1,000,000 and there are houses selling at over \$6,000,000.

(b) Housing prices in a country where 25% of the houses cost below \$300,000, 50% of the houses cost below \$600,000, 75% of the houses cost below \$900,000 and there no houses selling at over \$1,200,000.

(c) Number of alcoholic drinks consumed by college students in a given week.

(d) Annual salaries of the employees of a Fortune 500 company.

1.9.4 Considering categorical data

1.32 The table below shows the relationship between hair color and eye color for a group of 1,770 German men.

		Hair Color			
		Brown	Black	Red	Total
Eye	Brown	400	300	20	720
Color	Blue	800	200	50	1050
	Total	1200	500	70	1770

(a) What percentage of the men have black hair?

(b) What percentage of the men have blue eyes?

(c) What percentage of the men with black hair have blue eyes?

(d) Does it appear hair and eye color are independent?

1.33 Below is a bar plot of proportions and a pie chart showing the distribution of marital status in the data set on the smoking habits of UK residents introduced in Exercise 1.2.

(a) What features are apparent in the bar plot but not in the pie chart?

(b) What features are apparent in the pie chart but not in the bar plot?

(c) Which graph would you prefer to use for displaying these categorical data?

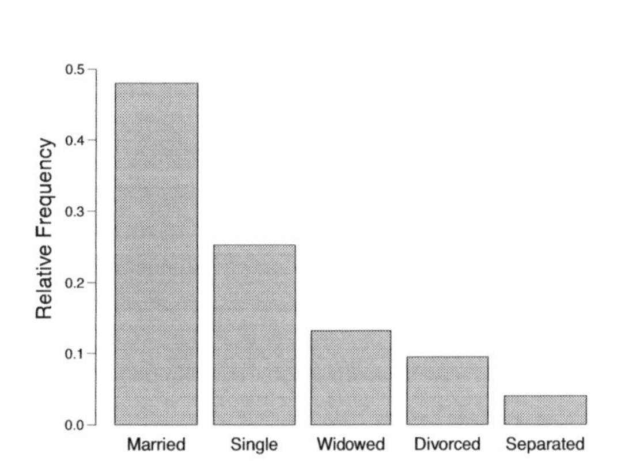

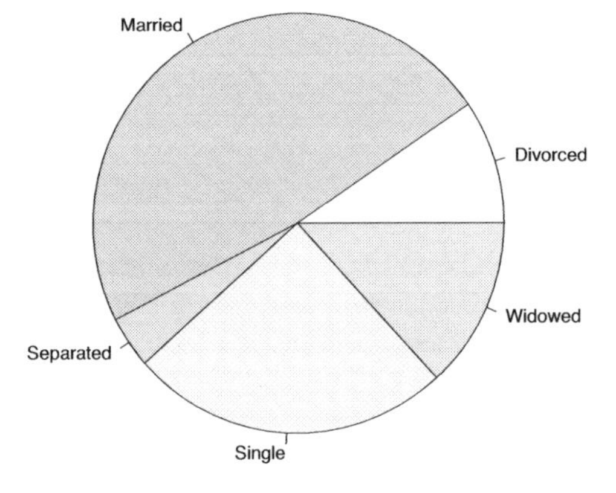

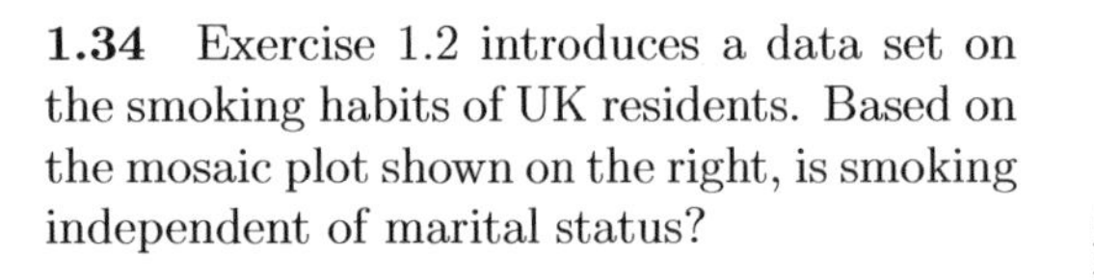

1.34 Exercise 1.2 introduces a data set on the smoking habits of UK residents. Based on the mosaic plot shown on the right, is smoking independent of marital status?

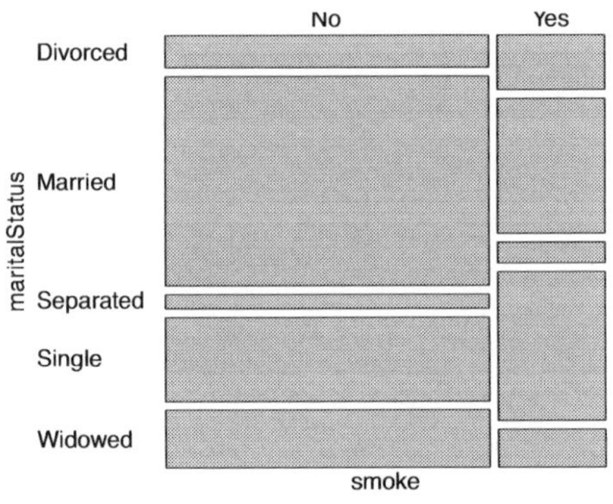

1.35 The Stanford University Heart Transplant Study was conducted to determine whether an experimental heart transplant program increased lifespan. Each patient entering the program was designated officially a heart transplant candidate, meaning that he was gravely ill and would most likely benefit from a new heart. Some patients got a transplant and some did not. The variable `transplant` indicates what group the patients were in; the treatment group got a transplant and control group did not. Another variable in the study, `survived`, indicates whether or not the patient was alive at the end of the study. [10]

(a) Based on the mosaic plot shown below on the left, is survival independent of whether or not the patient got a transplant? Explain your reasoning.

(b) What does the boxplots below suggest about the efficacy of transplants?

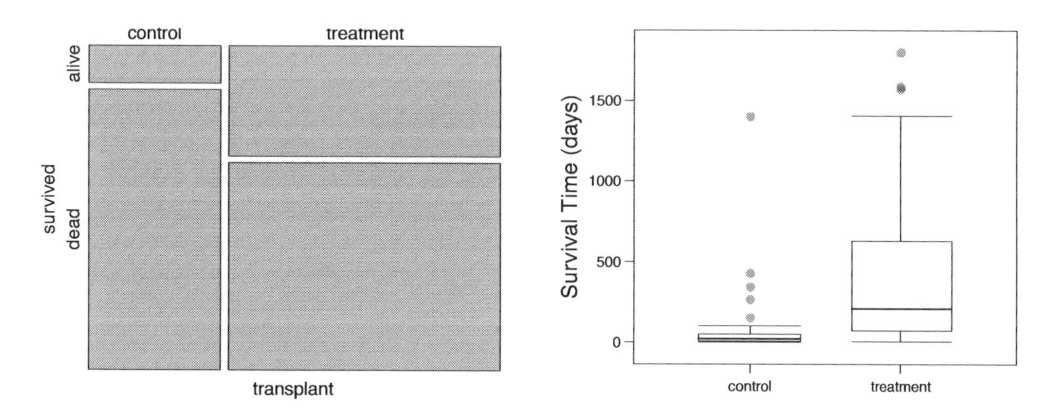

1.9.5 Overview of data collection principles

1.36 Identify the population and the sample in the following studies.

(a) Study about air pollution exposure and frequency of preterm births in Southern California described in Exercise 1.3.

(b) Study on the Buteyko method from Exercise 1.4.

1.37 Identify the population and the sample in the following studies.

(a) Study on the relationship between BMI and body fat percentage from Exercise 1.5.

(b) Study on the characteristics of California voters who do and do not support Proposition 23 from Exercise 1.6.

1.38 Below is a scatterplot displaying the relationship between the number of hours per week students watch TV and the grade they got in a statistics class (out of 100).

(a) What is the explanatory variable and what is the response variable?

(b) Is this an experiment or an observational study?

(c) Describe the relationship between the two variables.

(d) Can we conclude that watching longer hours of TV causes the students to get lower grades?

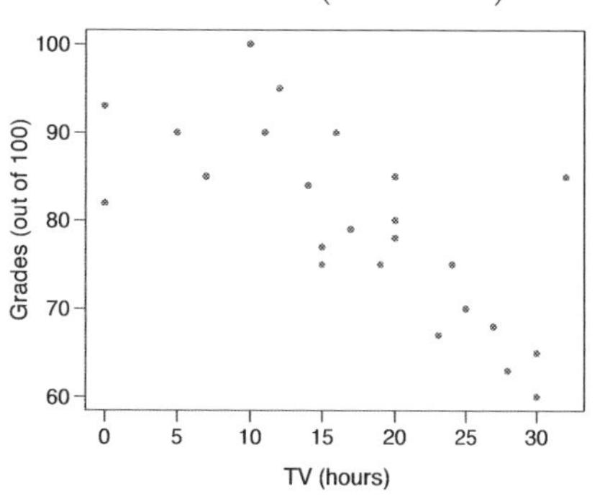

1.9.6 Observational studies and sampling strategies

1.39 A study shows that countries in which a higher percentage of the population have access to the Internet also tend to have higher average life expectancies.

(a) What type of study is this?

(b) State a possible lurking variable that might explain this relationship and describe its potential effect.

1.40 A large college class has 160 students. All 160 students attend the lectures together but are split into 4 groups of 40 for lab sections. The professor wants to conduct a survey about how satisfied the students are with the course so she decides to randomly sample 5 students from each lab section.

(a) What type of study is this?

(b) Which sampling method was used?

1.41 A Statistics student curious about the relationship between amount of time students spend on social networking sites and their performance at school decides to conduct a survey. Two research strategies for collecting data are described below. Name the sampling method used and any bias you might expect.

(a) He randomly samples 40 students, gives them the survey, asks them to fill it out and bring it back the next day.

(b) He gives out the survey only to his friends, and makes sure each one of them fills out the survey.

1.42 The Gallup Poll uses a procedure called random digit dialing (RDD) which creates phone numbers based on a list of all area codes in America, along with estimates of the number of residential households those exchanges have attached to them. This procedure is a lot more complicated than obtaining a list of all phone numbers from the phone book. Give a possible reason the Gallup Poll chooses to use RDD instead of picking phone numbers from the phone book?

1.43 Identify the flaw in reasoning in the following scenarios. Explain what the individuals in the study should have done differently if they wanted to make such strong conclusions.

(a) Students at an elementary school are given a questionnaire that they are required to return after their parents have completed it. One of the questions asked is, "Do you find that your work schedule makes it difficult for you to spend time with your kids after school?" Of the parents who replied, 85% said "no". Based on these results, the school officials conclude that a great majority of the parents have no difficulty spending time with their kids after school.

(b) A survey is conducted on a simple random sample of 1,000 women who recently gave birth, asking them about whether or not they smoked during pregnancy. A follow-up survey asking if the children have respiratory problems is conducted 3 years later, however only 567 of these women are reached at the same address. The researcher reports that these 567 women represent a simple random sample of all mothers.

(c) A orthopedist administers a questionnaire to 30 of his patients who do not have any joint problems and find that 20 of them regularly go running. He concludes that running decreases the risk of joint problems.

1.44 Suppose we want to estimate family size, where family is defined as parents and children. If we select students at random at an elementary school and ask them what their family size is, will our average be biased? If so, will it overestimate or underestimate the true value?

1.45 An article published in the *New York Times* called *Risks: Smokers Found More Prone to Dementia* states the following: [11]

> "Researchers analyzed the data of 23,123 health plan members who participated in a voluntary exam and health behavior survey from 1978 to 1985, when they were 50 to 60 years old.
>
> Twenty-three years later, about one-quarter of the group, or 5,367, had dementia, including 1,136 with Alzheimers disease and 416 with vascular dementia.
>
> After adjusting for other factors, the researchers concluded that pack-a-day smokers were 37 percent more likely than nonsmokers to develop dementia, and the risks went up sharply with increased smoking; 44 percent for one to two packs a day; and twice the risk for more than two packs."

Based on this study can we conclude that smokers are more likely to have dementia later in life? Explain your reasoning.

1.9.7 Experiments

1.46 In order to assess the effectiveness of a vitamin supplement, researchers prescribed a certain vitamin to patients. After 6 months, the researchers asked the patients whether or not they have been taking the vitamin. Then they divided the patients into two groups, those who took the pills, and those who did not and compared some health conditions between the two groups to measure the effectiveness of the vitamin.

(a) Was this an experiment or an observational study? Why?

(b) What are the explanatory and response variables in this study?

(c) Use the language of statistics to explain the flaw and how this affects the validity of the conclusion reached by the researchers.

(d) Were the patients blinded to their treatment? If not, explain how to make this a blinded study.

(e) Were the researchers blinded in this study? If not, explain how to make this a double blind study.

1.47 You would like to conduct an experiment in class to see if your classmates prefer the taste of regular Coke or Diet Coke. Briefly outline a design for this study.

1.48 Can chia seeds help you lose weight? Chia Pets - those terra-cotta figurines that sprout fuzzy green hair - made the chia plant a household name. But chia has gained an entirely new reputation as a diet supplement. In one study in 2009, a team of researchers randomly assigned half of 38 recruited men into a treatment group and the other half into a control group. They also recruited 38 women, and they randomly placed half of these participants into the treatment group and the other half into the control group. One group was given 25 grams of chia seeds twice a day, and the other was given a placebo. The subjects volunteered to be a part of the study. After 12 weeks, the scientists found no significant difference between the groups in appetite or weight loss.

(a) What type of study is this?

(b) What are the treatment and control groups in this this study?

(c) Has blocking been used in this study? If so, what is the blocking variable?

(d) Has blinding been used in this study?

(e) Comment on whether or not we can make a causal statement, and whether or not we can generalize the conclusion to the population at large.

1.49 A researcher is interested in the effects of exercise on mental health. She uses stratified random sampling to obtain representative proportions of 18-30, 31-40 and 41-55 year olds as in the population. She then randomly assigns half the subjects from each age group to exercise twice a week, and the rest are told to get no exercise. She conducts a mental health exam at the beginning and at the end of the study, and she compares the results.

(a) What type of study is this?

(b) What are the treatment and control groups in this study?

(c) Has blocking been used in this study? If so, what is the blocking variable?

(d) Has blinding been used in this study?

(e) Comment on whether or not we can make a causal statement, and whether or not we can generalize the conclusion to the population at large.

1.9.8 Case study: efficacy of sulphinpyrazone

1.50 Exercise 1.35 introduces the Stanford Heart Transplant Study. Of the 34 patients in the control group, 4 were alive at the end of the study. Of the 69 patients in the treatment group, 24 were alive. The contingency table below summarizes these results.

		Group		
		Control	Treatment	Total
Outcome	Alive	4	24	28
	Dead	30	45	75
	Total	34	69	103

(a) What proportion of all patients died?

(b) If outcome and treatment were independent (i.e. there was no difference between the success rates of two the groups), about how many deaths would we have expected in the treatment group?

Continue to part (c) on the next page.

(c) Using a randomization technique, a researcher investigated the relationship between outcome and treatment in this study. In order to simulate from the independence model, which states that the outcomes were independent of the treatment, she wrote whether or not each patient survived on cards, shuffled all the cards together, then dealt them into two groups of size 69 and 34. She repeated this simulation 250 times (using the help of a statistical software) and each time recorded the number of patients who died in the treatment group. Below is a histogram of these counts.

i. What are the claims being tested?

ii. Would more deaths or fewer deaths in the treatment group than the number calculated in part (b) provide support for the alternative hypothesis?

iii. What do the simulation results suggest about the effectiveness of the transplant program?

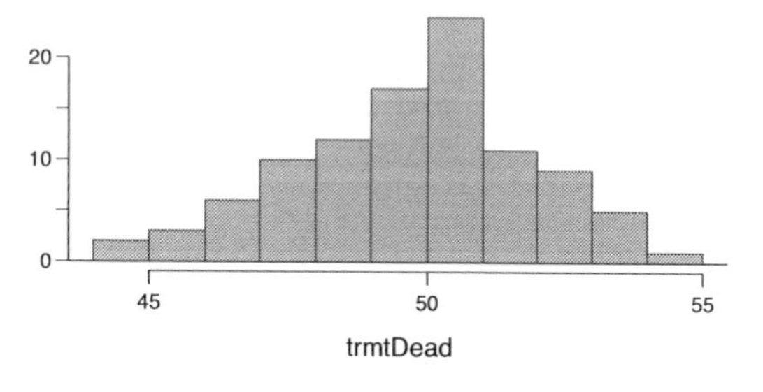

1.51 Rosiglitazone is the active ingredient in the controversial type 2 diabetes medicine Avandia. Rosiglitazone has been linked to an increased risk of serious cardiovascular problems such as acute myocardial infarction, stroke, heart failure, and death. A common alternative treatment is pioglitazone, the active ingredient in a diabetes medicine called Actos. In a nationwide retrospective observational study of a cohort of 227,571 Medicare beneficiaries aged 65 years or older, it was found that 2,593 of the 67,593 patients using rosiglitazone and 5,386 of the 159,978 using pioglitazone had serious cardiovascular problems. These data are summarized in the contingency table below. [12]

		Cardiovascular problems		
		Yes	No	Total
Treatment	Rosiglitazone	2,593	65,000	67,593
	Pioglitazone	5,386	154,592	159,978
	Total	7,979	219,592	227,571

Determine if each of the below statements is true or false. If false, explain why. Watch out: the reasoning may be wrong even if the statement's conclusion is correct. In such cases, the statement should be considered false.

(a) Since more patients on pioglitazone had cardiovascular problems (5,386 vs. 2,593) we can conclude that rate of cardiovascular problems for those on a pioglitazone treatment is higher.

(b) The data suggest that diabetic patients who are on a rosiglitazone treatment are more likely to have cardiovascular problems since the rate of incidence was (2,593 / 67,593 = 0.038) 3.8% for patients on this treatment, while it was only (5,386 / 159,978 = 0.034) 3.4% for patients on pioglitazone.

(c) The fact that the rate of incidence is higher for the rosiglitazone group proves that rosiglitazone causes serious cardiovascular problems.

(d) Based on the information provided so far, we cannot tell if the difference between the rates of incidences is due to a relationship between the two variables or due to chance.

1.52 Exercise 1.51 introduces a study that compares the rates of serious cardiovascular problems for diabetic patients on rosiglitazone and pioglitazone treatments. The study found that 2,593 of the 67,593 patients on rosiglitazone and 5,386 of the 159,978 patients on pioglitazone had cardiovascular problems.

(a) What proportion of all patients had cardiovascular problems?

(b) If having cardiovascular problems and treatment were independent, about how many patients with cardiovascular problems would we have expected in the rosiglitazone group?

(c) Using a randomization technique, a researcher investigated the relationship between taking Avandia and having cardiovascular problems in diabetic patients. In order to simulate from the independence model, which states that the outcomes were independent of the treatment, he wrote whether or not each patient had a cardiovascular problem on cards, shuffled all the cards together, then dealt them into two groups of size 67,593 and 159,978. He repeated this simulation 1,000 times (using the help of a statistical software) and each time recorded the number of people in the rosiglitazone group who had cardiovascular problems. Below is a relative frequency histogram of these counts.

   i. What are the claims being tested?

   ii. Would more or fewer patients with cardiovascular problems in the rosiglitazone group than the number calculated in part (b) provide support for the alternative hypothesis?

   iii. What do the simulation results suggest about the relationship between taking rosiglitazone and having cardiovascular problems in diabetic patients?

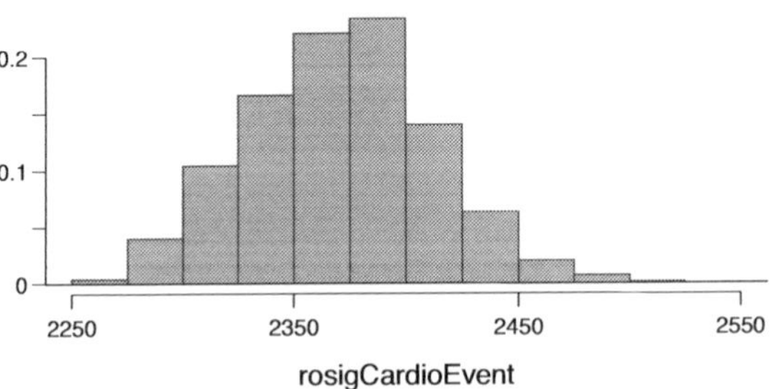

Chapter 2

Probability (special topic)

Probability forms a foundation for statistics. You may already be familiar with many aspects of probability, however, formalization of the concepts is new for most. This chapter aims to introduce probability on familiar terms using processes most people have seen before.

2.1 Defining probability (special topic)

● **Example 2.1** A "die", the singular of dice, is a cube with six faces numbered 1, 2, 3, 4, 5, and 6. What is the chance of getting 1 when rolling a die?

If the die is fair, then the chance of a 1 is as good as the chance of any other number. Since there are six outcomes, the chance must be 1-in-6 or, equivalently, $1/6$.

● **Example 2.2** What is the chance of getting a 1 or 2 in the next roll?

1 and 2 constitute two of the six equally likely possible outcomes, so the chance of getting one of these two outcomes must be $2/6 = 1/3$.

● **Example 2.3** What is the chance of getting either 1, 2, 3, 4, 5, or 6 on the next roll?

100%. The die must be one of these numbers.

● **Example 2.4** What is the chance of not rolling a 2?

Since the chance of rolling a 2 is $1/6$ or $16.\bar{6}\%$, the chance of not getting a 2 must be $100\% - 16.\bar{6}\% = 83.\bar{3}\%$ or $5/6$.

Alternatively, we could have noticed that not rolling a 2 is the same as getting a 1, 3, 4, 5, or 6, which makes up five of the six equally likely outcomes and has probability $5/6$.

● **Example 2.5** Consider rolling two dice. If $1/6^{th}$ of the time the first die is 1 and $1/6^{th}$ of those times the second die is a 1, what is the chance of getting two 1s?

If $16.\bar{6}\%$ of the time the first die is a 1 and $1/6^{th}$ of *those* times the second die is also a 1, then the chance both dice are 1 is $(1/6) * (1/6)$ or $1/36$.

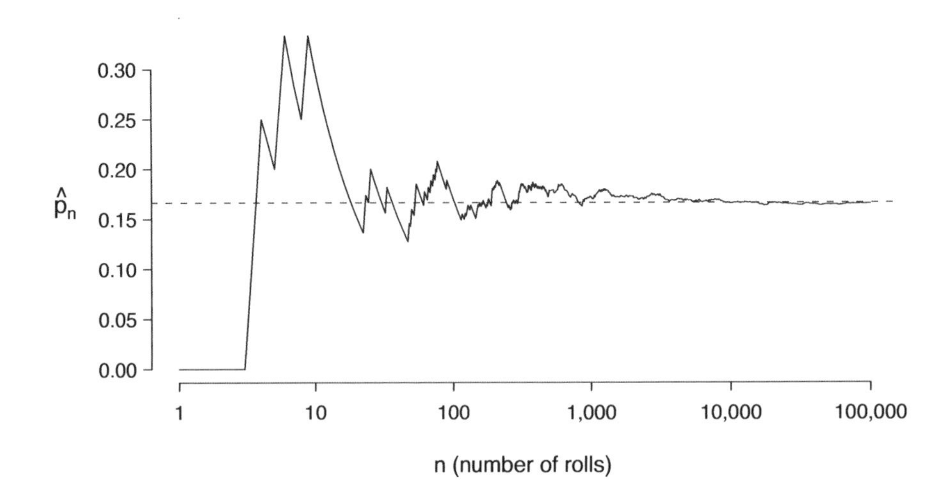

Figure 2.1: The fraction of die rolls that are 1 at each stage in a simulation. The proportion tends to get closer to the probability $1/6 \approx 0.167$ as the sample size gets large.

2.1.1 Probability

We use probability to build tools to describe and understand apparent randomness. We often frame probability in terms of a **random process** giving rise to an **outcome**.

Roll a die	$\rightarrow$	`1, 2, 3, 4, 5, or 6`
Flip a coin	$\rightarrow$	`H or T`

Rolling a die or flipping a coin is a seemingly random process and each gives rise to an outcome.

> **Probability**
> The **probability** of an outcome is the proportion of times the outcome would occur if we observed the random process an infinite number of times.

Probability is defined as a proportion, and it must always be between 0 and 1 (inclusively). It may also be displayed as a percentage between 0% and 100%.

Probability can be illustrated by rolling a die many times. Let $\hat{p}_n$ be the proportion of outcomes that are `1` after the first n rolls. As the number of rolls increases, $\hat{p}_n$ will converge to the probability of rolling `1`, $p = 1/6$. Figure 2.1 shows this convergence for 100,000 die rolls. The tendency of $\hat{p}_n$ to stabilize around p is described by the **Law of Large Numbers**.

> **Law of Large Numbers**
> As more observations are collected, the proportion $\hat{p}_n$ of occurrences with a particular outcome converges to the probability p of that outcome.

Occasionally the proportion will veer off from the probability and appear to defy the Law of Large Numbers, as $\hat{p}_n$ does many times in Figure 2.1. However, these deviations become smaller as the sample size becomes larger.

Above we write p as the probability of rolling a 1. We can also write this probability as

$$P(\texttt{rolling a 1})$$

$P(A)$
Probability of outcome A

As we become more comfortable with this notation, we will abbreviate it further. For instance, if it is clear the process is "rolling a die", we could abbreviate P(rolling a 1) as $P(1)$.

⊙ **Exercise 2.6** Random processes include rolling a die and flipping a coin. (a) Think of another random process. (b) Describe all the possible outcomes of that process. For instance, rolling a die is a random process with potential outcomes 1, 2, ..., 6. Several examples are in the footnote[1].

What we think of as random processes are not necessarily random, but they may just be too difficult to understand exactly. Example (iv) in Exercise 2.6 suggests whether your roommate does her dishes tonight is random. However, even if your roommate's behavior is not truly random, modeling her behavior as a random process can still be useful.

TIP: Modeling a process as random
It can be helpful to model a process as random even if it is not truly random.

2.1.2 Disjoint or mutually exclusive outcomes

Two outcomes are called **disjoint** or **mutually exclusive** if they cannot both happen. For instance, if we roll a die, the outcomes 1 and 2 are disjoint since they cannot both occur. On the other hand, the outcomes 1 and "rolling an odd number" are not disjoint since both occur if the die is a 1. The terms *disjoint* and *mutually exclusive* are equivalent and interchangeable.

Calculating the probability of disjoint outcomes is easy. When rolling a die, the outcomes 1 and 2 are disjoint, and we compute the probability that one of these outcomes will occur by adding their separate probabilities:

$$P(1 \text{ or } 2) = P(1) + P(2) = 1/6 + 1/6 = 1/3$$

What about the probability of rolling a 1, 2, 3, 4, 5, or 6? Here again, all of the outcomes are disjoint so we add the probabilities:

$$\begin{aligned}
&P(1 \text{ or } 2 \text{ or } 3 \text{ or } 4 \text{ or } 5 \text{ or } 6) \\
&\quad = P(1) + P(2) + P(3) + P(4) + P(5) + P(6) \\
&\quad = 1/6 + 1/6 + 1/6 + 1/6 + 1/6 + 1/6 = 1.
\end{aligned}$$

The **Addition Rule** guarantees the accuracy of this approach when the outcomes are disjoint.

[1](i) Whether someone gets sick in the next month or not is an apparently random process with outcomes `sick` and `not`. (ii) We can *generate* a random process by randomly picking a person and measuring that person's height. The outcome of this process will be a positive number. (iii) Whether the stock market goes up or down next week is a seemingly random process with possible outcomes `up`, `down`, and `noChange`. Alternatively, we could have used the percent change in the stock market as a numerical outcome. (iv) Whether your roommate cleans her dishes tonight probably seems like a random process with possible outcomes `cleansDishes` and `leavesDishes`.

Addition Rule of disjoint outcomes

If A_1 and A_2 represent two disjoint outcomes, then the probability that one of them occurs is given by

$$P(A_1 \text{ or } A_2) = P(A_1) + P(A_2)$$

If there are many disjoint outcomes A_1, ..., A_k, then the probability that one of these outcomes will occur is

$$P(A_1) + P(A_2) + \cdots + P(A_k) \qquad (2.7)$$

⊙ **Exercise 2.8** We are interested in the probability of rolling a `1`, `4`, or `5`. (a) Explain why the outcomes `1`, `4`, and `5` are disjoint. (b) Apply the Addition Rule for disjoint outcomes to determine $P($`1` or `4` or `5`$)$.

⊙ **Exercise 2.9** In the `cars` data set in Chapter 1, the `type` variable described the size of the vehicle: `small` (21 cars), `midsize` (22 cars), or `large` (11 cars). (a) Are the outcomes `small`, `midsize`, and `large` disjoint? Answer in the footnote[2]. (b) Determine the proportion of `midsize` and `large` cars separately. (c) Use the Addition Rule for disjoint outcomes to compute the probability a randomly selected car from this sample is either `midsize` or `large`.

Statisticians rarely work with individual outcomes and instead consider *sets* or *collections* of outcomes. Let A represent the event where a die roll results in `1` or `2` and B represent the event that the die roll is a `4` or a `6`. We write A as the set of outcomes {`1`, `2`} and B = {`4`, `6`}. These sets are commonly called **events**. Because A and B have no elements in common, they are disjoint events. A and B are represented in Figure 2.2.

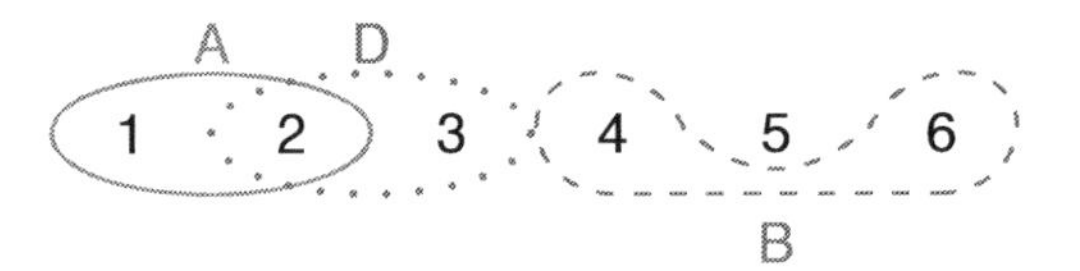

Figure 2.2: Three events, A, B, and D, consist of outcomes from rolling a die. A and B are disjoint since they do not have any outcomes in common.

⊙ **Exercise 2.10** Events A and B in Figure 2.2 are disjoint. What other pair of events in the figure are disjoint? Answer in the footnote[3].

The Addition Rule applies to both disjoint outcomes and disjoint events. The probability that one of the disjoint events A or B occurs is the sum of the separate probabilities:

$$P(A \text{ or } B) = P(A) + P(B) = 1/3 + 1/3 = 2/3$$

⊙ **Exercise 2.11** (a) Verify the probability of event A, $P(A)$, is $1/3$ using the Addition Rule for disjoint events. (b) Do the same for B.

[2]Yes. Each car is categorized in only one level of `type`.

[3]Sets B and D since they have no outcomes in common.

2♣	3♣	4♣	5♣	6♣	7♣	8♣	9♣	10♣	J♣	Q♣	K♣	A♣
2♢	3♢	4♢	5♢	6♢	7♢	8♢	9♢	10♢	J♢	Q♢	K♢	A♢
2♡	3♡	4♡	5♡	6♡	7♡	8♡	9♡	10♡	J♡	Q♡	K♡	A♡
2♠	3♠	4♠	5♠	6♠	7♠	8♠	9♠	10♠	J♠	Q♠	K♠	A♠

Table 2.3: Representations of the 52 unique cards in a deck.

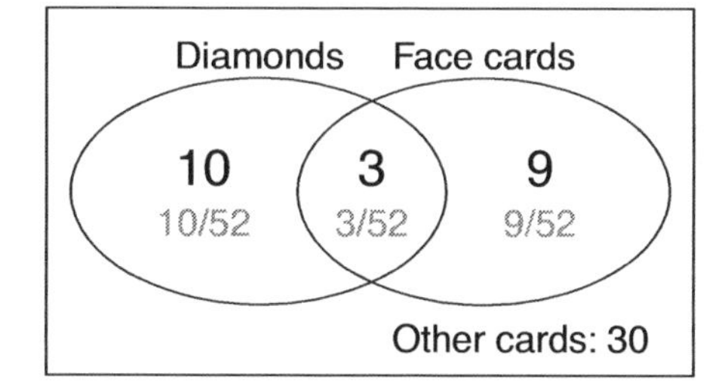

Figure 2.4: A Venn diagram for diamonds and face cards.

⊙ **Exercise 2.12** (a) Using Figure 2.2 as a reference, what outcomes are represented by the event D? (b) Are events B and D disjoint? (c) Are events A and D disjoint? Answer to part (c) is in the footnote[4].

⊙ **Exercise 2.13** In Exercise 2.12, you confirmed B and D from Figure 2.2 are disjoint. Compute the probability that either event B or event D occurs.

2.1.3 Probabilities when events are not disjoint

Let's consider calculations for two events that are not disjoint in the context of a regular deck of 52 cards, represented in Table 2.3. If you are unfamiliar with the cards in a regular deck, please see the footnote[5].

⊙ **Exercise 2.14** (a) What is the probability a randomly selected card is a diamond? (b) What is the probability a randomly selected card is a face card?

Venn diagrams are useful when outcomes can be categorized as "in" or "out" for two or three variables, attributes, or random processes. The Venn diagram in Figure 2.4 uses a circle to represent diamonds and another to represent face cards. If a card is both a diamond and a face card, it falls into the intersection of the circles. If it is a diamond but not a face card, it will be in part of the left circle that is not in the right circle (and so on). The total number of cards that are diamonds is given by the total number of cards in the diamonds circle: $10 + 3 = 13$. The probabilities are shown in gray.

⊙ **Exercise 2.15** Using the Venn diagram, verify $P(\text{face card}) = 12/52 = 3/13$.

---

[4] The events A and D share an outcome in common, 2, and so are not disjoint.

[5] The 52 cards are split into four **suits**: ♣ (club), ♢ (diamond), ♡ (heart), ♠ (spade). Each suit has its 13 cards labeled: 2, 3, ..., 10, J (jack), Q (queen), K (king), and A (ace). Thus, each card is a unique combination of a suit and a label, e.g. 4♡. The 12 cards represented by the jacks, queens, and kings are called `face cards`. The cards that are ♢ or ♡ are typically colored `red` while the other two suits are typically colored `black`.

Let A represent the event that a randomly selected card is a diamond and B represent the event it is a face card. How do we compute $P(A \text{ or } B)$? Events A and B are not disjoint – the cards $J\diamondsuit$, $Q\diamondsuit$, and $K\diamondsuit$ fall into both categories – so we cannot use the Addition Rule for disjoint events. Instead we use the Venn diagram. We start by adding the probabilities of the two events:

$$P(A) + P(B) = P(\diamondsuit) + P(\text{face card}) = 12/52 + 13/52$$

However, the three cards that are in both events were counted twice, once in each probability. We must correct this double counting:

$$\begin{aligned} P(A \text{ or } B) &= P(\text{face card or } \diamondsuit) \\ &= P(\text{face card}) + P(\diamondsuit) - P(\text{face card } \& \diamondsuit) \quad (2.16) \\ &= 12/52 + 13/52 - 3/52 \\ &= 22/52 = 11/26 \end{aligned}$$

Equation (2.16) is an example of the **General Addition Rule**.

General Addition Rule
If A and B are any two events, disjoint or not, then the probability that at least one of them will occur is

$$P(A) + P(B) - P(A \& B) \quad (2.17)$$

where $P(A \& B)$ is the probability that both events occur.

TIP: "or" is inclusive
When we write "or" in statistics, we mean "and/or" unless we explicitly state otherwise. Thus, A or B occurs means A, B, or both A and B occur.

⊙ **Exercise 2.18** (a) Describe why, if A and B are disjoint, then $P(A \& B) = 0$. (b) Using part (a), verify that the General Addition Rule simplifies to the Addition Rule for disjoint events if A and B are disjoint. Answers in the footnote[6].

⊙ **Exercise 2.19** In the `cars` data set with 54 vehicles, 22 were `midsize` cars, 16 had a capacity of `6` people, and 5 `midsize` cars had a capacity of `6` people. Create a Venn diagram for this setup. Answer in the footnote[7].

⊙ **Exercise 2.20** (a) Use your Venn diagram from Exercise 2.19 to determine the probability a random car from the `cars` data set is both a `midsize` vehicle and has a capacity of 6. (b) What is the probability the car is a `midsize` car or has a `6` person capacity?

[6](a) If A and B are disjoint, A and B can never occur simultaneously. (b) If A and B are disjoint, then the last term of Equation (2.17) is 0 (see part (a)) and we are left with the Addition Rule for disjoint events.

[7]
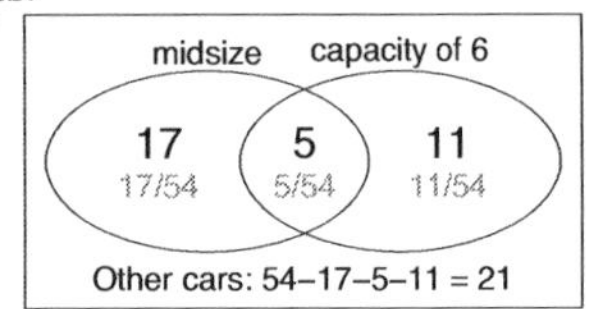

2.1.4 Probability distributions

The grocery receipt for a college student is shown in Table 2.5. Does anything seem odd about the total? The individual costs only add up to $23.20 while the total is written as $37.90. Where did the additional $14.70 come from?

Item	Cost
Spaghetti	$6.50
Carrots	$7.20
Apples	$3.10
Milk	$6.40
Tax	$0.00
Total	$37.90

Table 2.5: Grocery receipt with a total greater than the sum of the costs.

Table 2.6 shows another month of expenditures with a new problem. While the sum of the expenditures match up, the amount spent on milk is a negative.

Item	Cost
Spaghetti	$6.50
Carrots	$7.20
Apples	$5.10
Milk	-$4.40
Chocolate	$1.19
Tax	$0.11
Total	$15.70

Table 2.6: On this receipt, buying milk saves the customer money (!).

A **probability distribution** is a table of all disjoint outcomes and their associated probabilities. It is like a grocery bill, except that instead of foods there are outcomes, and instead of costs for each food item there are probabilities for each outcome. Table 2.7 shows the probability distribution for the sum of two dice.

Dice sum	2	3	4	5	6	7	8	9	10	11	12
Probability	$\frac{1}{36}$	$\frac{2}{36}$	$\frac{3}{36}$	$\frac{4}{36}$	$\frac{5}{36}$	$\frac{6}{36}$	$\frac{5}{36}$	$\frac{4}{36}$	$\frac{3}{36}$	$\frac{2}{36}$	$\frac{1}{36}$

Table 2.7: Probability distribution for the sum of two dice.

Probability distributions share the same structure as grocery receipts. However, probability distributions impose one special rule: the total must be 1.

Rules for probability distributions

A probability distribution is a list of the possible outcomes with corresponding probabilities that satisfies three rules:

1. The outcomes listed must be disjoint.
2. Each probability must be between 0 and 1.
3. The probabilities must total 1.

Income range ($1000s)	0-25	25-50	50-100	100+
(a)	0.18	0.39	0.33	0.16
(b)	0.38	-0.27	0.52	0.37
(c)	0.28	0.27	0.29	0.16

Table 2.8: Proposed distributions of US household incomes (Exercise 2.21).

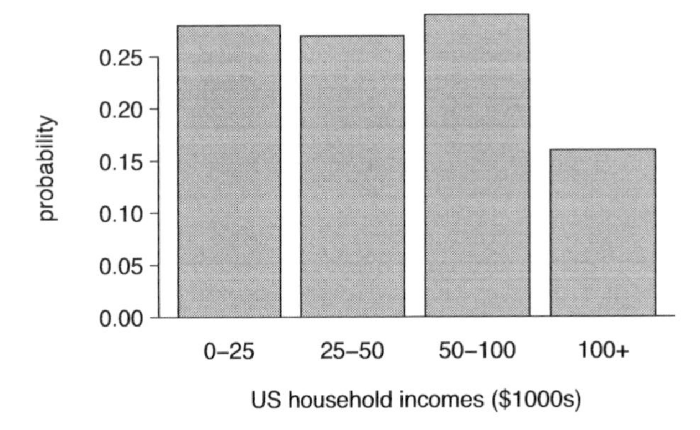

Figure 2.9: The probability distribution of US household income.

⊙ **Exercise 2.21** Table 2.8 suggests three distributions for household income in the United States. Only one is correct. Which one must it be? What is wrong with the other two? Answer in the footnote[8].

Chapter 1 emphasized the importance of plotting data to provide quick summaries. Probability distributions can also be summarized in a bar plot. For instance, the distribution of US household incomes is shown in Figure 2.9 as a bar plot[9]. The probability distribution for the sum of two dice is shown in Table 2.7 and plotted in Figure 2.10.

In these bar plots, the bar heights represent the outcome probabilities. If the outcomes are numerical and discrete, it is usually (visually) convenient to place the bars at their associated locations on the axis, as in the case of the sum of two dice. Another example of plotting the bars at their respective locations is shown in Figure 2.26 on page 84.

2.1.5 Complement of an event

Rolling a die produces a value in the set {1, 2, 3, 4, 5, 6}. This set of all possible outcomes is called the **sample space** (S) for rolling a die. We often use the sample space to examine the scenario where an event does not occur.

A^c
Complement of outcome A

Let $D = \{2, 3\}$ represent the event we roll a die and get 2 or 3. Then the **complement** of D represents all outcomes in our sample space that are not in D, which is denoted by $D^c = \{1, 4, 5, 6\}$. That is, D^c is the set of all possible outcomes not already included in D. Figure 2.11 shows the relationship between D, D^c, and the sample space S.

[8]The probabilities of (a) do not sum to 1. The second probability in (b) is negative. This leaves (c), which sure enough satisfies the requirements of a distribution. One of the three was said to be the actual distribution of US household incomes, so it must be (c).

[9]It is also possible to construct a distribution plot when income is not artificially binned into four groups. *Continuous* distributions are considered in Section 2.2.

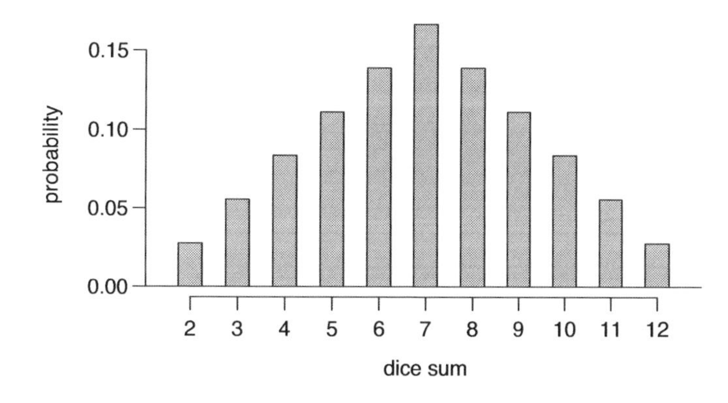

Figure 2.10: The probability distribution of the sum of two dice.

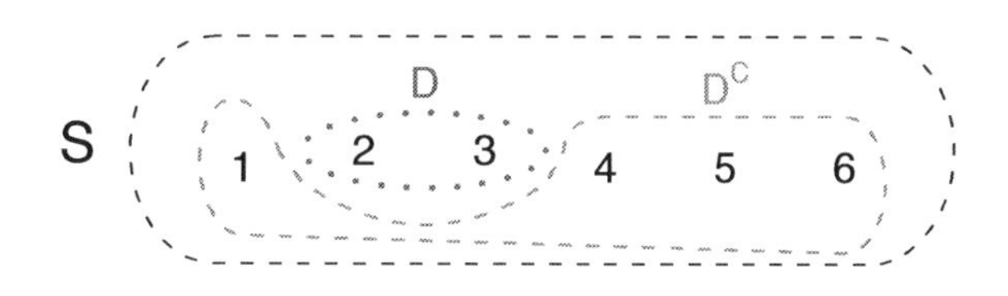

Figure 2.11: Event $D = \{2, 3\}$ and its complement, $D^c = \{1, 4, 5, 6\}$. S represents the sample space, which is the set of all possible events.

⊙ **Exercise 2.22** (a) Compute $P(D^c) = P(\text{rolling a 1, 4, 5, or 6})$. (b) What is $P(D) + P(D^c)$?

⊙ **Exercise 2.23** Events $A = \{1, 2\}$ and $B = \{4, 6\}$ are shown in Figure 2.2 on page 58. (a) Write out what A^c and B^c represent. (b) Compute $P(A^c)$ and $P(B^c)$. (c) Compute $P(A) + P(A^c)$ and $P(B) + P(B^c)$.

A complement of an event A is constructed to have two very important properties: (i) every possible outcome not in A is in A^c, and (ii) A and A^c are disjoint. Property (i) implies

$$P(A \text{ or } A^c) = 1 \tag{2.24}$$

That is, if the outcome is not in A, it must be represented in A^c. We use the Addition Rule for disjoint events to apply Property (ii):

$$P(A \text{ or } A^c) = P(A) + P(A^c) \tag{2.25}$$

Combining Equations (2.24) and (2.25) yields a very useful relationship between the probability of an event and its complement.

Complement
The complement of event A is denoted A^c, and A^c represents all outcomes not in A. A and A^c are mathematically related:

$$P(A) + P(A^c) = 1, \quad \text{i.e.} \quad P(A) = 1 - P(A^c) \tag{2.26}$$

In simple examples, computing A or A^c is feasible in a few steps. However, using the complement can save a lot of time as problems grow in complexity.

⊙ **Exercise 2.27** Let A represent the event where we roll two dice and their total is less than 12. (a) What does the event A^c represent? Answer in the footnote[10]. (b) Determine $P(A^c)$ from Table 2.7 on page 61. (c) Determine $P(A)$. Answer in the footnote[11].

⊙ **Exercise 2.28** Consider again the probabilities from Table 2.7 and rolling two dice. Find the following probabilities: (a) The sum of the dice is *not* 6. Hint in the footnote[12]. (b) The sum is at least 4. That is, determine the probability of the event $B = \{4, 5, ..., 12\}$. Answer in the footnote[13]. (c) The sum is no more than 10. That is, determine the probability of the event $D = \{2, 3, ..., 10\}$.

2.1.6 Independence

Just as variables and observations can be independent, random processes can be independent, too. Two processes are **independent** if knowing the outcome of one provides no useful information about the outcome of the other. For instance, flipping a coin and rolling a die are two independent processes – knowing the coin was H does not help determine what the die will be. On the other hand, stock prices usually move up or down together, so they are not independent.

Example 2.5 provides a basic example of two independent processes: rolling two dice. We want to determine the probability that both will be 1. Suppose one of the dice is red and the other white. If the outcome of the red die is a 1, it provides no information about the outcome of the white die. Example 2.5's argument (page 55) is as follows: $1/6^{th}$ of the time the red die is 1, and $1/6^{th}$ of *those* times the white die will also be 1. This is illustrated in Figure 2.12. Because the rolls are independent, the probabilities of the corresponding outcomes can be multiplied to get the final answer: $(1/6) * (1/6) = 1/36$. This can be generalized to many independent processes.

● **Example 2.29** What if there was also a blue die independent of the other two? What is the probability of rolling the three dice and getting all 1s?

The same logic applies from Example 2.5. If $1/36^{th}$ of the time the white and red dice are both 1, then $1/6^{th}$ of *those* times the blue die will also be 1, so multiply:

$$\begin{aligned} &P(white = 1 \text{ and } red = 1 \text{ and } blue = 1) \\ &= P(white = 1) * P(red = 1) * P(blue = 1) \\ &= (1/6) * (1/6) * (1/6) = (1/36) * (1/6) = 1/216 \end{aligned}$$

Examples 2.5 and 2.29 illustrate what is called the Product Rule for independent processes.

---

[10]The complement of A: when the total is equal to 12.

[11]Use the probability of the complement from part (b), $P(A^c) = 1/36$, and Equation (2.26): $P(\text{less than } 12) = 1 - P(12) = 1 - 1/36 = 35/36$.

[12]First find $P(6)$.

[13]We first find the complement, which requires much less effort: $P(2 \text{ or } 3) = 1/36 + 2/36 = 1/12$. Then we find $P(B) = 1 - P(B^c) = 1 - 1/12 = 11/12$.

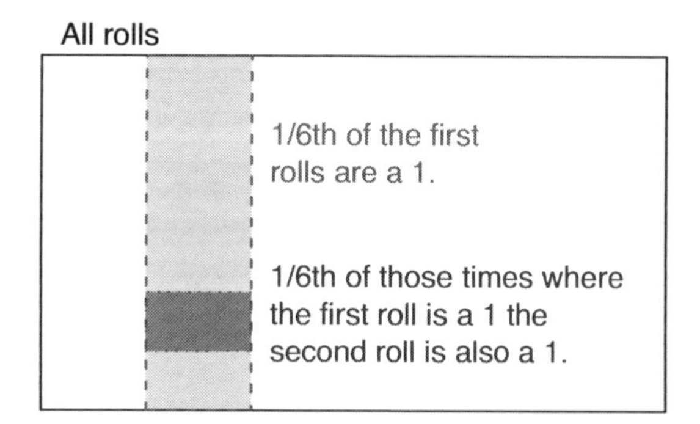

Figure 2.12: $1/6^{th}$ of the time, the first roll is a `1`. Then $1/6^{th}$ of *those* times, the second roll will also be a `1`.

Product Rule for independent processes
If A and B represent events from two different and independent processes, then the probability both A and B occur can be computed as the product of their separate probabilities:

$$P(A \text{ and } B) = P(A) * P(B) \tag{2.30}$$

Similarly, if there are k events A_1, ..., A_k from k independent processes, then the probability they all occur is

$$P(A_1) * P(A_2) * \cdots * P(A_k)$$

⊙ **Exercise 2.31** About 9% of people are `leftHanded`. Suppose 2 people are selected at random from the U.S. population. Because the sample of 2 is so small relative to the population, it is reasonable to assume these two people are independent. (a) What is the probability that both are `leftHanded`? (b) What is the probability that both are `rightHanded`? Answer to (a) and hint to (b) in the footnote[14].

⊙ **Exercise 2.32** Suppose 5 people are selected at random.

(a) What is the probability that all are `rightHanded`? Answer in the footnote[15].

(b) What is the probability that all are `leftHanded`?

(c) What is the probability that not all of the people are `rightHanded`? Hint in the footnote[16].

---

[14](a) The probability the first person is `leftHanded` is 0.09, which is the same for the second person. We apply the Product Rule for independent processes to determine the probability that both will be `leftHanded`: $0.09 * 0.09 = 0.0081$.
(b) It is reasonable to assume the proportion of people who are ambidextrous (both right and left handed) is nearly 0, which results in $P(\texttt{rightHanded}) = 1 - 0.09 = 0.91$.

[15]Since each are independent, we apply the Product Rule for independent processes:

$$\begin{aligned} &P(\text{all five are } \texttt{rightHanded}) \\ &= P(\text{first} = \texttt{rH}, \text{second} = \texttt{rH}, ..., \text{fifth} = \texttt{rH}) \\ &= P(\text{first} = \texttt{rH}) * P(\text{second} = \texttt{rH}) * \cdots * P(\text{fifth} = \texttt{rH}) \\ &= 0.91 * 0.91 * 0.91 * 0.91 * 0.91 = 0.624 \end{aligned}$$

[16]Use the complement, $P(\text{all five are } \texttt{rightHanded})$, to answer this question.

Suppose the variables `handedness` and `gender` are independent, i.e. knowing someone's `gender` provides no useful information about their `handedness` and vice-versa. We can compute whether a randomly selected person is `rightHanded` and `female`[17] using the Product Rule:

$$\begin{aligned} P(\texttt{rightHanded and female}) &= P(\texttt{rightHanded}) * P(\texttt{female}) \\ &= 0.91 * 0.50 = 0.455 \end{aligned}$$

⊙ **Exercise 2.33** Three people are selected at random. Answers to each part are in the footnote[18]. (a) What is the probability the first person is `male` and `rightHanded`? (b) What is the probability the first two people are `male` and `rightHanded`?. (c) What is the probability the third person is `female` and `leftHanded`? (d) What is the probability the first two people are `male` and `rightHanded` and the third person is `female` and `leftHanded`?

Sometimes we wonder if one outcome provides useful information about another outcome. The question we are asking is, are the occurrences of the two events independent? We say that two events A and B are independent if they satisfy Equation (2.30).

● **Example 2.34** If we shuffle up a deck of cards and draw one, is the event that the card is a heart independent of the event that the card is an ace?

The probability the card is a heart is 1/4 and the probability it is an ace is 1/13. The probability the card is the ace of hearts is 1/52. We check whether Equation 2.30 is satisfied:

$$\begin{aligned} P(\heartsuit) * P(\text{ace}) &= \frac{1}{4} * \frac{1}{13} \\ &= \frac{1}{52} \\ &= P(\heartsuit \ \& \ \text{ace}) \end{aligned}$$

Because the equation holds, the event that the card is a heart and the event that the card is an ace are independent events.

2.2 Continuous distributions (special topic)

● **Example 2.35** Figure 2.13 shows a few different hollow histograms of the variable `height` for 3 million US adults from the mid-90's[19]. How does changing the number of bins allow you to make different interpretations of the data?

Adding more bins provides greater detail. This sample is extremely large, which is why much smaller bins still work well. Usually we do not use so many bins with smaller sample sizes since small counts per bin mean the bin heights are very volatile.

[17] The actual proportion of the U.S. population that is `female` is about 50%, and so we use 0.5 for the probability of sampling a woman. However, this probability is different in other countries.

[18] This is the same as P(a randomly selected person is `male` and `rightHanded`) $= 0.455$. (b) 0.207. (c) 0.045. (d) 0.0093.

[19] This sample can be considered a simple random sample from the US population. It relies on the USDA Food Commodity Intake Database.

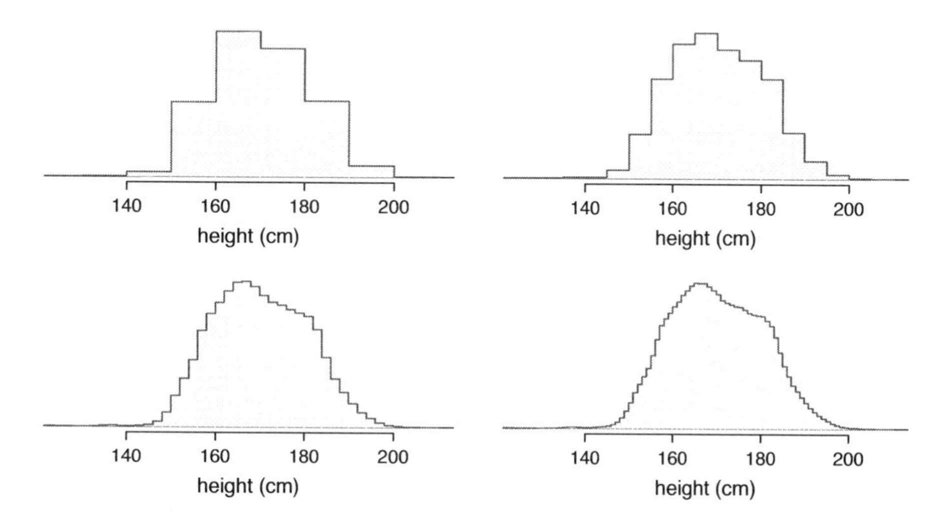

Figure 2.13: Four hollow histograms of US adults heights with varying bin widths.

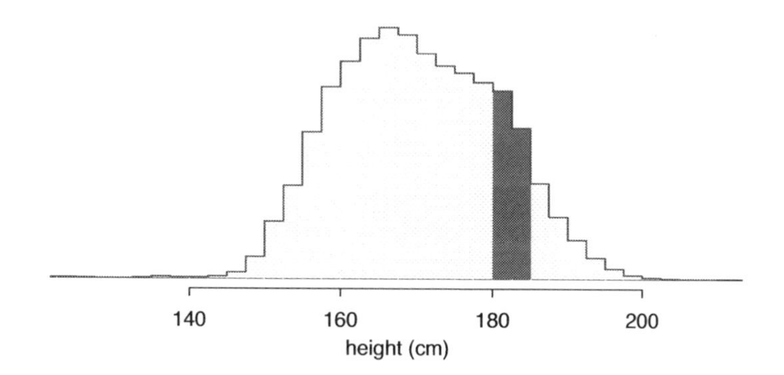

Figure 2.14: A histogram with bin sizes of 2.5 cm. The shaded region represents individuals with heights between 180 and 185 cm.

Example 2.36 What proportion of the sample is between 180 cm and 185 cm tall (about 5'11" to 6'1")?

The probability that a randomly selected person is between 180 cm and 185 cm can be estimated by finding the proportion of people in the sample who are between these two heights. This means adding up the heights of the bins in this range and dividing by the sample size. For instance, this can be done with the two shaded bins in Figure 2.14. The two bins in this region have counts of 195,307 and 156,239 people, resulting in the following estimate of the probability:

$$\frac{195307 + 156239}{3{,}000{,}000} = 0.1172$$

This fraction is the same as the proportion of the histogram's area that falls in the range 180 to 185 cm.

2.2.1 From histograms to continuous distributions

Examine the transition from a boxy hollow histogram in the top-left of Figure 2.13 to the much more smooth plot in the lower-right. In this last plot, the bins are so slim that the hollow histogram is starting to resemble a smooth curve. This suggests the population height as a *continuous* numerical variable might best be explained by a curve that represents the top of extremely slim bins.

This smooth curve represents a **probability density function** (also called a **density** or **distribution**), and such a curve is shown in Figure 2.15 overlaid on a histogram of the sample. A density has a special property: the total area under the density's curve is 1.

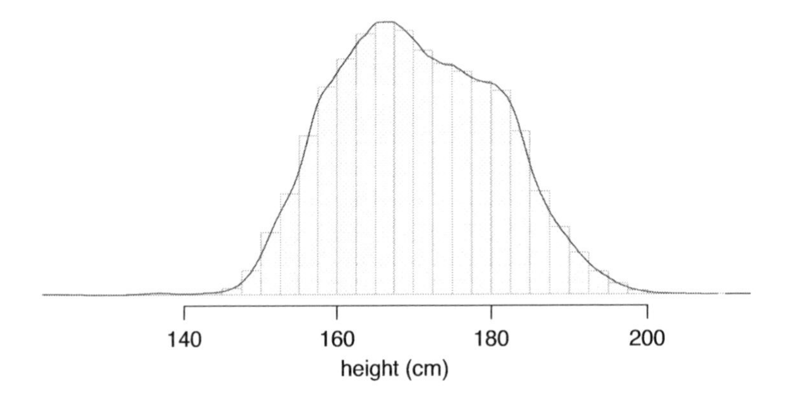

Figure 2.15: The continuous probability distribution of heights for US adults.

2.2.2 Probabilities from continuous distributions

We computed the proportion of individuals with heights 180 to 185 cm in Example 2.36 as a fraction:

$$\frac{\text{number of people between 180 and 185}}{\text{total sample size}}$$

We found the number of people with heights between 180 and 185 cm by determining the shaded boxes in this range, which represented the fraction of the box area in this region. Similarly, we can use the area in the shaded region under the curve to find a probability (with the help of a computer):

$$P(\texttt{height} \text{ between 180 and 185}) = \text{area between 180 and 185} = 0.1157$$

The probability that a randomly selected person is between 180 and 185 cm is 0.1157. This is very close to the estimate from Example 2.36: 0.1172.

⊙ **Exercise 2.37** Three US adults are randomly selected. The probability a single adult is between 180 and 185 cm is 0.1157. Short answers are in the footnote[20].

(a) What is the probability that all three are between 180 and 185 cm tall?

(b) What is the probability that none are between 180 and 185 cm?

[20](a) $0.1157 * 0.1157 * 0.1157 = 0.0015$. (b) $(1 - 0.1157)^3 = 0.692$

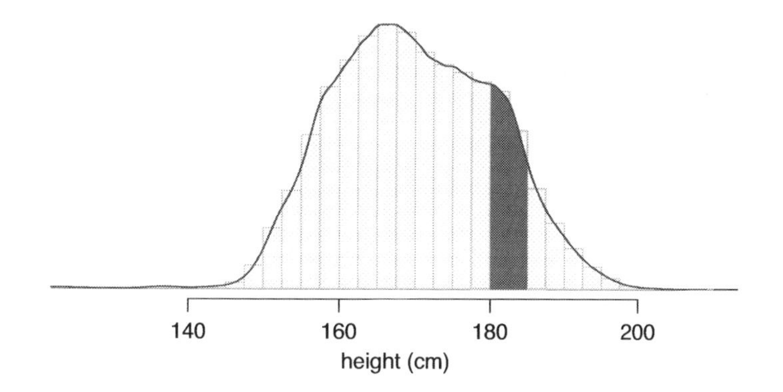

Figure 2.16: Density for heights in the US adult population with the area between 180 and 185 cm shaded. Compare this plot with Figure 2.14.

⊙ **Exercise 2.38** What is the probability a randomly selected person is **exactly** 180 cm? Assume you can measure perfectly. Answer in the footnote[21].

⊙ **Exercise 2.39** Suppose a person's height is rounded to the nearest centimeter. Is there a chance that a random person's **measured** height will be 180 cm? Answer in the footnote[22].

2.3 Conditional probability (special topic)

Are students more likely to use marijuana when their parents used drugs? The `drugUse`[23] data set contains a sample of 445 cases with two variables, `student` and `parents`, and is summarized in Table 2.17. The `student` variable is either `uses` or `not`, where a student `uses` if she has recently used marijuana. The `parents` variable takes the value `used` if at least one of the parents used drugs, including alcohol.

		parents		
		used	not	Total
student	uses	125	94	219
	not	85	141	226
	Total	210	235	445

Table 2.17: Contingency table summarizing the `drugUse` data set.

● **Example 2.40** If at least one parent used drugs, what is the chance their child (`student`) uses?

We will estimate this probability using the data. Of the 210 cases in this data set where `parents` = `used`, 125 represent cases where `student` = `uses`:

$$P(\texttt{student} = \texttt{uses} \text{ given } \texttt{parents} = \texttt{used}) = \frac{125}{210} = 0.60$$

[21]This probability is zero. A person might be close to 180 cm, but not exactly 180 cm tall. This also makes sense with the definition of probability as area; there is no area between 180 cm to 180 cm.

[22]This has positive probability. Anyone between 179.5 cm and 180.5 cm will have a *measured* height of 180 cm. This is probably a more realistic scenario to encounter in practice versus Exercise 2.38.

[23]Ellis GJ and Stone LH. 1979. Marijuana Use in College: An Evaluation of a Modeling Explanation. Youth and Society 10:323-334.

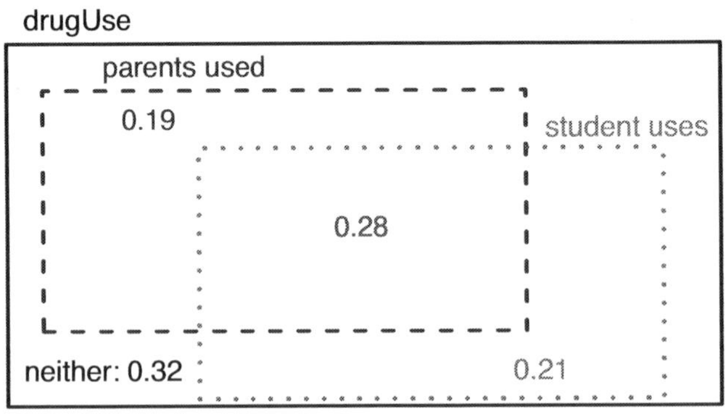

Figure 2.18: A Venn diagram using boxes for the **drugUse** data set.

	parents: used	parents: not	Total
student: uses	0.28	0.21	0.49
student: not	0.19	0.32	0.51
Total	0.47	0.53	1.000

Table 2.19: Probability table summarizing parental and student drug use.

● **Example 2.41** A student is randomly selected from the study and she does not use drugs. What is the probability at least one of her parents used?

If the student does not use drugs, then she is one of the 226 students in the second row. Of these 226 students, 85 had at least one parent who used drugs:

$$P(\texttt{parents} = \texttt{used} \text{ given } \texttt{student} = \texttt{not}) = \frac{85}{226} = 0.376$$

2.3.1 Marginal and joint probabilities

Table 2.17 includes row and column totals for the counts. These totals provide summary information about each variable separately. These totals represent **marginal probabilities** for the sample, which are the probabilities based on a single variable without conditioning on any other variables. For instance, a probability based solely on the student variable is a marginal probability:

$$P(\texttt{student} = \texttt{uses}) = \frac{219}{445} = 0.492$$

A probability of outcomes for two or more variables or processes is called a **joint probability**:

$$P(\texttt{student} = \texttt{uses} \ \& \ \texttt{parents} = \texttt{not}) = \frac{94}{445} = 0.21$$

It is common to substitute a comma for the ampersand (&) in a joint probability, although either is acceptable.

Joint outcome	Probability
`parents` = `used`, `student` = `uses`	0.28
`parents` = `used`, `student` = `not`	0.19
`parents` = `not`, `student` = `uses`	0.21
`parents` = `not`, `student` = `not`	0.32
Total	1.00

Table 2.20: A joint probability distribution for the `drugUse` data set.

Marginal and joint probabilities
If a probability is based on a single variable, it is a *marginal probability.* The probability of outcomes for two or more variables or processes is called a *joint probability.*

We use **table proportions** to summarize joint probabilities for the `drugUse` sample. These proportions are computed by dividing each count in Table 2.17 by 445 to obtain the proportions in Table 2.19. The joint probability distribution of the `parents` and `student` variables is shown in Table 2.20.

⊙ **Exercise 2.42** Verify Table 2.20 represents a probability distribution: events are disjoint, all probabilities are non-negative, and the probabilities sum to 1.

⊙ **Exercise 2.43** Which table do you find more useful, Table 2.19 or Table 2.20?

We can compute marginal probabilities using joint probabilities in simple cases. For example, the probability a random student from the study uses drugs is found by summing the outcomes from Table 2.20 where `student` = `uses`:

$$\begin{aligned} &P(\underline{\texttt{student} = \texttt{uses}}) \\ &\quad = P(\texttt{parents} = \texttt{used}, \underline{\texttt{student} = \texttt{uses}}) + \\ &\qquad\qquad P(\texttt{parents} = \texttt{not}, \underline{\texttt{student} = \texttt{uses}}) \\ &\quad = 0.28 + 0.21 = 0.49 \end{aligned}$$

2.3.2 Defining conditional probability

There is some connection between drug use of parents and of the student: drug use of one is associated with drug use of the other[24]. In this section, we discuss how to use information about associations between two variables to improve probability estimation.

The probability that a random student from the study uses drugs is 0.49. Could we update this probability if we knew that this student's parents used drugs? Absolutely. To do so, we limit our view to only those 210 cases where parents used drugs and look at the fraction where the student uses drugs:

$$P(\texttt{student} = \texttt{uses} \text{ given } \texttt{parents} = \texttt{used}) = \frac{125}{210} = 0.60$$

We call this a **conditional probability** because we computed the probability under a condition: `parents` = `used`. There are two parts to a conditional probability, **the outcome**

[24]This is an observational study and no causal conclusions may be reached.

of interest and the **condition**. It is useful to think of the condition as information we know to be true, and this information usually can be described as a known outcome or event.

We separate the text inside our probability notation into the outcome of interest and the condition:

$$\begin{aligned} &P(\texttt{student} = \texttt{uses given parents} = \texttt{used}) \\ &= P(\texttt{student} = \texttt{uses} \mid \texttt{parents} = \texttt{used}) = \frac{125}{210} = 0.60 \end{aligned} \quad (2.44)$$

$P(A|B)$
Probability of outcome A given B

The vertical bar "|" is read as *given*.

In Equation (2.44), we computed the probability a student uses based on the condition that at least one parent used as a fraction:

$$\begin{aligned} &P(\texttt{student} = \texttt{uses} \mid \texttt{parents} = \texttt{used}) \\ &= \frac{\#\text{ times } \texttt{student} = \texttt{uses} \;\&\; \texttt{parents} = \texttt{used}}{\#\text{ times } \texttt{parents} = \texttt{used}} \\ &= \frac{125}{210} = 0.60 \end{aligned} \quad (2.45)$$

We considered only those cases that met the condition, `parents` = `used`, and then we computed the ratio of those cases that satisfied our outcome of interest, that the student uses.

Counts are not always available for data, and instead only marginal and joint probabilities may be provided. Disease rates are commonly listed in percentages rather than in a count format. We would like to be able to compute conditional probabilities even when no counts are available, and we use Equation (2.45) as an example demonstrating this technique.

We considered only those cases that satisfied the condition, `parents` = `used`. Of these cases, the conditional probability was the fraction who represented the outcome of interest, `student` = `uses`. Suppose we were provided only the information in Table 2.19 on page 70, i.e. only probability data. Then if we took a sample of 1000 people, we would anticipate about 47% or $0.47 * 1000 = 470$ would meet our information criterion. Similarly, we would expect about 28% or $0.28 * 1000 = 280$ to meet both the information criterion and represent our outcome of interest. Thus, the conditional probability could be computed:

$$\begin{aligned} P(\texttt{student} = \texttt{uses} \mid \texttt{parents} = \texttt{used}) &= \frac{\#\,(\texttt{student} = \texttt{uses} \;\&\; \texttt{parents} = \texttt{used})}{\#\,(\texttt{parents} = \texttt{used})} \\ &= \frac{280}{470} = \frac{0.28}{0.47} = 0.60 \end{aligned} \quad (2.46)$$

where `S` and `parents` represent the `student` and `parents` variables. In Equation (2.46), we examine exactly the fraction of two probabilities, 0.28 and 0.47, which we can write as

$$P(\texttt{S} = \texttt{uses} \;\&\; \texttt{parents} = \texttt{used}) \quad \text{and} \quad P(\texttt{parents} = \texttt{used}).$$

The fraction of these probabilities represents our general formula for conditional probability.

Conditional Probability

The conditional probability of the outcome of interest A given condition B is computed as the following:

$$P(A|B) = \frac{P(A \;\&\; B)}{P(B)} \quad (2.47)$$

		inoculated		
		yes	no	Total
result	lived	238	5136	5374
	died	6	844	850
	Total	244	5980	6224

Table 2.21: Contingency table for the `smallpox` data set.

⊙ **Exercise 2.48** Answers in the footnote[25].

(a) Write out the following statement in conditional probability notation: "*The probability a random case has* `parents` = `not` *if it is known that* `student` = `not`". Notice that the condition is now based on the student, not the parent.

(b) Determine the probability from part (a). Table 2.19 on page 70 may be helpful.

⊙ **Exercise 2.49**

(a) Determine the probability that one of the parents had used drugs if it is known the student does not use drugs.

(b) Using the answers from part (a) and Exercise 2.48(b), compute

$$P(\texttt{parents} = \texttt{used}|\texttt{student} = \texttt{not}) + P(\texttt{parents} = \texttt{not}|\texttt{student} = \texttt{not})$$

(c) Provide an intuitive argument to explain why the sum in (b) is 1. Answer in the footnote[26].

⊙ **Exercise 2.50** The data indicate that drug use of parents and children are associated. Does this mean the drug use of parents causes the drug use of the students? Answer in the footnote[27].

2.3.3 Smallpox in Boston, 1721

The `smallpox` data set provides a sample of 6,224 individuals from the year 1721 who were exposed to smallpox in Boston[28]. Doctors at the time believed that inoculation, which involves exposing a person to the disease in a controlled form, could reduce the likelihood of death.

Each case represents one person with two variables: `inoculated` and `result`. The variable `inoculated` takes two levels: `yes` or `no`, indicating whether the person was inoculated or not. The variable `result` has outcomes `lived` or `died`. We summarize the data in Tables 2.21 and 2.22.

⊙ **Exercise 2.51** Write out, in formal notation, the probability a randomly selected person who was not inoculated died from smallpox, and find this probability. Brief answer in the footnote[29].

[25](a) $P(\texttt{parent} = \texttt{not}|\texttt{student} = \texttt{not})$. (b) Equation (2.47) for conditional probability suggests we should first find $P(\texttt{parents} = \texttt{not}\ \&\ \texttt{student} = \texttt{not}) = 0.32$ and $P(\texttt{student} = \texttt{not}) = 0.51$. Then the ratio represents the conditional probability: $0.32/0.51 = 0.63$.

[26]Under the condition the student does not use drugs, the parents must either use drugs or not. The complement still appears to work *when conditioning on the same information*.

[27]No. This was an observational study. Two potential lurking variables include `income` and `region`. Can you think of others?

[28]Fenner F. 1988. *Smallpox and Its Eradication (History of International Public Health, No. 6)*. Geneva: World Health Organization. ISBN 92-4-156110-6.

[29]$P(\texttt{result} = \texttt{died} \mid \texttt{inoculated} = \texttt{no}) = \frac{0.1356}{0.9608} = 0.1411$.

		inoculated		
		yes	no	Total
result	lived	0.0382	0.8252	0.8634
	died	0.0010	0.1356	0.1366
	Total	0.0392	0.9608	1.0000

Table 2.22: Table proportions for the `smallpox` data, computed by dividing each count by the table total, 6224.

⊙ **Exercise 2.52** Determine the probability an inoculated person died from smallpox. How does this result compare with the result of Exercise 2.51?

⊙ **Exercise 2.53** The people of Boston self-selected whether or not to be inoculated. (a) Is this study observational or experimental? (b) Can we infer any causal connection using these data? (c) What are some potential lurking variables that might influence whether someone `lived` or `died` and affect whether that person was inoculated or not? Answers in the footnote[30].

2.3.4 General multiplication rule

Section 2.1.6 introduced a multiplication rule for independent processes. Here we provide a General Multiplication Rule for events that might not be independent.

General Multiplication Rule
If A and B represent two outcomes or events, then

$$P(A \text{ and } B) = P(A|B) * P(B)$$

It is useful to think of A as the outcome of interest and B as the condition.

This General Multiplication Rule is simply a rearrangement of the definition for conditional probability in Equation (2.47) on page 72.

● **Example 2.54** Consider the `smallpox` data set. Suppose we knew that 85.88% of residents exposed to smallpox were not inoculated, 96.08% of exposed residents were not inoculated, and no other information was provided. How could we compute the probability a resident who was exposed to smallpox was not inoculated and also survived?

We will compute our answer and then verify it using Table 2.22. We want to determine

$$P(\texttt{result} = \texttt{lived} \,\&\, \texttt{inoculated} = \texttt{no})$$

and we are given that

$$P(\texttt{result} = \texttt{lived} \mid \texttt{inoculated} = \texttt{no}) = 0.8588$$
$$P(\texttt{inoculated} = \texttt{no}) = 0.9608$$

[30](a) Observational. (b) No! We cannot infer causation from this observational study. (c) Accessibility to the latest and best medical care. (There are other valid answers.)

Among the 96.08% of people who were not inoculated, 85.88% survived:

$$P(\texttt{result} = \texttt{lived \& inoculated} = \texttt{no}) = 0.8588 * 0.9608 = 0.8251$$

This is equivalent to the General Multiplication Rule. We can confirm this probability in Table 2.22 on the facing page at the intersection of `no` and `lived` (with a little rounding error).

⊙ **Exercise 2.55** Use $P(\texttt{inoculated} = \texttt{yes}) = 0.0392$ and $P(\texttt{result} = \texttt{lived} \mid \texttt{inoculated} = \texttt{yes}) = 0.9754$ to determine the probability a person was both inoculated and lived. Verify your answer using Table 2.22.

Sum of conditional probabilities
Let A_1, ..., A_k represent all the disjoint outcomes for a variable or process. Then if B is a condition for another variable or process, we have:

$$P(A_1|B) + \cdots + P(A_k|B) = 1$$

The rule for complements also holds when an event and its complement are conditioned on the same information:

$$P(A|B) = 1 - P(A^c|B)$$

⊙ **Exercise 2.56** If 97.54% of the people who were inoculated lived, what proportion of inoculated people must have died?

⊙ **Exercise 2.57** Do you think inoculation is effective at reducing the risk of death from smallpox? Use Table 2.21 to support your answer.

2.3.5 Independence considerations in conditional probability

If two processes are independent, then knowing the outcome of one should provide no information about the other. We can show this is mathematically true using conditional probabilities.

⊙ **Exercise 2.58** Let X and Y represent the outcomes of rolling two dice. (a) What is the probability the first die, X, is 1? (b) What is the probability both X and Y are 1? (c) Use the formula for conditional probability to compute $P(Y = 1 \mid X = 1)$. (d) When X was conditioned to be 1, did it alter the probability Y was 1 in part (c)?

We can show in Exercise 2.58(c) that the conditioning information has no influence by using the Multiplication Rule for independence processes:

$$\begin{aligned} P(Y = 1|X = 1) &= \frac{P(Y = 1 \ \& \ X = 1)}{P(X = 1)} \\ &= \frac{P(Y = 1) * P(X = 1)}{P(X = 1)} \\ &= P(Y = 1) \end{aligned}$$

⊙ **Exercise 2.59** Ron is watching a roulette table in a casino and notices that the last five outcomes were `black`. He figures that the chances of getting `black` six times in a row is very small (about 1/64) and puts his paycheck on red. What is wrong with his reasoning?

2.3.6 Tree diagrams

Tree diagrams are a tool to organize outcomes and probabilities around the structure of the data. They are most useful when two or more processes occur in a sequence and each process is conditioned on its predecessors.

The `smallpox` data fit this description. We see the population as split by `inoculation`: `yes` and `no`. Following this split, survival rates were observed for each group. We construct the tree diagram to follow the data structure. It splits the group by `inoculation` and then split those subgroups by `result`, shown in Figure 2.23. The first branch for `inoculation` is said to be the **primary** branch while the other branches are **secondary**.

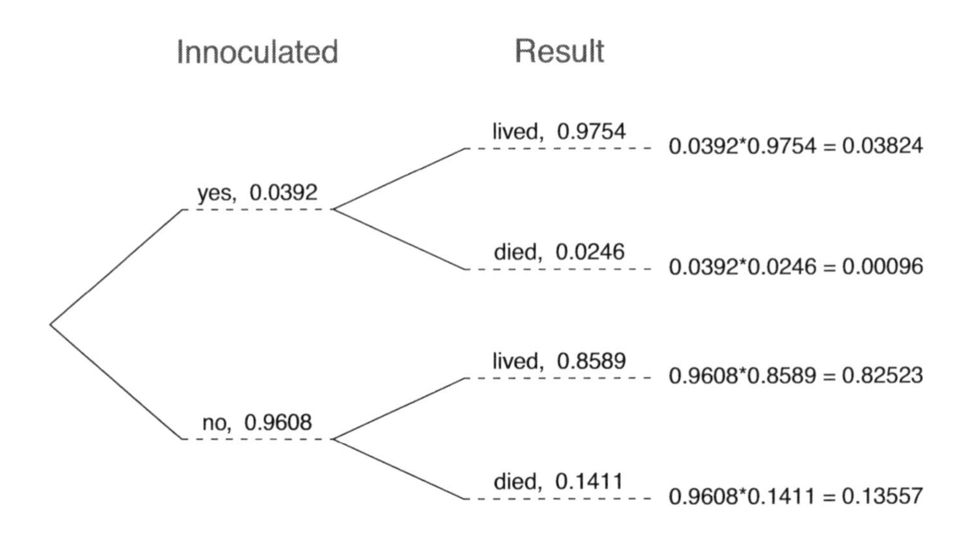

Figure 2.23: A tree diagram of the `smallpox` data set.

We annotate the tree diagram with marginal and conditional probabilities on the branches. We first split the data by `inoculation` into the `yes` and `no` groups with respective marginal probabilities 0.0392 and 0.9608. The second split is conditioned on the first, so we assign conditional probabilities to the branches. For example, the top branch in Figure 2.23 is the probability that `result` = `lived` conditioned on the information that `inoculated` = `yes`. We may (and usually do) construct joint probabilities at the end of each branch in our tree by multiplying the numbers we come across as we move from left to right. These joint probabilities are computed directly from the General Multiplication Rule:

$$\begin{aligned}
&P(\texttt{inoculated} = \texttt{yes} \ \& \ \texttt{result} = \texttt{lived}) \\
&\quad = P(\texttt{inoculated} = \texttt{yes}) * P(\texttt{result} = \texttt{lived} | \texttt{inoculated} = \texttt{yes}) \\
&\quad = 0.0392 * 0.9754 = 0.0382
\end{aligned}$$

● **Example 2.60** Consider the midterm and final for a statistics class. Suppose 13% of students earn an A on the midterm. Of those students who earned an A on the midterm, 47% got an A on the final, and 11% of the students who got lower than an A on the midterm got an A on the final. You randomly pick up a final exam and notice the student got an A. What is the probability this student got an A on the midterm?

It is not obvious how to solve this problem. However, we can start by organizing our information into a tree diagram. First split the students based on the midterm and then split those primary branches into secondary branches based on the final. We associate the marginal probability with the primary branches and the conditional probabilities with the secondary branches. The result is shown in Figure 2.24.

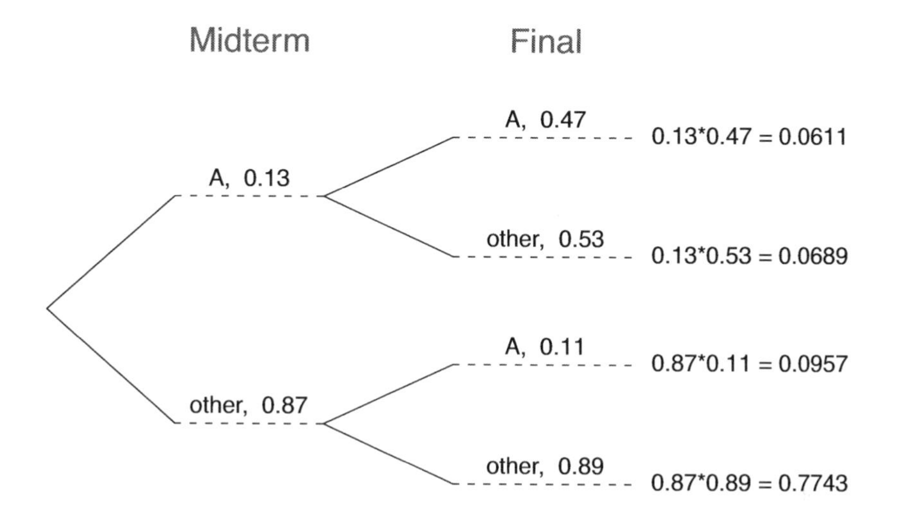

Figure 2.24: A tree diagram describing the `midterm` and `final` variables.

The end-goal is to find $P(\texttt{midterm} = \texttt{A}|\texttt{final} = \texttt{A})$. We can start by finding the two probabilities associated with this conditional probability using the tree diagram:

$$
\begin{aligned}
&P(\texttt{midterm} = \texttt{A}\ \&\ \texttt{final} = \texttt{A}) = 0.0611\\
&P(\texttt{final} = \texttt{A})\\
&\quad = P(\texttt{midterm} = \texttt{other}\ \&\ \texttt{final} = \texttt{A}) + P(\texttt{midterm} = \texttt{A}\ \&\ \texttt{final} = \texttt{A})\\
&\quad = 0.0611 + 0.0957 = 0.1568
\end{aligned}
$$

Then we take the ratio of these two probabilities:

$$
\begin{aligned}
P(\texttt{midterm} = \texttt{A}|\texttt{final} = \texttt{A}) &= \frac{P(\texttt{midterm} = \texttt{A}\ \&\ \texttt{final} = \texttt{A})}{P(\texttt{final} = \texttt{A})}\\
&= \frac{0.0611}{0.1568} = 0.3897
\end{aligned}
$$

The probability the student also got an A on the midterm is about 0.39.

⊙ **Exercise 2.61** After an introductory statistics course, 78% of students can successfully construct tree diagrams. Of those who can construct tree diagrams, 97% passed, while only 57% of those students who could not construct tree diagrams passed. (a) Organize this information into a tree diagram. (b) What is the probability a randomly selected student passed? (c) Compute the probability a student is able to construct a tree diagram if it is known that she passed. Hints plus a short answer to (c) in the footnote[31].

2.3.7 Bayes' Theorem

In many instances, we are given a conditional probability of the form

$$P(\text{statement about variable 1} \mid \text{statement about variable 2})$$

but we would really like to know the inverted conditional probability:

$$P(\text{statement about variable 2} \mid \text{statement about variable 1})$$

Tree diagrams can sometimes be used to find the second conditional probability when given the first. However, sometimes it is not always possible to draw the scenario in a tree diagram. In these cases, we can apply a very useful and general formula: Bayes' Theorem.

We first take a critical look at a new example of inverting conditional probabilities where we still apply a tree diagram.

● **Example 2.62** In the US, 0.5% of the population has Lupus. Suppose we consider a new test for Lupus, but this test is not perfectly accurate. In 1% of patients with Lupus, the test gives a *false negative*: it indicates a person does not have Lupus when she does have Lupus. Similarly, the test gives a *false positive* in 2% of patients who do not have Lupus: it indicates these patients have Lupus when they actually do not. If we tested a random person for Lupus and the test came back positive – that is, the test indicated the patient has Lupus – what is the probability the person actually has Lupus?

This is a conditional probability, and it can be broken into two pieces:

$$P(\text{patient has Lupus } | \text{pos. test}) = \frac{P(\text{patient has Lupus and the test is positive})}{P(\text{pos. test})}$$

A tree diagram is useful for identifying each probability and is shown in Figure 2.25. The probability the patient has Lupus and has a positive test result is

$$\begin{aligned} P(\text{pos. test \& has Lupus}) &= P(\text{pos. test} \mid \text{has Lupus})P(\text{has Lupus}) \\ &= 0.99 * 0.005 = 0.00495 \end{aligned}$$

The probability of a positive test result is the sum of the two corresponding scenarios:

$$\begin{aligned} P(\text{pos. test}) &= P(\text{pos. test \& no Lupus}) + P(\text{pos. test \& has Lupus}) \\ &= P(\text{pos. test} \mid \text{no Lupus})P(\text{no Lupus}) \\ &\quad + P(\text{pos. test} \mid \text{has Lupus})P(\text{has Lupus}) \\ &= 0.02 * 0.995 + 0.99 * 0.005 = 0.02485 \end{aligned}$$

[31](a) The first branch should be based on the variable that directly splits the data into groups: whether students can construct tree diagrams. (b) Identify which two joint probabilities represent students who passed. (c) Use the definition of conditional probability. Your answer to part (b) may be useful in part (c). The solution to (c) is 0.86.

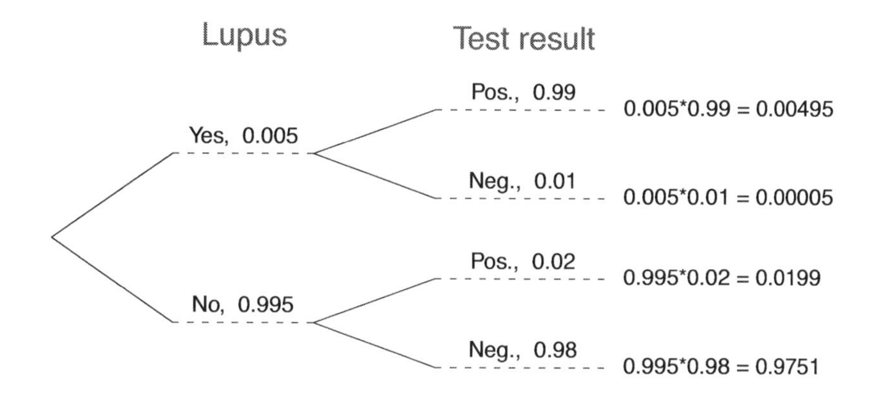

Figure 2.25: Tree diagram for Example 2.62, computing the probability a random patient who tests positive for Lupus actually has Lupus.

Then the probability the patient has Lupus conditioned on the test result being positive is

$$P(\text{patient has Lupus }|\text{pos. test}) = \frac{0.00495}{0.02485} \approx 0.1992$$

That is, even if a patient has a positive test result, there is still only a 20% chance she has Lupus.

Consider the conditional probability of Example 2.62:

$$P(\text{patient has Lupus }|\text{pos. test}) = \frac{P(\text{patient has Lupus and the test is positive})}{P(\text{the test is positive})}$$

Using the tree diagram, we can see that the numerator (the top of the fraction) is equal to the following product:

$$P(\text{pos. test \& has Lupus}) = P(\text{pos. test } | \text{ has Lupus})P(\text{has Lupus})$$

The denominator – the probability of a positive test result – is equal to all the different ways the test could be positive:

$$P(\text{pos. test}) = P(\text{pos. test \& no Lupus}) + P(\text{pos. test \& has Lupus})$$

Each of these probabilities can be broken down into a product of a conditional and marginal probability, just like what was done in the numerator:

$$\begin{aligned} P(\text{pos. test}) &= P(\text{pos. test \& no Lupus}) + P(\text{pos. test \& has Lupus}) \\ &= P(\text{pos. test } | \text{ no Lupus})P(\text{no Lupus}) \\ &\qquad + P(\text{pos. test } | \text{ has Lupus})P(\text{has Lupus}) \end{aligned}$$

This example, where the numerator and the denominator are broken down into their pieces, actually shows a particular case of Bayes' Theorem.

$$\begin{aligned} &P(\text{patient has Lupus }|\text{pos. test}) \\ &= \frac{P(\text{pos. test } | \text{ Lupus})P(\text{Lupus})}{P(\text{pos. test } | \text{ no Lupus})P(\text{no Lupus}) + P(\text{pos. test } | \text{ Lupus})P(\text{Lupus})} \end{aligned}$$

Bayes' Theorem: inverting probabilities
Consider the following conditional probability for variable 1 and variable 2:

$$P(\text{outcome } A_1 \text{ of variable 1} \mid \text{outcome } B \text{ of variable 2})$$

Bayes' Theorem states that this conditional probability can be identified as the following fraction:

$$\frac{P(B|A_1) * P(A_1)}{P(B|A_1) * P(A_1) + P(B|A_2) * P(A_2) + \cdots + P(B|A_k) * P(A_k)} \quad (2.63)$$

where A_2, A_3, ..., and A_k represent all other possible outcomes of the first variable.

Bayes' Theorem is just a generalization of what we have done using tree diagrams. The numerator (top of the fraction) identifies the probability of getting both A_1 and B. The denominator is the marginal probability of getting B. This second part looks long and complicated since we have to add up probabilities from all the different ways to get B. We always had to complete this step when using tree diagrams. However, we usually did it in a separate step so it didn't seem as complex.

To apply Bayes' Theorem correctly, there are two preparatory steps:

(1) First identify the marginal probabilities of each possible outcome of the first variable: $P(A_1)$, $P(A_2)$, ..., $P(A_k)$.

(2) Then identify the probability of the outcome B, conditioned on each possible scenario for the first variable: $P(B|A_1)$, $P(B|A_2)$, ..., $P(B|A_k)$.

Once each of these probabilities is identified, they can be applied directly within the formula.

TIP: Only use Bayes' Theorem when tree diagrams are difficult
Drawing a tree diagram makes it easier to understand how two variables are connected. Use Bayes' Theorem only when there are so many scenarios that drawing a tree diagram would be complex.

⊙ **Exercise 2.64** Jose visits campus every Thursday evening. However, some days the parking garage is full, often due to events. There are academic events on 35% of evenings and sporting events on 20% of evenings. When there is an academic event, the garage fills up about 25% of the time, and it fills up 70% of evenings with sporting events. On evenings when there are no events, it only fills up about 5% of the time. If Jose comes to campus and finds the garage full, what is the probability that there is a sporting event? Solve this problem using a tree diagram. Hint about the diagram in the footnote[32].

● **Example 2.65** Here we solve Exercise 2.64 using Bayes' Theorem.

The outcome of interest is whether there is a sporting event (call this A_1), and the condition is that the lot is full (B). Let A_2 represent an academic event and A_3 represent there being no event on campus. Then the given probabilities can be written as

[32]This tree diagram will have three primary branches: there is an academic event, sporting event, or no event on campus that evening.

$$P(A_1) = 0.2 \qquad P(A_2) = 0.35 \qquad P(A_3) = 1 - 0.2 - 0.35 = 0.45$$
$$P(B|A_1) = 0.7 \qquad P(B|A_2) = 0.25 \qquad P(B|A_3) = 0.05$$

Bayes' Theorem can be used to compute the probability of a sporting event (A_1) under the condition that the parking lot is full (B):

$$\begin{aligned} P(A_1|B) &= \frac{P(B|A_1) * P(A_1)}{P(B|A_1) * P(A_1) + P(B|A_2) * P(A_2) + P(B|A_3) * P(A_3)} \\ &= \frac{(0.7)(0.2)}{(0.7)(0.2) + (0.25)(0.35) + (0.05)(0.45)} \\ &= 0.56 \end{aligned}$$

While we have greater reason to believe there is a sporting event, we are by no means certain.

⊙ **Exercise 2.66** Using the probabilities given in Exercise 2.64, verify the probability that there is an academic event conditioned on the parking lot being full is 0.35.

⊙ **Exercise 2.67** In Exercise 2.64 and 2.66, you found that if the parking lot is full, the probability a sporting event is 0.56 and the probability there is an academic event is 0.35. Using this information, compute P(no event | the lot is full)?

The last several exercises offered a way to update how much we might believe there is a sporting event, academic event, or no event going on at the school based on the information that the parking lot was full. This strategy of *updating beliefs* using Bayes' Theorem is actually the foundation of an entire section of statistics called **Bayesian statistics**. While Bayesian statistics is very important and useful, we will not have time to cover much more of it in this book.

2.4 Sampling from a small population (special topic)

● **Example 2.68** Professors sometimes select a student at random to answer a question. If the selection is truly random, and there are 15 people in your class, what is the chance that she will pick you for the next question?

If there are 15 people to ask and none are skipping class (for once), then the probability is $1/15$, or about 0.067.

● **Example 2.69** If the professor asks 3 questions, what is the probability you will not be selected? Assume that she will not pick the same person twice in a given lecture.

For the first question, she will pick someone else with probability $14/15$. When she asks the second question, she only has 14 people who have not yet been asked. Thus, if you were not picked on the first question, the probability you are again not picked is $13/14$. Similarly, the probability you are again not picked on the third question is $12/13$, and the probability of not being picked for any of the three questions is

$$\begin{aligned} & P(\text{not picked in 3 questions}) \\ &= P(\texttt{Q1} = \texttt{notPicked}, \texttt{Q2} = \texttt{notPicked}, \texttt{Q3} = \texttt{notPicked}.) \\ &= \frac{14}{15} \times \frac{13}{14} \times \frac{12}{13} = \frac{12}{15} = 0.80 \end{aligned}$$

⊙ **Exercise 2.70** What rule permitted us to multiply the probabilities in Example 2.69? Answer in the footnote[33].

● **Example 2.71** Suppose the professor randomly picks without regard to who she already selected, i.e. students can be picked more than once. What is the probability you will not be picked for any of the three questions?

Each pick is independent, and the probability of `notPicked` on any one question is 14/15. Thus, we can use the Product Rule for independent events.

$$\begin{aligned} &P(\text{not picked in 3 questions}) \\ &\quad = P(\texttt{Q1} = \texttt{notPicked}, \texttt{Q2} = \texttt{notPicked}, \texttt{Q3} = \texttt{notPicked}.) \\ &\quad = \frac{14}{15} \times \frac{14}{15} \times \frac{14}{15} = 0.813 \end{aligned}$$

You have a slightly higher chance of not being picked compared to when she picked a new person for each question. However, you now may be picked more than once.

⊙ **Exercise 2.72** Under the setup of Example 2.71, what is the probability of being picked to answer all three questions?

If we sample from a small population **without replacement**, we no longer have independence between our observations. In Example 2.69, the probability of `notPicked` for the second question was conditioned on the event that you were `notPicked` for the first question. In Example 2.71, the professor sampled her students **with replacement**: she repeatedly sampled the entire class without regard to who she already picked.

⊙ **Exercise 2.73** Your department is holding a raffle. They sell 30 tickets and offer seven prizes. (a) They place the tickets in a hat and draw one for each prize. The tickets are sampled without replacement, i.e. the selected tickets are not placed back in the hat. What is the probability of winning a prize if you buy one ticket? (b) What if the tickets are sampled with replacement? Answers in the footnote[34].

⊙ **Exercise 2.74** Compare your answers in Exercise 2.73. How much influence does the sampling method have on your chances of winning a prize?

Had we repeated Exercise 2.73 with 300 tickets instead of 30, we would have found something interesting: the results would be nearly identical. The probability would be

[33]The three probabilities we computed were actually one marginal probability, $P(\texttt{Q1}{=}\texttt{notPicked})$, and two conditional probabilities:

$$\begin{aligned} &P(\texttt{Q2} = \texttt{notPicked}|\texttt{Q1} = \texttt{notPicked}) \\ &P(\texttt{Q3} = \texttt{notPicked}|\texttt{Q1} = \texttt{notPicked}, \texttt{Q2} = \texttt{notPicked}) \end{aligned}$$

Using the General Multiplication Rule, the product of these three probabilities is the probability of not being picked in 3 questions.

[34](a) First determine the probability of not winning. The tickets are sampled without replacement, which means the probability you do not win on the first draw is 29/30, 28/29 for the second, ..., and 23/24 for the seventh. The probability you win no prize is the product of these separate probabilities: 23/30. That is, the probability of winning a prize is $7/30 = 0.233$. (b) When the tickets are sampled with replacement, they are seven independent draws. Again we first find the probability of not winning a prize: $(29/30)^7 = 0.789$. Thus, the probability of winning (at least) one prize when drawing with replacement is 0.211.

0.0233 without replacement and 0.0231 with replacement. When the sample size is only a small fraction of the population (under 10%), observations are nearly independent even when sampling without replacement.

2.5 Random variables (special topic)

● **Example 2.75** Two books are assigned for a statistics class: a textbook and its corresponding study guide. The student bookstore determined 20% of enrolled students do not buy either book, 55% buy the textbook, and 25% buy both books, and these percentages are relatively constant from one term to another. If there are 100 students enrolled, how many books should the bookstore expect to sell to this class?

Around 20 students will not buy either book (0 books total), about 55 will buy one book (55 total), and approximately 25 will buy two books (totaling 50 books for these 25 students). The bookstore should expect to sell about 105 books for this class.

⊙ **Exercise 2.76** Would you be surprised if the bookstore sold slightly more or less than 105 books?

● **Example 2.77** The textbook costs \$137 and the study guide \$33. How much revenue should the bookstore expect from this class of 100 students?

About 55 students will just buy a textbook, providing revenue of

$$\$137 * 55 = \$7,535$$

The roughly 25 students who buy both the textbook and a study guide would pay a total of

$$(\$137 + \$33) * 25 = \$170 * 25 = \$4,250$$

Thus, the bookstore should expect to generate about \$11,785 from these 100 students for this one class. However, there might be some *sampling variability* so the actual amount may differ by a little bit.

● **Example 2.78** What is the average revenue per student for this course?

The expected total revenue is \$11,785, and there are 100 students. Therefore the expected revenue per student is $\$11,785/100 = \$117.85$.

2.5.1 Expectation

We call a variable or process with a numerical outcome a **random variable**, and usually represent this random variable with a capital letter such as X, Y, or Z. The amount of money a single student will spend on her statistics books is a random variable, and we represent it by X.

Random variable
A random process or variable with a numerical outcome.

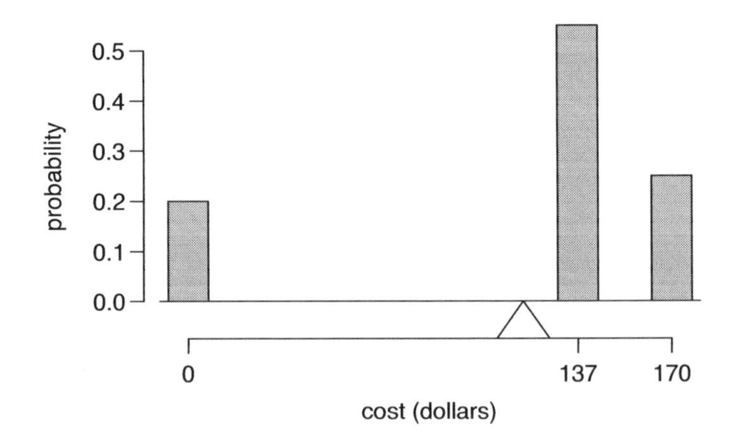

Figure 2.26: Probability distribution for the bookstore's revenue from a single student. The distribution balances on a pyramid representing the average revenue per student.

i	1	2	3	Total
x_i	\$0	\$137	\$170	–
$P(X = x_i)$	0.20	0.55	0.25	1.00

Table 2.27: The probability distribution for the random variable X, representing the bookstore's revenue from a single student.

The possible outcomes of X are labeled with a corresponding lower case letter and subscripts. For example, we write $x_1 = \$0$, $x_2 = \$137$, and $x_3 = \$170$, which occur with probabilities 0.20, 0.55, and 0.25. The distribution of X is summarized in Figure 2.26 and Table 2.27.

$E(X)$
Expected value of X

We computed the average outcome of X as \$117.85 in Example 2.78. We call this average the **expected value** of X, denoted by $E(X)$. The expected value of a random variable is computed by adding each outcome weighted by its probability:

$$\begin{aligned} E(X) &= \$0 * P(X = \$0) + \$137 * P(X = \$137) + \$170 * P(X = \$170) \\ &= \$0 * 0.20 + \$137 * 0.55 + \$170 * 0.25 = \$117.85 \end{aligned}$$

Expected value of a Discrete Random Variable
If X takes outcomes x_1, ..., x_k with probabilities $P(X = x_1)$, ..., $P(X = x_k)$, the expected value of X is the sum of each outcome multiplied by its corresponding probability:

$$\begin{aligned} E(X) &= x_1 * P(X = x_1) + \cdots + x_k * P(X = x_k) \\ &= \sum_{i=1}^{k} x_i P(X = x_i) \end{aligned} \tag{2.79}$$

The notation μ may be used in place of $E(X)$.

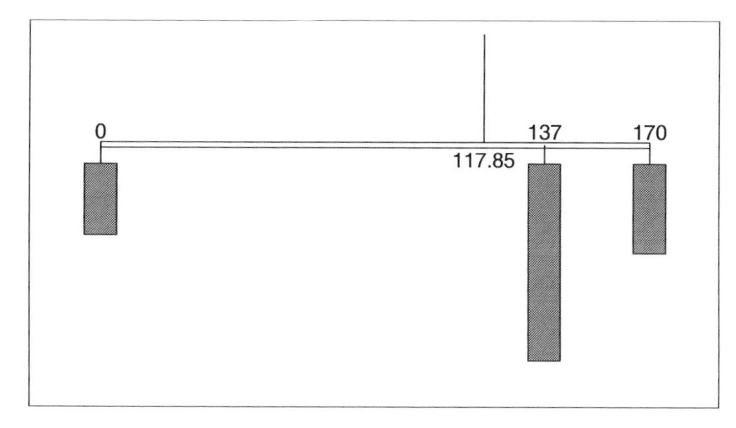

Figure 2.28: A weight system representing the probability distribution for X. The string holds the distribution at the mean to keep the system balanced.

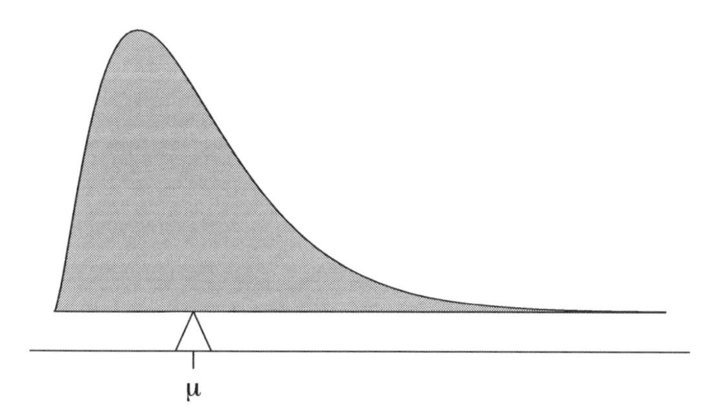

Figure 2.29: A continuous distribution can also be balanced at its mean.

The expected value for a random variable represents the average outcome. For example, $E(X) = 117.85$ represents the average amount the bookstore expects to make from a single student, which we could write as $\mu = 117.85$.

It is also possible to compute the expected value of a continuous random variable. However, it requires a little calculus and we save it for a later class[35].

In physics, the expectation holds the same meaning as the center of gravity. The distribution can be represented by a series of weights at each outcome, and the mean represents the balancing point. This is represented in Figures 2.26 and 2.28. The idea of a center of gravity also expands to a continuous probability distribution. Figure 2.29 shows a continuous probability distribution balanced atop a wedge placed at the mean.

2.5.2 Variability in random variables

Suppose you ran the university bookstore. Besides how much revenue you expect to generate, you might also want to know the volatility (variability) in your revenue.

[35] $\mu = \int xf(x)dx$ where $f(x)$ represents a function for the density curve.

The variance and standard deviation can be used to describe the variability of a random variable. Section 1.3.5 introduced a method for finding the variance and standard deviation for a data set. We first computed deviations from the mean ($x_i - \mu$), squared those deviations, and took an average to get the variance. In the case of a random variable, we again compute squared deviations. However, we take their sum weighted by their corresponding probabilities, just like we did for the expectation. This weighted sum of squared deviations equals the variance, and we calculate the standard deviation by taking the square root of the variance, just as we did in Section 1.3.5.

$Var(X)$
Variance of X

General variance formula
If X takes outcomes x_1, ..., x_k with probabilities $P(X = x_1)$, ..., $P(X = x_k)$ and expected value $\mu = E(X)$, then the variance of X, denoted by $Var(X)$ or the symbol σ^2, is

$$\begin{aligned} \sigma^2 &= (x_1 - \mu)^2 * P(X = x_1) + \cdots \\ &\qquad \cdots + (x_k - \mu)^2 * P(X = x_k) \\ &= \sum_{j=1}^{k} (x_j - \mu)^2 P(X = x_j) \end{aligned} \tag{2.80}$$

The standard deviation of X (σ) is the square root of the variance.

Example 2.81 Compute the expected value, variance, and standard deviation of X, the revenue of a single statistics student for the bookstore.

It is useful to construct a table that holds computations for each outcome separately, then add up the results.

i	1	2	3	Total
x_i	\$0	\$137	\$170	
$P(X = x_i)$	0.20	0.55	0.25	
$x_i * P(X = x_i)$	0	75.35	42.50	117.85

Thus, the expected value is $\mu = 117.35$, which we computed earlier. The variance can be constructed by extending this table:

i	1	2	3	Total
x_i	\$0	\$137	\$170	
$P(X = x_i)$	0.20	0.55	0.25	
$x_i * P(X = x_i)$	0	75.35	42.50	117.85
$x_i - \mu$	-117.85	19.15	52.15	
$(x_i - \mu)^2$	13888.62	366.72	2719.62	
$(x_i - \mu)^2 * P(X = x_i)$	2777.7	201.7	679.9	3659.3

The variance of X is $\sigma^2 = 3659.3$, which means the standard deviation is $\sigma = \sqrt{3659.3} = \$60.49$.

⊙ **Exercise 2.82** The bookstore also offers a chemistry textbook for \$159 and a book supplement for \$41. From past experience, they know about 25% of chemistry students just buy the textbook while 60% buy both the textbook and supplement. Answers for each part below are provided in the footnote[36].

(a) What proportion of students don't buy either book? Again assume no students buy the supplement without the textbook.

(b) Let Y represent the revenue from a single student. Write out the probability distribution of Y, i.e. a table for each outcome and its associated probability.

(c) Compute the expected revenue from a single chemistry student.

(d) Find the standard deviation to describe the variability associated with the revenue from a single student.

2.5.3 Linear combinations of random variables

So far, we have thought of each variable as being a complete story in and of itself. Sometimes it is more appropriate to use a combination of variables. For instance, the amount of time a person spends commuting to work each week can be broken down into several daily commutes. Similarly, the total gain or loss in a stock portfolio is only the sum of the gains and losses in its components.

● **Example 2.83** John travels to work five days a week. We will use X_1 to represent his travel time on Monday, X_2 to represent his travel time on Tuesday, and so on. Write an equation using X_1, ..., X_5 that represents his travel time for the week, denoted by W.

His total weekly travel time is the sum of the five daily values:

$$W = X_1 + X_2 + X_3 + X_4 + X_5$$

Breaking the weekly travel time into pieces provides a framework for understanding each source of randomness and how best to model them.

● **Example 2.84** It takes John an average of 18 minutes each day to commute. What would you expect his average commute time to be for the week?

We were told that the average (i.e. expected value) of the commute time is 18 minutes per day: $E(X_i) = 18$. To get the expected time for the sum of the five days, we can

[36] (a) 100% - 25% - 60% = 15% of students do not buy any books for the class. Part (b) is represented by the first two lines in the table below. The expectation for part (c) is given as the total on the line $y_i * P(Y = y_i)$. The result of part (d) is the square-root of the variance listed on in the total on the last line: $\sigma = \sqrt{Var(Y)} = \$69.28$.

i (scenario)	1 (noBook)	2 (textbook)	3 (both)	Total
y_i	0.00	159.00	200.00	
$P(Y = y_i)$	0.15	0.25	0.60	
$y_i * P(Y = y_i)$	0.00	39.75	120.00	$E(Y) = \$159.75$
$y_i - E(Y)$	-159.75	-0.75	40.25	
$(y_i - E(Y))^2$	25520.06	0.56	1620.06	
$(y_i - E(Y))^2 * P(Y)$	3828.0	0.1	972.0	$Var(Y) \approx 4800$

add up the expected time for each individual day:

$$\begin{aligned} E(W) &= E(X_1 + X_2 + X_3 + X_4 + X_5) \\ &= E(X_1) + E(X_2) + E(X_3) + E(X_4) + E(X_5) \\ &= 18 + 18 + 18 + 18 + 18 = 90 \text{ minutes} \end{aligned}$$

The expectation of the total time is equal to the sum of the expected individual times. More generally, the expectation of a sum of random variables is always the sum of the expectation for each random variable.

⊙ **Exercise 2.85** Elena is selling a TV at a cash auction and also intends to buy a toaster oven in the auction. If X represents the profit for selling the TV and Y represents the cost of the toaster oven, write an equation that represents the net change in Elena's cash. Answer in the footnote[37].

⊙ **Exercise 2.86** Based on past auctions, Elena figures she should expect to make about \$175 on the TV and pay about \$23 for the toaster oven. In total, how much should she expect to make or spend? Answer in the footnote[38].

⊙ **Exercise 2.87** Would you be surprised if John's weekly commute wasn't exactly 90 minutes or if Elena didn't make exactly \$152? Explain.

Two important concepts concerning combinations of random variables have so far been introduced. First, a final value can sometimes be described as the sum of its parts in an equation. Second, intuition suggests that putting the individual average values into this equation gives the average value we would expect in total. This second point needs clarification – it is guaranteed to be true in what are called *linear combinations of random variables*.

A **linear combination** of two random variables X and Y is a fancy phrase to describe a combination

$$aX + bY$$

where a and b are some fixed and known numbers. For John's commute time, there were five random variables – one for each work day – and each random variable could be written as having a fixed coefficient of 1:

$$1 * X_1 + 1 * X_2 + 1 * X_3 + 1 * X_4 + 1 * X_5$$

For Elena's net gain or loss, the X random variable had a coefficient of +1 and the Y random variable had a coefficient of -1.

When considering the average of a linear combination of random variables, it is safe to plug in the mean of each random variable and then compute the final result. For a few examples of nonlinear combinations of random variables – cases where we cannot simply plug in the means – are provided in the footnote[39].

[37] She will make X dollars on the TV but spend Y dollars on the toaster oven: $X - Y$.

[38] $E(X - Y) = E(X) - E(Y) = 175 - 23 = \$152$. She should expect to make about \$152.

[39] If X and Y are random variables, consider the following combinations: X^{1+Y}, $X * Y$, X/Y. In such cases, plugging in the average value for each random variable and computing the result will not generally lead to an accurate average value for the end result.

Linear combinations of random variables and the average result
If X and Y are random variables, then a linear combination of the random variables is given by

$$aX + bY \tag{2.88}$$

where a and b are some fixed numbers. To compute the average value of a linear combination of random variables, plug in the average of each individual random variable and compute the result:

$$a\mu_X + b\mu_Y$$

Recall that the expected value is the same as the mean (e.g. $E(X) = \mu_X$).

● **Example 2.89** Leonard has invested \$6000 in Google Inc. (stock ticker: GOOG) and \$2000 in Exxon Mobil Corp. (XOM). If X represents the change in Google's stock next month and Y represents the change in Exxon Mobil stock next month, write an equation that describes how much money will be made or lost in Leonard's stocks for the month.

For simplicity, we will suppose X and Y are not in percents but are in decimal form (e.g. if Google's stock increases 1%, then $X = 0.01$; or if it loses 1%, then $X = -0.01$). Then we can write an equation for Leonard's gain as

$$\$6000 * X + \$2000 * Y$$

If we plug in the change in the stock value for X and Y, this equation gives the change in value of Leonard's stock portfolio for the month. A positive value represents a gain, and a negative value represents a loss.

⊙ **Exercise 2.90** Suppose Google and Exxon Mobil stocks have recently been rising 1.1% and 0.56% per month, respectively. Compute the expected change in Leonard's stock portfolio for next month.

⊙ **Exercise 2.91** You should have found that Leonard expects a positive gain in Exercise 2.90. However, would you be surprised if he actually had a loss this month?

2.5.4 Variability in linear combinations of random variables

Quantifying the average outcome from a linear combination of random variables is helpful, but it is also important to have some sense of the uncertainty associated with the total outcome of that combination of random variables. The expected net gain or loss of Leonard's stock portfolio was considered in Exercise 2.90. However, there was no quantitative discussion of the volatility of this portfolio. For instance, while the average monthly gain might be about \$75 according to the data, that gain is not guaranteed. Figure 2.30 shows the monthly changes in a portfolio like Leonard's during the 36 months from 2008 to 2010. The gains and losses vary widely, and quantifying these fluctuations is important when investing in stocks.

Just as we have done in many previous cases, we use the variance and standard deviation to describe the uncertainty associated with Leonard's monthly returns. To do

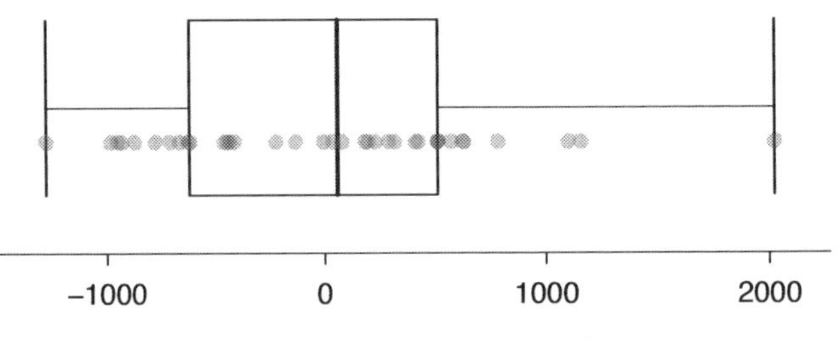

Figure 2.30: The change in a portfolio like Leonard's for the 36 months from 2008 to 2010, where \$6000 is in Google's stock and \$2000 is in Exxon Mobil's.

	Mean ($\bar{x}$)	Standard deviation (s)	Variance (s^2)
GOOG	0.0110	0.1050	0.0110
XOM	0.0056	0.0503	0.0025

Table 2.31: The mean, standard deviation, and variance of the GOOG and XOM stocks. These statistics were estimated from historical stock data, so notation used for sample statistics has been used.

so, the variances of each stock's monthly return will be useful, and these are shown in Table 2.31. The stocks' returns are nearly independent.

Here we use an equation from probability theory to describe the uncertainty of Leonard's monthly returns; we leave the proof of this method to a dedicated probability course. The variance of a linear combination of random variables can be computed by plugging in the variances of the individual random variables and squaring the coefficients of the random variables:

$$Var(aX + bY) = a^2 * Var(X) + b^2 * Var(Y)$$

It is important to note that this equality assumes the random variables are independent; if independence doesn't hold, then more advanced methods are necessary. This equation can be used to compute the variance of Leonard's monthly return:

$$\begin{aligned} Var(6000 * X + 2000 * Y) &= 6000^2 * Var(X) + 2000^2 * Var(Y) \\ &= 36,000,000 * 0.0110 + 4,000,000 * 0.0025 \\ &= 405,796 \end{aligned}$$

To get the standard deviation of the monthly variability, take the square root of the variance: $\sqrt{405,796} = \$637$. While an average monthly return of \$79 on a \$6000 investment is nothing to scoff at, the monthly returns are so volatile that Leonard should not expect this income to be very stable.

Variability of linear combinations of random variables
The variance of a linear combination of random variables may be computed by squaring the constants, substituting in the variances for the random variables, and computing the result:

$$Var(aX + bY) = a^2 * Var(X) + b^2 * Var(Y)$$

This equation is valid as long as the random variables are independent of each other. The standard deviation of the linear combination may be found by taking the square root of the variance result.

● **Example 2.92** Suppose John's daily commute has a standard deviation of 4 minutes. What is the uncertainty in his total commute time for the week?

The expression for John's commute time was

$$X_1 + X_2 + X_3 + X_4 + X_5$$

Each coefficient is 1, and the variance of each day's time is $4^2 = 16$. Thus, the variance of the total weekly commute time is

$$\begin{aligned} \text{variance} &= 1^2 * 16 + 1^2 * 16 + 1^2 * 16 + 1^2 * 16 + 1^2 * 16 = 5 * 16 = 80 \\ \text{standard deviation} &= \sqrt{\text{variance}} = \sqrt{80} = 8.94 \end{aligned}$$

The standard deviation for John's weekly work commute time is about 9 minutes.

⊙ **Exercise 2.93** The computation in Example 2.92 relied on an important assumption: the commute time for each day is independent of the time on other days of that week. Do you think this is valid? Explain.

⊙ **Exercise 2.94** Consider Elena's two auction's from Exercise 2.85 on page 88. Suppose these auctions are approximately independent and the variability in auction prices associated with the TV and toaster oven can be described using standard deviations of $25 and $8. Compute the standard deviation of Elena's net gain. Answer in the footnote[40].

Consider again Exercise 2.94. The negative coefficient for Y in the linear combination was eliminated when we squared the coefficients. This generally holds true: negatives in a linear combination will have no impact on the variability computed for a linear combination, but they do impact the expected value computations.

[40] The equation for Elena can be written as

$$(1) * X + (-1) * Y$$

The variances of X and Y are 625 and 64. We square the coefficients and plug in the variances:

$$(1)^2 * Var(X) + (-1)^2 * Var(Y) = 1 * 625 + 1 * 64 = 689$$

The variance of the linear combination is 689, and the standard deviation is the square root of 689: about $26.25.

2.6 Exercises

2.6.1 Defining probability

2.1 True or false: If a fair coin is tossed many times and the last eight tosses are all heads, then the chance that the next toss will be heads is somewhat less than 50%.

2.2 Determine if the statements below are true or false, and explain your reasoning.

(a) Drawing a face card (jack, queen, or king) and drawing a red card from a full deck of playing cards are mutually exclusive events.

(b) Drawing a face card and drawing an ace from a full deck of playing cards are mutually exclusive events.

2.3 Below are four versions of the same game. Your archnemisis gets to pick the version of the game, and then you get to choose how many times to flip a coin: 10 times or 100 times. Identify how many coin flips you should choose for each version of the game. Explain your reasoning.

(a) If the proportion of heads is larger than 0.60, you win $1.

(b) If the proportion of heads is larger than 0.40, you win $1.

(c) If the proportion of heads is between 0.40 and 0.60, you win $1.

(d) If the proportion of heads is smaller than 0.30, you win $1.

2.4 Backgammon is a board game for two players in which the playing pieces are moved according to the roll of two dice. Players win by removing all of their pieces from the board, so it is usually good to roll high numbers. You are playing backgammon with a friend and you roll two sixes in your first roll and two sixes in your second roll. Your friend rolls two threes in his first roll and again in his second row. Your friend claims that you are cheating, because rolling double sixes twice in a row is very unlikely. Using probability, how could you argue that your rolls were just as likely as his?

2.5 If you flip a fair coin 10 times, what is the probability of

(a) getting all tails? (b) getting all heads? (c) getting at least one tails?

2.6 If you roll a pair of fair dice, what is the probability of

(a) getting a sum of 1? (b) getting a sum of 5? (c) getting a sum of 12?

2.7 An online survey shows that 25% of bloggers own a DSLR camera, 90% own a point-and-shoot camera, and 20% own both types of cameras.

(a) Draw a Venn diagram summarizing these probabilities.

(b) What percent of bloggers own a DSLR but not a point-and-shoot camera?

(c) What percent of bloggers own a point-and-shoot camera but not a DSLR?

(d) What percent of bloggers own a point-and-shoot camera or a DSLR or both?

(e) What percent of bloggers own neither a DSLR nor a point-and-shoot camera?

(f) Are owning a DSLR and point-and-shoot camera mutually exclusive?

2.8 In a class where everyone's native language is English, 70% of students speak Spanish, 45% speak French as a second language and 20% speak both.

(a) Draw a Venn diagram summarizing these probabilities.

(b) What percent of students speak Spanish but not French?

(c) What percent of students speak French but not Spanish?

(d) What percent of students speak French or Spanish?

(e) What percent of the students do not speak French or Spanish?

(f) Are speaking French and Spanish mutually exclusive?

2.9 In parts (a) and (b), identify whether the events are mutually exclusive, independent, or neither (they cannot be both).

(a) You and a randomly selected student from your class both earn A's in this course.

(b) You and a friend you study with from your class both earn A's in this course.

(c) If two events can occur at the same time must they be dependent?

2.10 In a multiple choice exam, there are 5 questions and 4 choices for each question (a, b, c, d). Nancy has not studied for the exam at all and decided to randomly guess the answers. What is the probability that:

(a) the first question she gets right is the 5th question?

(b) she gets all questions right?

(c) she gets at least one question right?

2.11 The US Census is conducted every 10 years and collects demographic information from the residents of United States. The table below shows the distribution of education level attained by US residents by gender. Answer the following questions based on this table.

		Gender	
		Male	Female
	Less than high school	0.19	0.19
	High school graduate	0.28	0.30
Highest	Some college	0.27	0.28
education	Bachelor's degree	0.16	0.15
attained	Master's degree	0.06	0.06
	Professional school degree	0.03	0.01
	Doctorate degree	0.01	0.01

(a) What is the probability that a randomly chosen man has at least a Bachelor's degree?

(b) What is the probability that a randomly chosen woman has at least a Bachelor's degree?

(c) What is the probability that a man and a woman getting married both have at least a Bachelor's degree? Note any assumptions you make.

(d) If you made an assumption in part (c), do you think it was reasonable? If you didn't make an assumption, double check your earlier answer.

2.12 A middle school estimates that 20% of its students miss exactly one day of school per semester due to sickness, 13% miss exactly two days, and 5% miss three or more days.

(a) What is the probability that a student chosen at random doesn't miss any days of school due to sickness?

(b) What is the probability that a student chosen at random misses no more than one day?

(c) What is the probability that a student chosen at random misses at least one day?

(d) If a parent has two kids at this middle school, what is the probability that neither kid will miss any school? Note any assumption you make.

(e) If a parent has two kids at this middle school, what is the probability that that both kids will miss some school, i.e. at least one day? Note any assumption you make.

(f) If a parent has two kids at this middle school, what is the probability that that at least one kid will miss some school? Note any assumption you make.

(g) If you made an assumption in parts (c) through (e), do you think it was reasonable? If you didn't make an assumption, double check your earlier answers.

2.13 Indicate what, if anything, is wrong with the following probability distributions for the final grades within a Statistics.

	Grades				
	A	B	C	D	F
(a)	0.3	0.3	0.3	0.2	0.1
(b)	0	0	1	0	0
(c)	0.3	0.3	0.3	0	0
(d)	0.3	0.5	0.2	0.1	-0.1
(e)	0.2	0.4	0.2	0.1	0.1
(f)	0	-0.1	1.1	0	0

2.14 The table below shows the relationship between hair color and eye color for a group of 1,770 German men. You might recognize this table from Chapter 1 exercises. Here we will take another look at it from a probability perspective.

		Hair Color			
		Brown	Black	Red	Total
Eye	Brown	400	300	20	720
Color	Blue	800	200	50	1050
	Total	1200	500	70	1770

(a) If we draw one man at random, what is the probability that he has brown hair and blue eyes?

(b) If we draw one man at random, what is the probability that he has brown hair or blue eyes?

2.6.2 Continuous distributions

2.15 Below is a histogram of body weights (in *kg*) of 144 male and female cats.

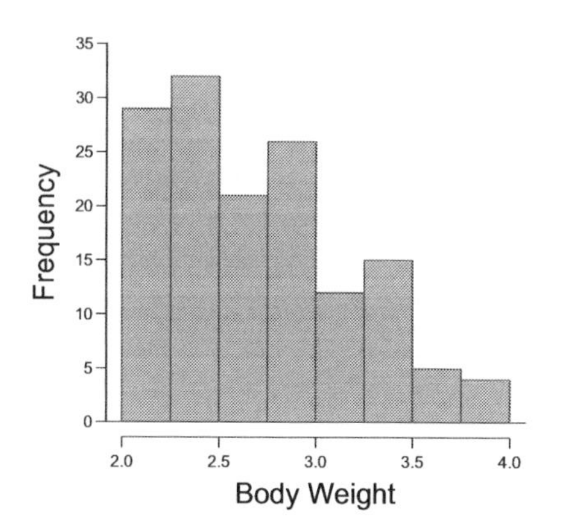

(a) What is the probability that a randomly chosen cat weighs less than 2.5 *kg*?

(b) What is the probability that a randomly chosen cat weighs between 2.5 and 2.75 *kg*?

(c) What is the probability that a randomly chosen cat weighs between 2.75 and 3.5 *kg*?

(d) What is the probability that a randomly chosen cat weighs more than 3.5 *kg*? How quickly did you solve this part of the problem?

2.16 Exercise 2.15 introduces a data set of 144 cats' body weights. The histogram below shows the distribution of these cats' body weights by gender. There are 47 female and 97 male cats in the data set.

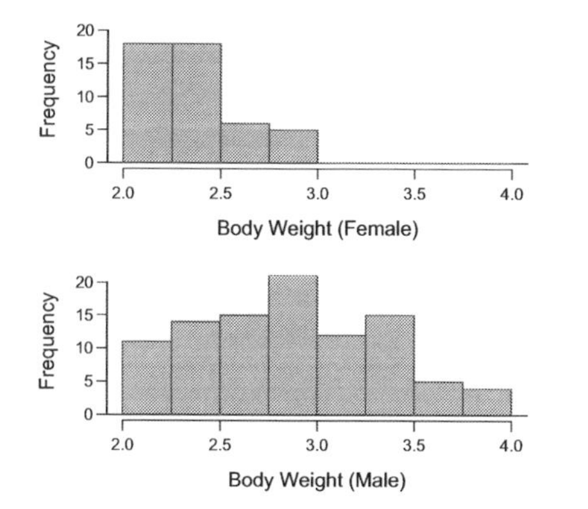

(a) What is the probability that a randomly chosen female cat weighs less than 2.5 *kg*?

(b) What is the probability that a randomly chosen male cat weighs less than 2.5 *kg*?

2.17 The American Community Survey (ACS) is an ongoing statistical survey by the U.S. Census Bureau, sent to approximately 250,000 addresses monthly (or 3 million per year). The relative frequency table below displays the distribution of annual total personal income (in 2009 inflation-adjusted dollars) for a random sample of 96,420,486 Americans. These data come from 2005-2009 ACS 5-year estimates. This sample is comprised of 59% males and 41% females. [13]

(a) Describe the distribution of total personal income.

(b) What is the probability that a randomly chosen person makes less than \$50,000 per year?

(c) What is the probability that a randomly chosen person makes less than \$50,000 per year and is female? Note any assumptions you make.

(d) The same data source indicates that 71.8% of females make less than \$50,000 per year. Use this value to determine whether or not the assumption you made in part (c) is valid.

Income	*Total*
\$1 to \$9,999 or loss	2.2%
\$10,000 to \$14,999	4.7%
\$15,000 to \$24,999	15.8%
\$25,000 to \$34,999	18.3%
\$35,000 to \$49,999	21.2%
\$50,000 to \$64,999	13.9%
\$65,000 to \$74,999	5.8%
\$75,000 to \$99,999	8.4%
\$100,000 or more	9.7%

2.6.3 Conditional probability

2.18 P(A) = 0.3, P(B) = 0.7

(a) Can you compute P(A and B) if you only know P(A) and P(B)?

(b) Assuming that events A and B arise from independent random processes,

(i) what is P(A and B)?
(ii) what is P(A or B)?
(iii) what is P(A|B)?

(c) If we are given that P(A and B) = 0.1, are the random variables giving rise to events A and B independent?

(d) If we are given that P(A and B) = 0.1, what is P(A|B)?

2.19 Based on the probabilities provided in Exercise 2.7, are the variables owning a DSLR and owning a point-and-shoot camera independent?

2.20 Based on the probabilities provided in Exercise 2.8, are the variables speaking French and speaking Spanish independent?

2.21 Exercise 2.14 introduces the contingency table summarizing the relationship between hair color and eye color for a group of 1,770 German men. Answer the following questions based on this table.

		Hair Color			
		Brown	Black	Red	Total
Eye	Brown	400	300	20	720
Color	Blue	800	200	50	1050
	Total	1200	500	70	1770

(a) What is the probability that a randomly chosen man has black hair?

(b) What is the probability that a randomly chosen man has black hair given that he has blue eyes?

(c) What is the probability that a randomly chosen man has black hair given that he has brown eyes?

(d) Are hair color and eye color independent?

2.22 The contingency table below shows two variables: `type` and `drivetrain` from the `cars` data set introduced in Chapter 1. Answer the following questions based on the probabilities you can estimate from the data in this table.

		drivetrain		
		4WD	front	rear
type	large	0	7	4
	midsize	0	17	5
	small	2	19	0

(a) What is the probability that a randomly chosen car is midsize?

(b) What is the probability that a randomly chosen car has front wheel drive?

(c) What is the probability that a randomly chosen car has front wheel drive and is midsize?

(d) What is the probability that a randomly chosen car has front wheel drive given that it is midsize?

(e) What is the marginal distribution of drivetrain?

(f) What is the conditional distribution of drivetrain for midsize cars, i.e. what are the conditional probabilities of 4WD, front wheel drive, and rear wheel drive when given that the car is midsize?

(g) Are type and drivetrain independent? Would you be comfortable generalizing your conclusion to all cars?

2.23 A survey conducted in 2010 by Survey USA for KABC-TV Los Angeles asked 500 Los Angeles residents "What is the best hamburger place in Southern California? Five Guys Burgers? In-N-Out Burger? Fat Burger? Tommy's Hamburgers? Umami Burger? Or somewhere else?" Consider the results of this survey, shown in the table below, in each of the following questions.

		Gender		
		Male	Female	Total
Best hamburger place	Five Guys Burgers	5	6	11
	In-N-Out Burger	162	181	343
	Fat Burger	10	12	22
	Tommy's Hamburgers	27	27	54
	Umami Burger	5	1	6
	Other	26	20	46
	Not Sure	13	5	18
	Total	248	252	500

(a) What is the probability that a randomly chosen male likes In-N-Out the best?

(b) What is the probability that a randomly chosen female likes In-N-Out the best?

(c) What is the probability that a man and a woman who are dating both like In-N-Out the best? Note any assumption you make and evaluate whether you think that assumption is reasonable.

(d) What is the probability that a randomly chosen person likes Umami best, or that person is female?

2.24 A Pew Research poll conducted in October 2010 asked respondents "From what you've read and heard, is there solid evidence that the average temperature on earth has been getting warmer over the past few decades, or not?". The contingency table below shows the distribution of responses by party and ideology. [14]

		Response			
		Earth is warming	Not warming	Don't Know Refuse	Total
Party and Ideology	Conservative Republican	146	265	31	442
	Mod/Lib Republican	77	68	16	161
	Mod/Cons Democrat	325	84	31	440
	Liberal Democrat	234	18	11	263
	Total	782	435	89	1306

(a) What is the probability that a randomly chosen respondent believes the earth is warming?

(b) What is the probability that a randomly chosen respondent is a liberal Democrat?

(c) What is the probability that a randomly chosen respondent believes the earth is warming and is a liberal Democrat?

(d) What is the probability that a randomly chosen respondent believes the earth is warming or is a liberal Democrat?

(e) What is the probability that a randomly chosen respondent believes the earth is warming given that s/he is a liberal Democrat?

(f) What is the probability that a randomly chosen respondent believes the earth is warming given that s/he is a conservative Republican?

(g) Does it appear that whether or not a respondent believes the earth is warming is independent of his/her party and ideology? Explain your reasoning.

(h) What is the probability that a randomly chosen respondent is a moderate/liberal Republican given that s/he does not believe that the earth is warming?

2.25 Exercise 2.15 introduces a data set of 144 cats' body weights, and Exercise 2.16 shows the distribution of gender in this sample of 144 cats. If a randomly chosen cat weighs less than 2.5 *kg*, is it more likely to be male or female?

2.26 Suppose that a student is about to take a multiple choice test that covers algebra and trigonometry. 40% of the questions are algebra and 60% are trigonometry. There is a 90% chance that she will get an algebra question right, and a 35% chance that she will get a trigonometry question wrong.

(a) If we choose a question at random from the exam, what is the probability that she will get it right?

(b) If we choose a question at random from the exam, what is the probability that she will get it wrong?

(c) If we choose a question at random from the exam that she got wrong, what is the probability it is an algebra question?

2.27 A genetic test is used to determine if people have a predisposition for *thrombosis*, which is the formation of a blood clot inside a blood vessel that obstructs the flow of blood through the circulatory system. It is believed that 3% of the people actually have this predisposition. This genetic test is 99% accurate if a person actually has the predisposition, meaning that the probability of a positive test result when a person actually has the predisposition is 0.99. The test is 98% accurate if a person does not have the predisposition. What is the probability that a randomly selected person who tests positive for the predisposition by the test actually has the predisposition?

2.28 Lupus is a medical phenomenon where antibodies that are supposed to attack foreign cells to prevent infections instead see plasma proteins as foreign bodies, leading to a high risk of blood clotting. It is believed that 2% of the population suffer from this disease.

The test for lupus is very accurate if the person actually has lupus, however is very inaccurate if the person does not. More specifically, the test is 98% accurate if a person actually has the disease. The test is 74% accurate if a person does not have the disease.

There is a line from the Fox television show *House*, often used after a patient tests positive for lupus: "It's never lupus." Do you think there is truth to this statement? Use appropriate probabilities to support your answer.

2.29 In November 2009, the US Preventive Services Task Force changed its recommendations for breast cancer screening. One of the reasons for this change was the high rate of false positives. The following quotation is from an opinion piece published in the *New York Times*, which tried to explain how false positive rates are calculated. [15, 16]

> "Assume there is a screening test for a certain cancer that is 95 percent accurate; that is, if someone has the cancer, the test will be positive 95 percent of the time. Let's also assume that if someone doesn't have the cancer, the test will be positive just 1 percent of the time. Assume further that 0.5 percent - one out of 200 people - actually have this type of cancer. Now imagine that you've taken the test and that your doctor somberly intones that you've tested positive. Does this mean you're likely to have the cancer? Surprisingly, the answer is no."

(a) Construct a tree diagram of this scenario.

(b) Calculate the probability that a person who tests positive does not actually have cancer.

(c) Calculate the probability that a person who tests positive actually has cancer.

(d) Suppose you are a doctor and a patient had a positive mammogram. How would you explain to the patient that this positive test is not conclusive and that more tests are necessary?

2.30 After an introductory statistics course, 80% of students can successfully construct box plots. Of those who can construct box plots, 86% passed, while only 65% of those students who could not construct box plots passed.

(a) Construct a tree diagram of this scenario.

(b) Calculate the probability that a student is able to construct a box plot if it is known that she passed.

2.6.4 Sampling from a small population

2.31 Imagine you have an urn with 5 red, 3 blue and 2 orange marbles in it.

(a) What is the probability that the first marble you draw is blue?

(b) Suppose you drew a blue marble in the first draw. If drawing with replacement, what is the probability of drawing a blue marble in the second draw?

(c) Suppose you instead drew an orange marble in the first draw. If drawing with replacement, what is the probability of drawing a blue marble in the second draw?

(d) If drawing with replacement, what is the probability of drawing two blue marbles in a row?

(e) When drawing with replacement, are the draws independent? Explain.

2.32 Suppose the same urn from Exercise 2.31 with 5 red, 3 blue and 2 orange marbles in it. You just drew one marble from the urn.

(a) Suppose the marble is blue. If drawing without replacement, what is the probability the next is also blue?

(b) Suppose the first marble is orange and you draw a second marble without replacement. What is the probability this second marble is blue?

(c) If drawing without replacement, what is the probability of drawing two blue marbles in a row?

(d) When drawing without replacement, are the draws independent? Explain.

2.33 In your sock drawer you have 4 blue, 5 gray, and 3 black socks. Half asleep one morning you grab 2 socks at random and put them on. Find the probability you end up wearing

(a) 2 blue socks

(b) no gray socks

(c) at least 1 black sock

(d) a green sock

(e) matching socks

2.34 The table below shows the distribution of books on a bookcase based on whether they are non-fiction or fiction and hardcover or paperback. Suppose you randomly draw two books without replacement. Calculate the probability that the books are as follows.

		Format		
		Hardcover	Paperback	Total
Type	Fiction	13	59	72
	Non-fiction	15	8	23
	Total	28	67	95

(a) Both books are fiction.

(b) The first book you draw is a hardcover fiction and the second is a paperback fiction.

(c) The first book you draw is a hardcover and the second is a paperback.

(d) The first book is a hardcover or a fiction book. The second book can be anything.

(e) At least one of the books is a paperback.

2.35 In a classroom with 24 students, 7 students are wearing jeans, 4 are wearing shorts, 8 are wearing skirts, and the rest are wearing leggings. If we randomly select 3 students without replacement, what is the probability that one of the selected students is wearing leggings and the other two are wearing jeans? Note that these are mutually exclusive clothing options.

2.36 Suppose we pick three people at random. For each of the following questions, ignore the special case where someone might be born on February 29th.

(a) What is the probability that the first two people share a birthday?

(b) What is the probability that at least two people share a birthday?

2.6.5 Random variables

2.37 At a university, 13% of students smoke.

(a) Calculate the expected number of smokers in a random sample of 100 students from this university.

(b) The university gym opens at 9am on Saturday mornings. On a Saturday morning at 8:55am there are 27 students outside the gym waiting for it to open. Should you use the same approach from part (a) to calculate the expected number of smokers among these 27 students?

2.38 Consider the following card game with a well-shuffled deck of cards. If you draw a red card, you win nothing. If you get a spade, you win \$5. For any club, you win \$10 plus an extra \$20 for the ace of clubs.

(a) Create a probability model for the amount you win at this game. Also, find the expected winnings for a single game and the standard deviation of the winnings.

(b) What is the maximum amount you would be willing to pay to play this game? Explain.

2.39 In a card game you start with a well-shuffled full deck and draw 3 cards without replacement. If you draw 3 hearts in a row, you win \$50. If you draw 3 black cards, you win \$25. For any other draws you win nothing.

(a) Create a probability model for the amount you win at this game and find the expected winnings. Also compute the standard deviation of this distribution.

(b) If the game costs \$5 to play, what would be the expected value and standard deviation of net profit (or loss)? *(Hint: profit = winnings − cost;* $X - 5$*)*

(c) If the game costs \$5 to play, should you play this game? Explain.

2.40 The game of roulette involves spinning a wheel with 38 slots: 18 red, 18 black, and 2 green. A ball is spun onto the wheel and will eventually land in a slot, where each slot has an equal chance of capturing the ball. Gamblers can place bets on red or black. If the ball lands on their color, they double their money. If it lands on another color, they lose their money. Suppose you bet \$1 on red.

(a) Let X represent the amount you win or lose on a single spin. Write a probability model for X.

(b) What's the expected value and standard deviation of your winnings?

2.41 Exercise 2.40 describes winnings on a game of roulette.

(a) Suppose you play roulette and bet $3 on a single round. What is the expected value and standard deviation of your total winnings?

(b) Suppose you bet $1 in three different rounds. What is the expected value and standard deviation of your total winnings?

(c) How do your answers to parts (a) and (b) compare? What does this say about the riskiness of the two games?

2.42 American Airlines charges the following baggage fees: $25 for the first bag and $35 for the second. Suppose 54% of passengers have no checked luggage, 34% have one piece of checked luggage and 12% have two pieces. We suppose a negligible portion of people check more than two bags.

(a) Build a probability model, compute the average revenue per passenger, and compute the corresponding standard deviation.

(b) About how much revenue should they expect for a flight of 120 passengers? With what standard deviation? Note any assumptions you make and if you think they are justified.

2.43 Andy is always looking for ways to make money fast. Lately, he has been trying to make money by gambling. Here is the game he is considering playing:

The game costs $2 to play. He draws a card from a deck. If he gets a number card (2-10), he wins nothing. For any face card (jack, queen or king) he wins $3. For any ace he wins $5, and he wins an *extra* $20 if he draws the ace of clubs.

(a) Create a probability model for the amount he profits at this game and find his expected profits per game.

(b) Would you recommend this game to Andy as a good way to make money? Explain.

2.44 A portfolio's value increases by 18% during a financial boom and by 9% during normal times. It decreases by 12% during a recession. What is the expected return on this portfolio if each scenario is equally likely?

2.45 You and your friend decide to bet on the Major League Baseball game happening one evening between the Los Angeles Dodgers and the San Diego Padres. Current standing statistics indicate that the Dodgers have 0.46 probability of winning this game against the Padres. If your friend bets you $5 that the Dodgers will win, how much would you need to bet on the Padres to make this a fair game?

2.46 Marcie has been tracking the following two items on Ebay:

- A textbook that sells for an average of $110 with a standard deviation of $4.
- Mario Kart for the Nintendo Wii, which sells for an average of $38 with a standard deviation of $5.

(a) Marcie wants to sell the video game and buy the textbook. How much net money (profits - losses) would she expect to make or spend? Also compute the standard deviation of how much she would make or spend.

(b) Lucy is selling the textbook on Ebay for a friend, and her friend is giving her a 10% commission (Lucy keeps 10% of the revenue). How much money should she expect to make? With what standard deviation?

2.47 Sally gets a cup of coffee and a muffin every day for breakfast from one of the many coffee shops in her neighborhood. She picks a coffee shop each morning at random and independent of previous days. The average price of a cup of coffee is \$1.40 with a standard deviation of 30¢. The average price of a muffin is \$2.50 with a standard deviation of 15¢.

(a) What is the mean and standard deviation of the amount she spends on breakfast daily?

(b) What is the mean and standard deviation of the amount she spends on breakfast weekly (7 days)?

2.48 Ice cream usually comes in 1.5 quart boxes (48 fluid ounces) and ice cream scoops hold about 2 ounces. However, there is some variability in the amount of ice cream in a box as well as the amount of ice cream scooped out. We represent the amount of ice cream in the box as X and the amount scooped out as Y, and these random variables have the following mean, standard deviations, and variances:

	mean	SD	variance
X	48	1	1
Y	2	0.25	0.0625

(a) An entire box of ice cream, plus 3 scoops from a second box is served at a party. How much ice cream do you expect to have been served at this party? What is the standard deviation of the amount of ice cream served?

(b) How much ice cream would you expect to be left in the box after scooping out one scoop of ice cream? That is find the expected value of $X - Y$. What is the standard deviation of the amount left in the box?

(c) Based on this example, can you explain why, when we subtract two random variables, we still add their variances?

Chapter 3

Distributions of random variables

3.1 Normal distribution

Among all the distributions we see in practice, one is overwhelmingly the most common. The symmetric, unimodal, bell curve is ubiquitous throughout statistics. Indeed it is so common, that people often know it as the **normal curve** or **normal distribution**[1], shown in Figure 3.1. Variables such as SAT scores and heights of US adult males closely follow the normal distribution.

> **Normal distribution facts**
> Many variables are nearly normal, but none are exactly normal. Thus the normal distribution, while not perfect for any single problem, is very useful for a variety of problems. We will use it in data exploration and to solve important problems in statistics.

[1] It is also introduced as the Gaussian distribution after Frederic Gauss, the first person to formalize its mathematical expression.

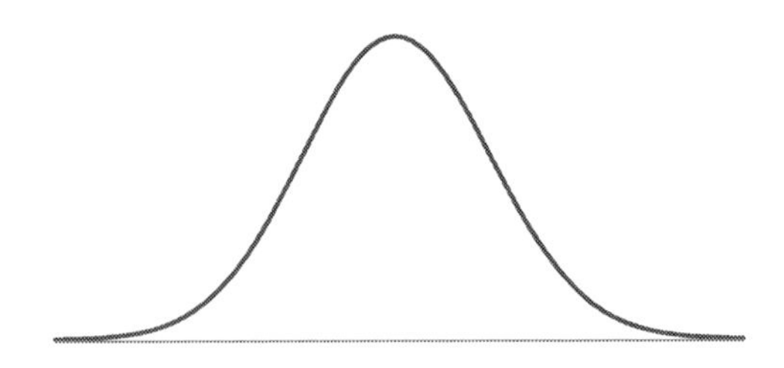

Figure 3.1: A normal curve.

3.1.1 Normal distribution model

The normal distribution model always describes a symmetric, unimodal, bell shaped curve. However, these curves can look different depending on the details of the model. Specifically, the normal distribution model can be adjusted using two parameters: mean and standard deviation. As you can probably guess, changing the mean shifts the bell curve to the left or right, while changing the standard deviation stretches or constricts the curve. Figure 3.2 shows the normal distribution with mean 0 and standard deviation 1 in the left panel and the normal distributions with mean 19 and standard deviation 4 in the right panel. Figure 3.3 shows these distributions on the same axis.

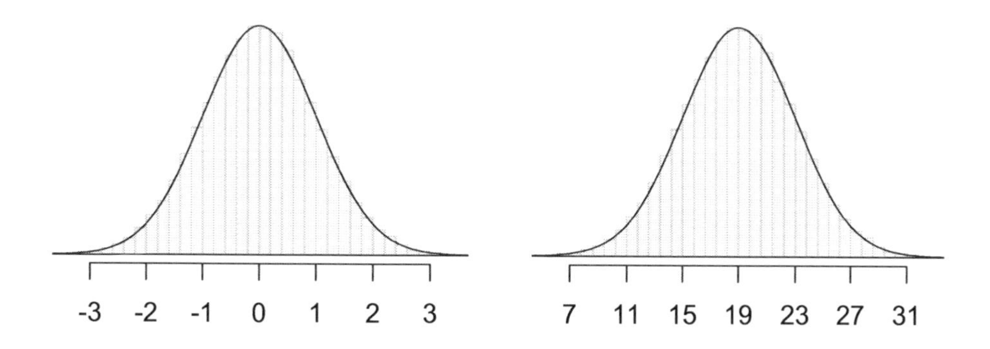

Figure 3.2: Both curves represent the normal distribution, however, they differ in their center and spread.

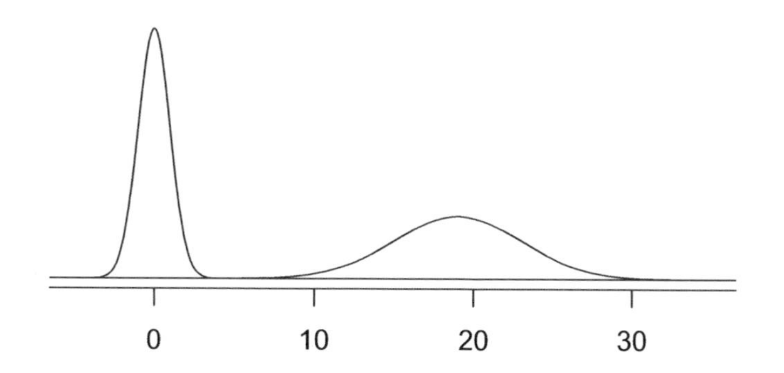

Figure 3.3: The normal models shown in Figure 3.2 but plotted together.

If a normal distribution has mean μ and standard deviation σ, we may write the distribution as $N(\mu,\sigma)$. The two distributions in Figure 3.3 can be written as

$$N(\mu = 0, \sigma = 1) \quad \text{and} \quad N(\mu = 19, \sigma = 4)$$

$N(\mu,\sigma)$
Normal dist. with mean μ & st. dev. σ

Because the mean and standard deviation describe a normal distribution exactly, they are called the distribution's **parameters**.

⊙ **Exercise 3.1** Write down the short-hand for a normal distribution with (a) mean 5 and standard deviation 3, (b) mean -100 and standard deviation 10, and (c) mean 2 and standard deviation 9. The answers for (a) and (b) are in the footnote[2].

[2] $N(\mu = 5, \sigma = 3)$ and $N(\mu = -100, \sigma = 10)$.

	SAT	ACT
Mean	1500	21
SD	300	5

Table 3.4: Mean and standard deviation for the SAT and ACT.

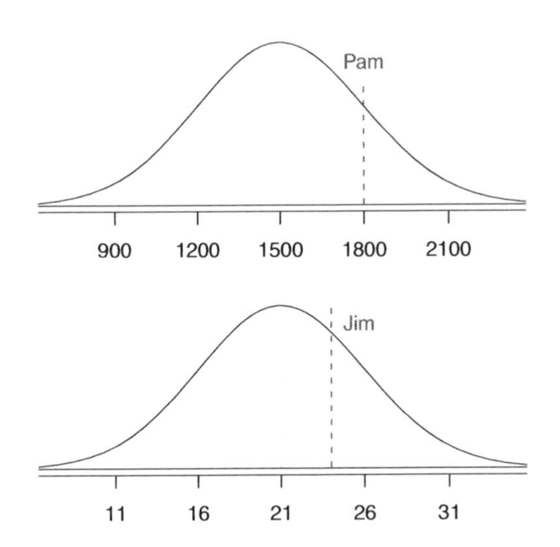

Figure 3.5: Pam's and Jim's scores shown with the distributions of SAT and ACT scores.

3.1.2 Standardizing with Z scores

● **Example 3.2** Table 3.4 shows the mean and standard deviation for total scores on the SAT and ACT. The distribution of SAT and ACT scores are both nearly normal. Suppose Pam earned an 1800 on her SAT and Jim obtained a 24 on his ACT. Which of them did better?

We use the standard deviation as a guide. Pam is 1 standard deviation above average on the SAT: $1500 + 300 = 1800$. Jim is 0.6 standard deviations above the mean: $21 + 0.6 * 5 = 24$. In Figure 3.5, we can see that Pam tends to do better with respect to everyone else than Jim did, so her score was better.

Z
Z score, the standardized observation

Example 3.2 used a standardization technique called a Z score. The **Z score** of an observation is defined as the number of standard deviations it falls above or below the mean. If the observation is one standard deviation above the mean, its Z score is 1. If it is 1.5 standard deviations *below* the mean, then its Z score is -1.5. If x is an observation from a distribution[3] $N(\mu,\sigma)$, we define the Z score mathematically as

$$Z = \frac{x - \mu}{\sigma}$$

Using $\mu_{SAT} = 1500$, $\sigma = 300$, and $x_{Pam} = 1800$, we find Pam's Z score:

$$Z_{Pam} = \frac{x_{Pam} - \mu_{SAT}}{\sigma_{SAT}} = \frac{1800 - 1500}{300} = 1$$

[3]It is still reasonable to use a Z score to describe an observation even when x is not nearly normal.

> **The Z score**
> The Z score of an observation is the number of standard deviations it falls above or below the mean. We compute the Z score for an observation x that follows a distribution with mean μ and standard deviation σ using
>
> $$Z = \frac{x - \mu}{\sigma}$$

⊙ **Exercise 3.3** Use Jim's ACT score, 24, along with the ACT mean and standard deviation to verify his Z score is 0.6.

Observations above the mean always have positive Z scores while those below the mean have negative Z scores. If an observation is equal to the mean (e.g. SAT score of 1500), then the Z score is 0.

⊙ **Exercise 3.4** Let X represent a random variable from $N(\mu = 3, \sigma = 2)$, and suppose we observe $x = 5.19$. (a) Find the Z score of x. (b) Use the Z score to determine how many standard deviations above or below the mean x falls. Answers in the footnote[4].

⊙ **Exercise 3.5** The variable `headL` from the `possum` data set is nearly normal with mean 92.6 mm and standard deviation 3.6 mm. Identify the Z scores for $\texttt{headL}_{14} =$ 95.4 mm and $\texttt{headL}_{79} = 85.8$, which correspond to the 14th and 79th cases in the data set.

We can use Z scores to identify which observations are more unusual than others. One observation x_1 is said to be more unusual than another observation x_2 if the absolute value of its Z score is larger than the absolute value of the other observations Z score: $|Z_1| > |Z_2|$.

⊙ **Exercise 3.6** Which of the observations in Exercise 3.5 is more unusual? Answer in the footnote[5].

3.1.3 Normal probability table

● **Example 3.7** Pam from Example 3.2 earned a score of 1800 on her SAT with a corresponding $Z = 1$. She would like to know what percentile she falls in for all SAT test-takers.

Pam's **percentile** is the percentage of people who earned a lower SAT score than Pam. We shade the area representing those individuals in Figure 3.6. The total area under the normal curve is always equal to 1, and the proportion of people who scored below Pam on the SAT is equal to the *area* shaded in Figure 3.6: 0.8413. In other words, Pam is in the 84th percentile of SAT takers.

We can use the normal model to find percentiles. A **normal probability table**, which lists Z scores and corresponding percentiles, can be used to identify a percentile based on the Z score (and vice versa). Statistical software can also be used.

---

[4](a) Its Z score is given by $Z = \frac{x-\mu}{\sigma} = \frac{5.19-3}{2} = 2.19/2 = 1.095$. (b) The observation x is 1.095 standard deviations *above* the mean. We know it must be above the mean since Z is positive.

[5]In Exercise 3.5, you should have found $Z_{14} = 0.78$ and $Z_{79} = -1.89$. Because the *absolute value* of Z_{79} is larger than Z_{14}, case 79 appears to have a more unusual head length.

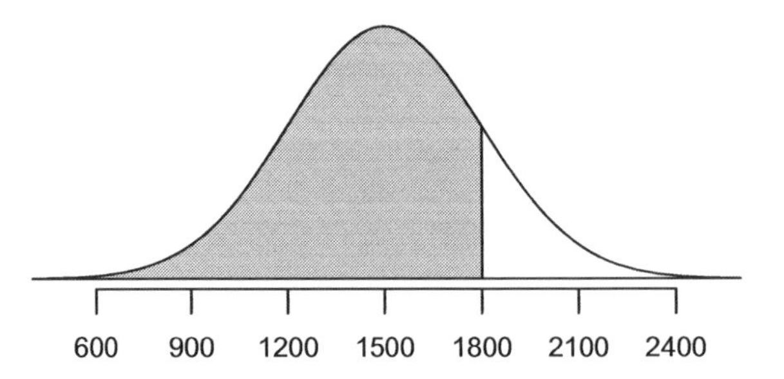

Figure 3.6: The normal model for SAT scores, shading the area of those individuals who scored below Pam.

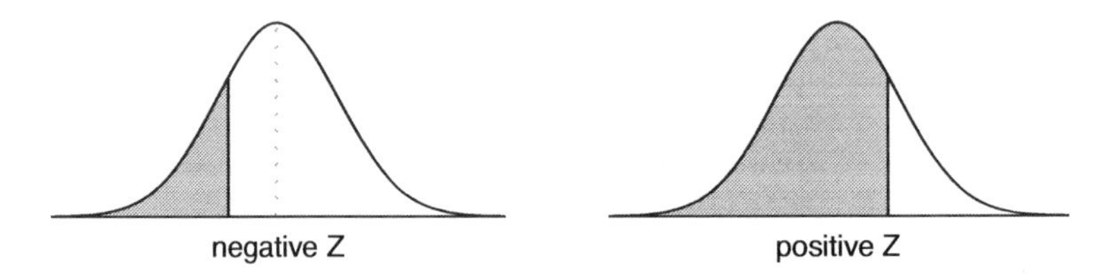

Figure 3.7: The area to the left of Z represents the percentile of the observation.

A normal probability table is given in Appendix C.1 on page 361 and abbreviated in Table 3.8. We use this table to identify the percentile corresponding to any particular Z score. For instance, the percentile of $Z = 0.43$ is shown in row 0.4 and column 0.03 in Table 3.8: 0.6664 or the 66.64^{th} percentile. Generally, we take Z rounded to two decimals, identify the proper row in the normal probability table up through the first decimal, and then determine the column representing the second decimal value. The intersection of this row and column is the percentile of the observation.

We can also find the Z score associated with a percentile. For example, to identify Z for the 80th percentile, we look for the value closest to 0.8000 in the middle portion of the table: 0.7995. We determine the Z score for the 80th percentile by combining the row and column Z values: 0.84.

⊙ **Exercise 3.8** Determine the proportion of SAT test takers who scored better than Pam on the SAT. Hint in the footnote[6].

3.1.4 Normal probability examples

Cumulative SAT scores are approximated well by a normal model, $N(\mu = 1500, \sigma = 300)$.

● **Example 3.9** Shannon is a randomly selected SAT taker, and nothing is known about Shannon's SAT aptitude. What is the probability Shannon scores at least 1630 on her SATs?

First, always draw and label a picture of the normal distribution. (Drawings need not be exact to be useful.) We are interested in the chance she scores above 1630, so we shade this upper tail:

[6]If 84% had lower scores than Pam, how many had better scores? Generally ties are ignored when the normal model is used.

	Second decimal place of Z									
Z	0.00	0.01	0.02	**0.03**	*0.04*	0.05	0.06	0.07	0.08	0.09
0.0	0.5000	0.5040	0.5080	0.5120	0.5160	0.5199	0.5239	0.5279	0.5319	0.5359
0.1	0.5398	0.5438	0.5478	0.5517	0.5557	0.5596	0.5636	0.5675	0.5714	0.5753
0.2	0.5793	0.5832	0.5871	0.5910	0.5948	0.5987	0.6026	0.6064	0.6103	0.6141
0.3	0.6179	0.6217	0.6255	0.6293	0.6331	0.6368	0.6406	0.6443	0.6480	0.6517
0.4	0.6554	0.6591	0.6628	**0.6664**	0.6700	0.6736	0.6772	0.6808	0.6844	0.6879
0.5	0.6915	0.6950	0.6985	0.7019	0.7054	0.7088	0.7123	0.7157	0.7190	0.7224
0.6	0.7257	0.7291	0.7324	0.7357	0.7389	0.7422	0.7454	0.7486	0.7517	0.7549
0.7	0.7580	0.7611	0.7642	0.7673	0.7704	0.7734	0.7764	0.7794	0.7823	0.7852
0.8	0.7881	0.7910	0.7939	0.7967	*0.7995*	0.8023	0.8051	0.8078	0.8106	0.8133
0.9	0.8159	0.8186	0.8212	0.8238	0.8264	0.8289	0.8315	0.8340	0.8365	0.8389
1.0	0.8413	0.8438	0.8461	0.8485	0.8508	0.8531	0.8554	0.8577	0.8599	0.8621
1.1	0.8643	0.8665	0.8686	0.8708	0.8729	0.8749	0.8770	0.8790	0.8810	0.8830
⋮	⋮	⋮	⋮	⋮	⋮	⋮	⋮	⋮	⋮	⋮

Table 3.8: A section of the normal probability table. The percentile for a normal random variable with $Z = 0.43$ has been **highlighted**, and the percentile closest to 0.8000 has also been ***highlighted***.

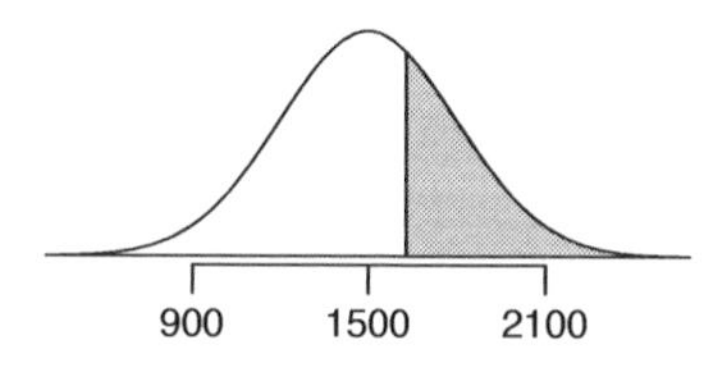

The picture shows the mean and the values at 2 standard deviations above and below the mean. To find areas under the normal curve, we will always need the Z score of the cutoff. With $\mu = 1500$, $\sigma = 300$, and the cutoff value $x = 1630$, the Z score is given by

$$Z = \frac{x - \mu}{\sigma} = \frac{1630 - 1500}{300} = 130/300 = 0.43$$

We look the percentile of $Z = 0.43$ in the normal probability table shown in Table 3.8 or in Appendix C.1 on page 361, which yields 0.6664. However, the percentile describes those who had a Z score *lower* than 0.43. To find the area *above* $Z = 0.43$, we compute one minus the area of the lower tail:

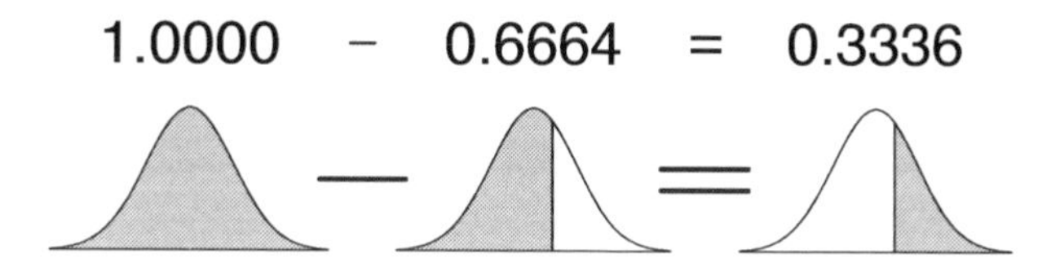

The probability Shannon scores at least 1630 on the SAT is 0.3336.

TIP: always draw a picture first
For any normal probability situation, *always always always* draw and label the normal curve and shade the area of interest first. The picture will provide an estimate of the probability.

TIP: find the Z score second
After drawing a figure to represent the situation, identify the Z score for the observation of interest.

⊙ **Exercise 3.10** If the probability of Shannon getting at least 1630 is 0.3336, then what is the probability she gets less than 1630? Draw the normal curve representing this exercise, shading the lower region instead of the upper one. Hint in the footnote[7].

● **Example 3.11** Edward earned a 1400 on his SAT. What is his percentile?

First, a picture is needed. Edward's percentile is the proportion of people who do not get as high as a 1400. These are the scores to the left of 1400.

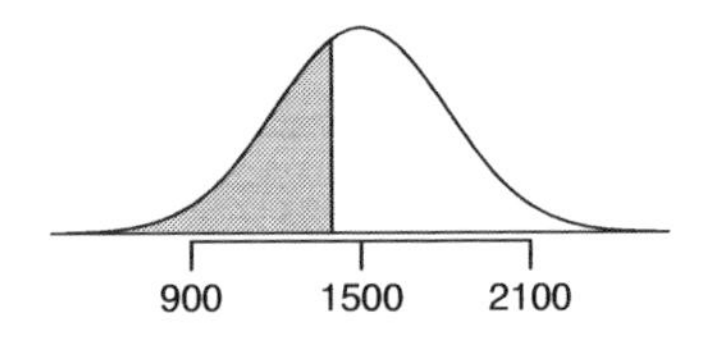

Identifying the mean $\mu = 1500$, the standard deviation $\sigma = 300$, and the cutoff for the tail area $x = 1400$ makes it easy to compute the Z score:

$$Z = \frac{x - \mu}{\sigma} = \frac{1400 - 1500}{300} = -0.33$$

Using the normal probability table, identify the row of -0.3 and column of 0.03, which corresponds to the probability 0.3707. Edward is at the 37th percentile.

⊙ **Exercise 3.12** Use the results of Example 3.11 to compute the proportion of SAT takers who did better than Edward. Also draw a new picture.

TIP: areas to the right
The normal probability table in most books gives the area to the left. If you would like the area to the right, first find the area to the left and then subtract this amount from one.

⊙ **Exercise 3.13** Stuart earned an SAT score of 2100. Draw a picture for each part. (a) What is his percentile? (b) What percent of SAT takers did better than Stuart? Short answers in the footnote[8].

[7]We found the probability in Example 3.9.
[8](a) 0.9772. (b) 0.0228.

Based on a sample of 100 men[9], the heights of male adults between the ages 20 and 62 in the US is nearly normal with mean 70.0" and standard deviation 3.3".

⊙ **Exercise 3.14** Mike is 5'7" and Jim is 6'4". (a) What is Mike's height percentile? (b) What is Jim's height percentile? Also draw one picture for each part.

The last several problems have focused on finding the probability or percentile for a particular observation. What if you would like to know the observation corresponding to a particular percentile?

● **Example 3.15** Erik's height is at the 40th percentile. How tall is he?

As always, first draw the picture.

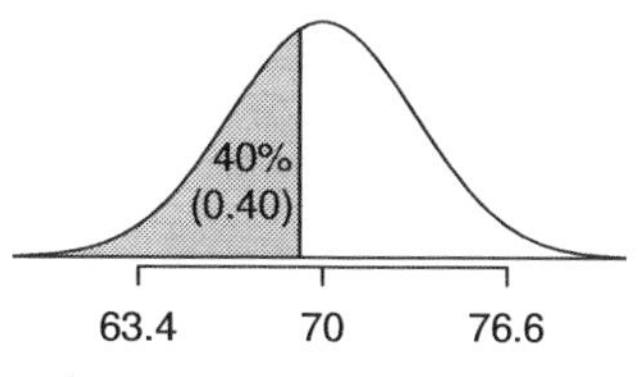

In this case, the lower tail probability is known (0.40), which can be shaded on the diagram. We want to find the observation that corresponds to this value. As a first step in this direction, we determine the Z score associated with the 40th percentile.

Because the percentile is below 50%, we know Z will be negative. Looking in the negative part of the normal probability table, we search for the probability *inside* the table closest to 0.4000. We find that 0.4000 falls in row -0.2 and between columns 0.05 and 0.06. Since it falls closer to 0.05, we take this one: $Z = -0.25$.

Knowing Erik's Z score $Z = -0.25$, $\mu = 70$ inches, and $\sigma = 3.3$ inches, the Z score formula can be setup to determine Erik's unknown height, labeled x:

$$-0.25 = Z = \frac{x - \mu}{\sigma} = \frac{x - 70}{3.3}$$

Solving for x yields the height 69.18 inches[10]. That is, Erik is about 5'9".

● **Example 3.16** What is the adult male height at the 82nd percentile?

Again, we draw the figure first.

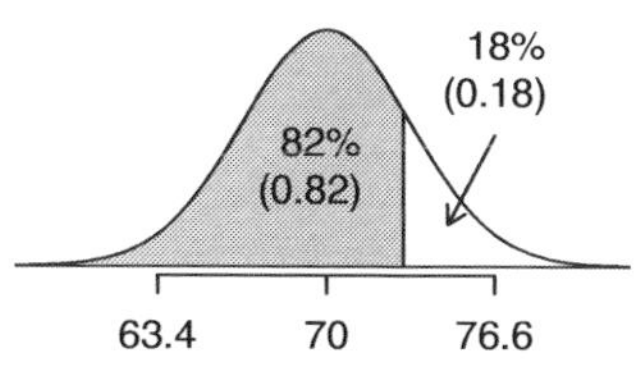

Next, we want to find the Z score at the 82nd percentile, which will be a positive value. Looking in the Z table, we find Z falls in row 0.9 and the nearest column is 0.02, i.e. $Z = 0.92$. Finally, the height x is found using the Z score formula with the known mean μ, standard deviation σ, and Z score $Z = 0.92$:

$$0.92 = Z = \frac{x - \mu}{\sigma} = \frac{x - 70}{3.3}$$

This yields 73.04 inches or about 6'1" as the height at the 82nd percentile.

---

[9]This sample was taken from the USDA Food Commodity Intake Database.

[10]To solve for x, first multiply by 3.3 and then add 70 to each side.

⊙ **Exercise 3.17** (a) What is the 95th percentile for SAT scores? (b) What is the 97.5^{th} percentile of the male heights? As always with normal probability problems, first draw a picture. Answers in the footnote[11].

⊙ **Exercise 3.18** (a) What is the probability that a randomly selected male adult is at least 6'2" (74")? (b) What is the probability that a male adult is shorter than 5'9" (69")? Short answers in the footnote[12].

● **Example 3.19** What is the probability that a random adult male is between 5'9" and 6'2"?

First, draw the figure. The area of interest is no longer an upper or lower tail.

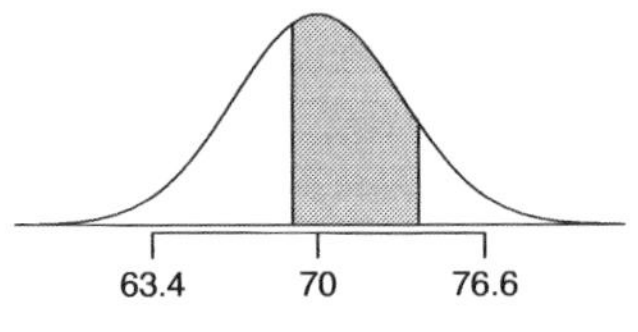

Because the total area under the curve is 1, the area of the two tails that are not shaded can be found (Exercise 3.18): 0.3821 and 0.1131. Then, the middle area is given by

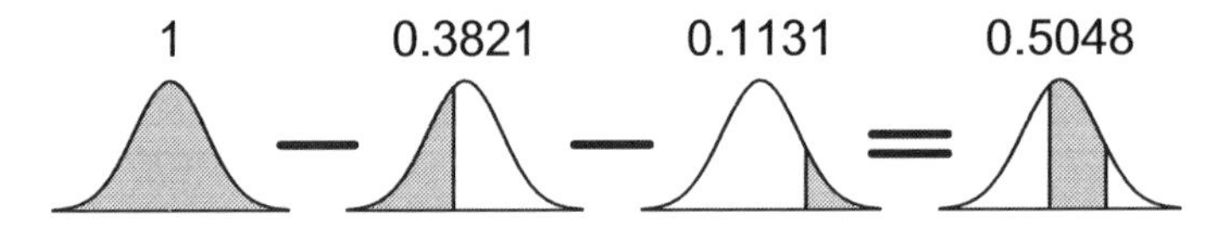

That is, the probability of being between 5'9" and 6'2" is 0.5048.

⊙ **Exercise 3.20** What percent of SAT takers get between 1500 and 2000? Hint in the footnote[13].

⊙ **Exercise 3.21** What percent of adult males are between 5'5" (65") and 5'7" (67")?

3.1.5 68-95-99.7 rule

Here, we present a useful rule of thumb for the probability of falling within 1, 2, and 3 standard deviations of the mean in the normal distribution. This will be useful in a wide range of practical settings, especially when trying to make a quick estimate without a calculator or Z table.

[11]Remember: draw a picture first, then find the Z score. (We leave the pictures to you.) The Z score can be found by using the percentiles and the normal probability table. (a) We look for 0.95 in the probability portion (middle part) of the normal probability table, which leads us to row 1.6 and (about) column 0.05, i.e. $Z_{95} = 1.65$. Knowing $Z_{95} = 1.65$, $\mu = 1500$, and $\sigma = 300$, we setup the Z score formula: $1.65 = \frac{x_{95} - 1500}{300}$. We solve for x_{95}: $x_{95} = 1995$. (b) Similarly, we find $Z_{97.5} = 1.96$, again setup the Z score formula for the heights, and calculate $x_{97.5} = 76.5$.

[12](a) 0.1131. (b) 0.3821.

[13]First find the percent who get below 1500 and the percent that get above 2000.

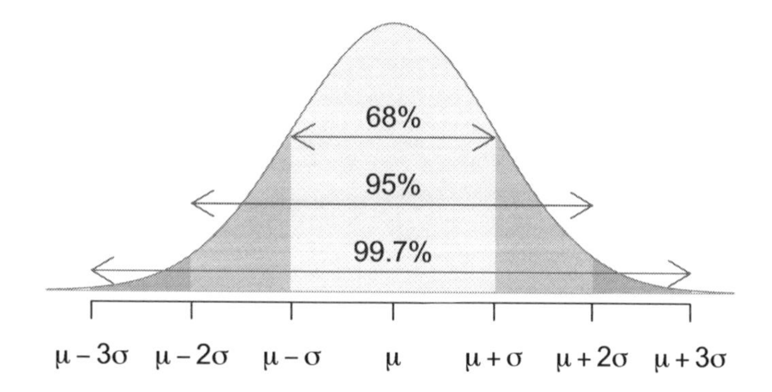

Figure 3.9: Probabilities for falling within 1, 2, and 3 standard deviations of the mean in a normal distribution.

⊙ **Exercise 3.22** Use the Z table to confirm that about 68%, 95%, and 99.7% of observations fall within 1, 2, and 3, standard deviations of the mean in the normal distribution, respectively. For instance, first find the area that falls between $Z = -1$ and $Z = 1$, which should have an area of about 0.68. Similarly there should be an area of about 0.95 between $Z = -2$ and $Z = 2$.

It is possible for a normal random variable to fall 4, 5, or even more standard deviations from the mean. However, these occurrences are very rare if the data are nearly normal. The probability of being further than 4 standard deviations from the mean is about 1-in-30,000. For 5 and 6 standard deviations, it is about 1-in-3.5 million and 1-in-1 billion, respectively.

⊙ **Exercise 3.23** SAT scores closely follow the normal model with mean $\mu = 1500$ and standard deviation $\sigma = 300$. (a) About what percent of test takers score 900 to 2100? (b) Can you determine how many score 1500 to 2100? Answer in the footnote[14].

3.2 Evaluating the normal approximation

Many processes can be well approximated by the normal distribution. We have already seen two good examples: SAT scores and the heights of US adult males. While using a normal model can be extremely convenient and useful, it is important to remember normality is always an approximation. Testing the appropriateness of the normal assumption is a key step in practical data analysis.

3.2.1 Normal probability plot

Example 3.15 suggests the distribution of heights of US males might be well approximated by the normal model. We are interested in proceeding under the assumption that the data are normally distributed, but first we must check to see if this is reasonable.

There are two visual methods for checking the assumption of normality, which can be implemented and interpreted quickly. The first is a simple histogram with the best fitting

[14](a) 900 and 2100 represent two standard deviations above and below the mean, which means about 95% of test takers will score between 900 and 2100. (b) Since the normal model is symmetric, then half of the test takers from part (a) ($95\%/2 = 47.5\%$ of all test takers) will score 900 to 1500 while 47.5% score between 1500 and 2100.

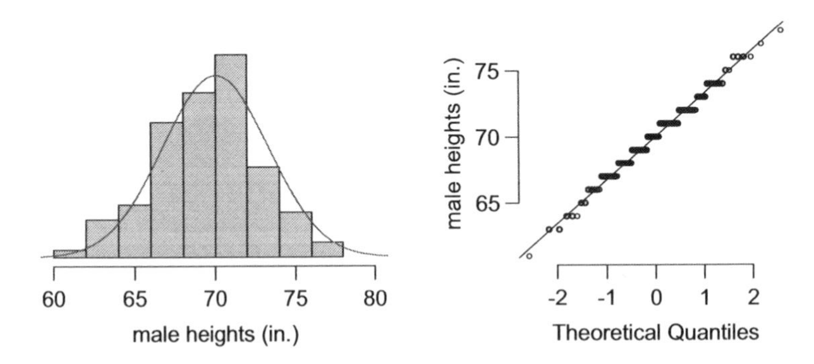

Figure 3.10: A sample of 100 male heights. The observations are rounded to the nearest whole inch, explaining why the points appear to jump in increments in the normal probability plot.

normal curve overlaid on the plot, as shown in the left panel of Figure 3.10. The sample mean $\bar{x}$ and standard deviation s are used as the parameters of the best fitting normal curve. The closer this curve fits the histogram, the more reasonable the normal model assumption. Another more common method is examining a **normal probability plot**[15], shown in the right panel of Figure 3.10. The closer the points are to a perfect straight line, the more confident we can be that the data follow the normal model. We outline the construction of the normal probability plot in Section 3.2.2.

● **Example 3.24** Three data sets of 40, 100, and 400 samples were simulated from a normal distribution, and the histograms and normal probability plots of the data sets are shown in Figure 3.11. These will provide a benchmark for what to look for in plots of real data.

The left panels show the histogram (top) and normal probability plot (bottom) for the simulated data set with 40 observations. The data set is too small to really see clear structure in the histogram. The normal probability plot also reflects this, where there are some deviations from the line. However, these deviations are not strong.

The middle panels show diagnostic plots for the data set with 100 simulated observations. The histogram shows more normality and the normal probability plot shows a better fit. While there is one observation that deviates noticeably from the line, it is not particularly extreme.

The data set with 400 observations has a histogram that greatly resembles the normal distribution, while the normal probability plot is nearly a perfect straight line. Again in the normal probability plot there is one observation (the largest) that deviates slightly from the line. If that observation had deviated 3 times further from the line, it would be of much greater concern in a real data set. Apparent outliers can occur in normally distributed data but they are rare and may be grounds to reject the normality assumption.

Notice the histograms look more normal as the sample size increases, and the normal probability plot becomes straighter and more stable. This is generally true when sample size increases.

[15] Also commonly called a **quantile-quantile plot**.

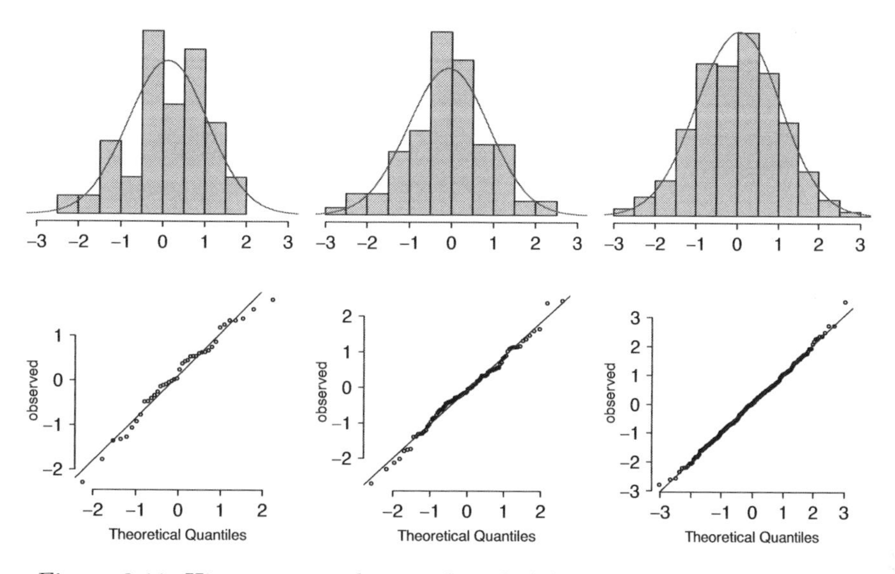

Figure 3.11: Histograms and normal probability plots for three simulated normal data sets; $n = 40$ (left), $n = 100$ (middle), $n = 400$ (right).

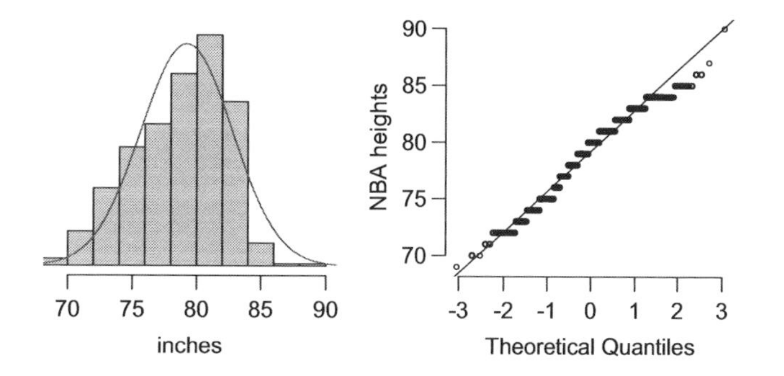

Figure 3.12: Histogram and normal probability plot for the NBA heights from the 2008-9 season.

● **Example 3.25** Are NBA player heights normally distributed? Consider all 435 NBA players[16] from the 2008-9 season presented in Figure 3.12.

We first create a histogram and normal probability plot of the NBA player heights. The histogram in the left panel is slightly left-skewed, which contrasts with the symmetric normal distribution. The points in the normal probability plot do not appear to closely follow a straight line but show what appears to be a "wave". We can compare these characteristics to the sample of 400 normally distributed observations in Example 3.24 and see that they represent much stronger deviations from the normal model. NBA player heights do not appear to come from a normal distribution.

[16]These data were collected from nba.com.

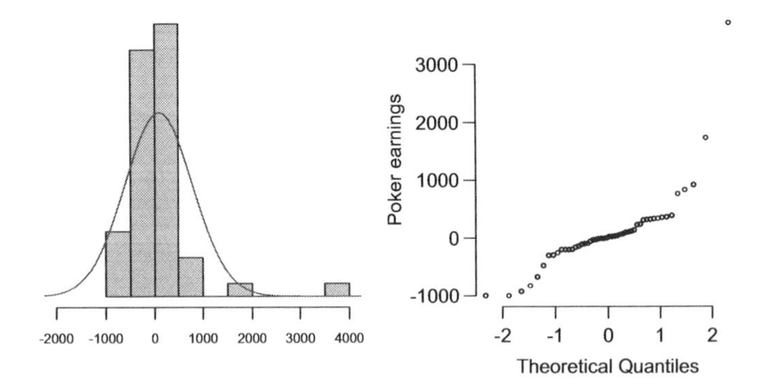

Figure 3.13: A histogram of poker data with the best fitting normal plot and a normal probability plot.

● **Example 3.26** Can we approximate poker winnings by a normal distribution? We consider the poker winnings of an individual over 50 days. A histogram and normal probability plot of these data are shown in Figure 3.13.

The data appear to be strongly right skewed in the histogram, which corresponds to the very strong deviations on the upper right component of the normal probability plot. If we compare these results to the sample of 40 normal observations in Example 3.24, it is apparent that these data set shows very strong deviations from the normal model.

⊙ **Exercise 3.27** Determine which data sets in of the normal probability plots in Figure 3.14 plausibly come from a nearly normal distribution. Are you confident in all of your conclusions? There are 100 (top left), 50 (top right), 500 (bottom left), and 15 points (bottom right) in the four plots. The authors' interpretations are in the footnote[17].

3.2.2 Constructing a normal probability plot (special topic)

We construct the plot as follows:

(1) Order the observations.

(2) Determine the percentile of each observation in the ordered data set.

(3) Identify the Z score corresponding to each percentile.

(4) Create a scatterplot of the observations (vertical) against the Z scores (horizontal).

If the observations are normally distributed, then their Z scores will approximately correspond to their percentiles and thus to the z_i in Table 3.15.

[17]The top-left plot appears show some deviations in the smallest values in the data set; specifically, the left tail of the data set probably has some outliers we should be wary of. The top-right and bottom-left plots do not show any obvious or extreme deviations from the lines for their respective sample sizes, so a normal model would be reasonable for these data sets. The bottom-right plot has a consistent curvature that suggests it is not from the normal distribution. If we examine just the vertical coordinates of these observations, we see that there is a lot of data between -20 and 0, and then about five observations scattered between 0 and 70. This describes a distribution that has a strong right skew.

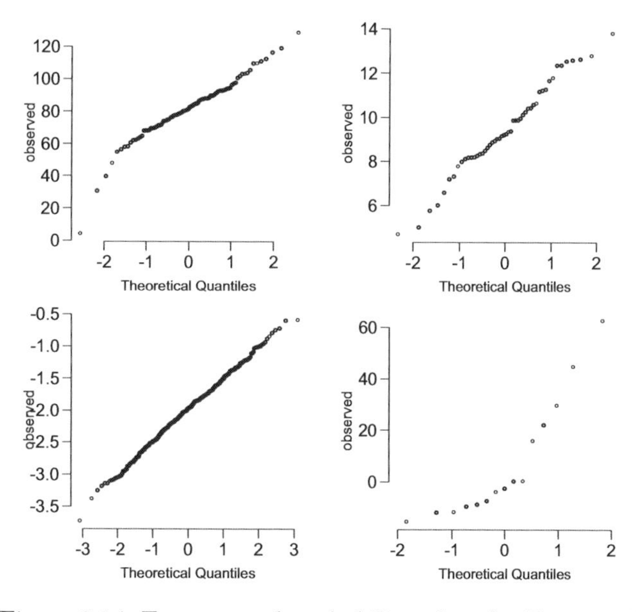

Figure 3.14: Four normal probability plots for Exercise 3.27.

Observation i	1	2	3	$\cdots$	100
x_i	61	63	63	$\cdots$	78
Percentile	0.99%	1.98%	2.97%	$\cdots$	99.01%
z_i	-2.33	-2.06	-1.89	$\cdots$	2.33

Table 3.15: Construction of the normal probability plot for the NBA players. The first observation is assumed to be at the 0.99^{th} percentile, and the z_i corresponding to a lower tail of 0.0099 is -2.33. To create the plot based on this table, plot each pair of points, (z_i, x_i).

Caution: z_i correspond to percentiles
The z_i in Table 3.15 are *not* the Z scores of the observations but only correspond to the percentiles of the observations.

Because of the complexity of these calculations, normal probability plots are generally created using statistical software.

3.3 Geometric distribution (special topic)

How long should we expect to flip a coin until it turns up `heads`? Or how many times should we expect to roll a die until we get a `1`? These questions can be answered using the geometric distribution. We first formalize each trial – such as a single coin flip or die toss – using the Bernoulli distribution, and then we combine these with our tools from probability (Chapter 2) to construct the geometric distribution.

3.3.1 Bernoulli distribution

Stanley Milgram began a series of experiments in 1963 to estimate what proportion of people would willingly obey an authority and give severe shocks to a stranger. Milgram found that about 65% of people would obey the authority and give such shocks. Over the years, additional research suggested this number is approximately consistent across communities and time[18].

Each person in Milgram's experiment can be thought of as a **trial**. We label a person a **success** if she refuses to administer the worst shock. A person is labeled a **failure** if she administers the worst shock. Because only 35% of individuals refused to administer the most severe shock, we denote the **probability of a success** with $p = 0.35$. The probability of a failure is sometimes denoted with $q = 1 - p$.

Thus, a `success` or `failure` is recorded for each person in the study. When an individual trial only has two possible outcomes, it is called a **Bernoulli random variable**.

> **Bernoulli random variable, descriptive**
> A Bernoulli random variable has exactly two possible outcomes. We typically label one of these outcomes a "success" and the other outcome a "failure". We may also denote a success by `1` and a failure by `0`.

> **TIP: "success" need not be something positive**
> We chose to label a person who refuses to administer the worst shock a "success" and all others as "failures". However, we could just as easily have reversed these labels. The mathematical framework we will build does not depend on which outcome is labeled a success and which a failure, as long as we are consistent.

Bernoulli random variables are often denoted as `1` for a success and `0` for a failure. In addition to being convenient in entering data, it is also mathematically handy. Suppose we observe ten trials:

`0 1 1 1 1 0 1 1 0 0`

Then the **sample proportion**, $\hat{p}$, is the sample mean of these observations:

$$\hat{p} = \frac{\text{\# of successes}}{\text{\# of trials}} = \frac{0+1+1+1+1+0+1+1+0+0}{10} = 0.6$$

This mathematical inquiry of Bernoulli random variables can be extended even further. Because `0` and `1` are numerical outcomes, we can define the mean and standard deviation of a Bernoulli random variable[19].

[18] Find further information on Milgram's experiment at http://www.cnr.berkeley.edu/ucce50/ag-labor/7article/article35.htm.

[19] If p is the true probability of a success, then the mean of a Bernoulli random variable X is given by

$$\begin{aligned}\mu = E[X] &= P(X=0)*0 + P(X=1)*1 \\ &= (1-p)*0 + p*1 = 0 + p = p\end{aligned}$$

Similarly, the variance of X can be computed:

$$\begin{aligned}\sigma^2 &= P(X=0)(0-p)^2 + P(X=1)(1-p)^2 \\ &= (1-p)p^2 + p(1-p)^2 = p(1-p)\end{aligned}$$

The standard deviation is $\sigma = \sqrt{p(1-p)}$.

Bernoulli random variable, mathematical
If X is a random variable that takes value `1` with probability of success p and `0` with probability $1 - p$, then X is a Bernoulli random variable with mean and standard deviation

$$\mu = p \qquad \sigma = \sqrt{p(1-p)}$$

In general, it is useful to think about a Bernoulli random variable as a random process with only two outcomes: a success or failure. Then we build our mathematical framework using the numerical labels `1` and `0` for successes and failures, respectively.

3.3.2 Geometric distribution

● **Example 3.28** Dr. Smith wants to repeat Milgram's experiments but she only wants to sample people until she finds someone who will not inflict the worst shock[20]. If the probability a person will *not* give the most severe shock is still 0.35 and the people are independent, what are the chances that she will stop the study after the first person? The second person? The third? What about if it takes her $n-1$ individuals who will administer the worst shock before finding her first success, i.e. the first success is on the nth person? (If the first success is the fifth person, then we say $n = 5$.)

The probability of stopping after the first person is just the chance the first person will not administer the worst shock: $1 - 0.65 = 0.35$. The probability it will be the second person is

$$\begin{aligned} &P(\text{second person is the first to not administer the worst shock}) \\ &\quad = P(\text{the first will, the second won't}) = (0.65)(0.35) = 0.228 \end{aligned}$$

Likewise, the probability it will be the third person is $(0.65)(0.65)(0.35) = 0.148$.

If the first success is on the nth person, then there are $n-1$ failures and finally 1 success, which corresponds to the probability $(0.65)^{n-1}(0.35)$. This is the same as $(1-0.35)^{n-1}(0.35)$.

Example 3.28 illustrates what is called the geometric distribution, which describes the waiting time until a success for **independent and identically distributed (iid)** Bernoulli random variables. In this case, the *independence* aspect just means the individuals in the example don't affect each other, and *identical* means they each have the same probability of success.

The geometric distribution from Example 3.28 is shown in Figure 3.16. In general, the probabilities for a geometric distribution decrease **exponentially** fast.

While this text will not derive the formulas for the mean (expected) number of trials needed to find the first success or the standard deviation or variance of this distribution, we present general formulas for each.

[20]This is hypothetical since, in reality, this sort of study probably would not be permitted any longer under current ethical standards.

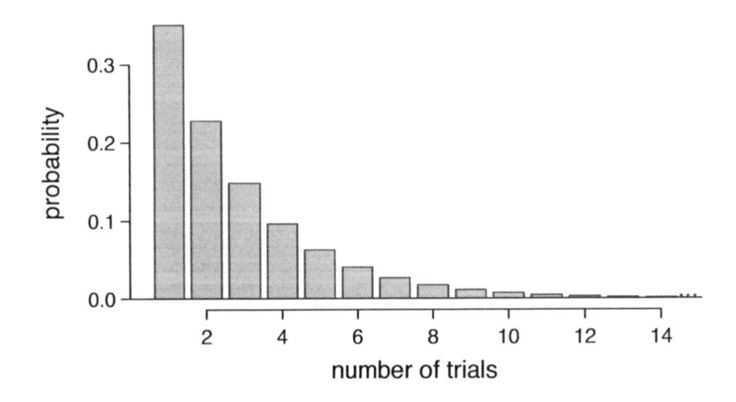

Figure 3.16: The geometric distribution when the probability of success is $p = 0.35$.

Geometric Distribution
If the probability of a success in one trial is p and the probability of a failure is $1 - p$, then the probability of finding the first success in the nth trial is given by

$$(1 - p)^{n-1}p \tag{3.29}$$

The mean (i.e. expected value), variance, and standard deviation of this wait time are given by

$$\mu = \frac{1}{p} \qquad \sigma^2 = \frac{1-p}{p^2} \qquad \sigma = \sqrt{\frac{1-p}{p^2}} \tag{3.30}$$

It is no accident that we use the symbol μ for both the mean and expected value. The mean and the expected value are one and the same.

The left side of Equation (3.30) says that, on average, it takes $1/p$ trials to get a success. This mathematical result is consistent with what we would expect intuitively. If the probability of a success is high (e.g. 0.8), then we don't wait very long for a success: $1/0.8 = 1.25$ trials on average. If the probability of a success is low (e.g. 0.1), then we would expect to view many trials before we see a success: $1/0.1 = 10$ trials.

⊙ **Exercise 3.31** The probability an individual would refuse to administer the worst shock is said to be about 0.35. If we were to examine individuals until we found one that did not administer the shock, how many people should we expect to check? The first expression in Equation (3.30) may be useful. The answer is in the footnote[21].

● **Example 3.32** What is the chance that Dr. Smith will find the first success within the first 4 people?

This is the chance it is the first ($n = 1$), second ($n = 2$), third ($n = 3$), or fourth ($n = 4$) person as the first success, which are four disjoint outcomes. Because the individuals in the sample are randomly sampled from a large population, they are

[21] We would expect to see about $1/0.35 = 2.86$ individuals to find the first success.

independent. We compute the probability of each case and add the separate results:

$$\begin{aligned}
&P(n = 1, 2, 3, \text{ or } 4) \\
&\quad = P(n = 1) + P(n = 2) + P(n = 3) + P(n = 4) \\
&\quad = (0.65)^{1-1}(0.35) + (0.65)^{2-1}(0.35) + (0.65)^{3-1}(0.35) + (0.65)^{4-1}(0.35) \\
&\quad = 0.82
\end{aligned}$$

She has about an 82% chance of ending the study within 4 people.

⊙ **Exercise 3.33** Determine a more clever way to solve Example 3.32. Show that you get the same result. Answer in the footnote[22].

● **Example 3.34** Suppose in one region it was found that the proportion of people who would administer the worst shock was "only" 55%. If people were randomly selected from this region, what is the expected number of people who must be checked before one was found that would be deemed a success? What is the standard deviation of this waiting time?

A success is when someone will **not** inflict the worst shock, which has probability $p = 1 - 0.55 = 0.45$ for this region. The expected number of people to be checked is $1/p = 1/0.45 = 2.22$ and the standard deviation is $\sqrt{(1-p)/p^2} = 1.65$.

⊙ **Exercise 3.35** Using the results from Example 3.34, $\mu = 2.22$ and $\sigma = 1.65$, would it be appropriate to use the normal model to find what proportion of experiments would end in 3 or fewer trials? Answer in the footnote[23].

⊙ **Exercise 3.36** The independence assumption is crucial to the geometric distribution's accurate description of a scenario. Why? Answer in the footnote[24].

3.4 Binomial distribution (special topic)

● **Example 3.37** Suppose we randomly selected four individuals to participate in the "shock" study. What is the chance exactly one of them will be a success? Let's call the four people Allen (A), Brittany (B), Caroline (C), and Damian (D) for convenience. Also, suppose 35% of people are successes as in the previous version of this example.

Let's consider a scenario where one person refuses:

$$\begin{aligned}
&P(A = \texttt{refuse},\ B = \texttt{shock},\ C = \texttt{shock},\ D = \texttt{shock}) \\
&\quad = P(A = \texttt{refuse}) * P(B = \texttt{shock}) * P(C = \texttt{shock}) * P(D = \texttt{shock}) \\
&\quad = (0.35) * (0.65) * (0.65) * (0.65) = (0.35)^1(0.65)^3 = 0.096
\end{aligned}$$

But there are three other scenarios: Brittany, Caroline, or Damian could have been the one to refuse. In each of these cases, the probability is again $(0.35)^1(0.65)^3$.

---

[22] Use the complement: P(there is no success in the first four observations). Compute one minus this probability.

[23] No. The geometric distribution is always right skewed and can never be well-approximated by the normal model.

[24] Independence simplified the situation. Mathematically, we can see that to construct the probability of the success on the nth trial, we had to use the Multiplication Rule for Independent processes. It is no simple task to generalize the geometric model for dependent trials.

These four scenarios exhaust all the possible ways that exactly one of these four people could refuse to administer the most severe shock, so the total probability is $4 * (0.35)^1(0.65)^3 = 0.38$.

⊙ **Exercise 3.38** Verify that the scenario where Brittany is the only one to refuse to give the most severe shock has probability $(0.35)^1(0.65)^3$.

3.4.1 The binomial distribution

The scenario outlined in Example 3.37 is a special case of what is called the binomial distribution. The **binomial distribution** describes the probability of having exactly k successes in n independent Bernoulli trials with probability of a success p (in Example 3.37, $n = 4$, $k = 1$, $p = 0.35$). We would like to determine the probabilities associated with the binomial distribution more generally, i.e. we want a formula where we can use n, k, and p to obtain the probability. To do this, we reexamine each part of the example.

There were four individuals who could have been the one to refuse, and each of these four scenarios had the same probability. Thus, we could identify the final probability as

$$[\# \text{ of scenarios}] * P(\text{single scenario}) \tag{3.39}$$

The first component of this equation is the number of ways to arrange the $k = 1$ successes among the $n = 4$ trials. The second component is the probability of any of the four (equally probable) scenarios.

Consider P(single scenario) under the general case of k successes and $n - k$ failures in the n trials. In any such scenario, we apply the Product Rule for independent events:

$$p^k(1-p)^{n-k}$$

This is our general formula for P(single scenario).

Secondly, we introduce a general formula for the number of ways to choose k successes in n trials, i.e. arrange k successes and $n - k$ failures:

$$\binom{n}{k} = \frac{n!}{k!(n-k)!}$$

The quantity $\binom{n}{k}$ is read **n choose k** [25]. The exclamation point notation (e.g. $k!$) denotes a **factorial** expression.

$$\begin{aligned}
0! &= 1 \\
1! &= 1 \\
2! &= 2 * 1 = 2 \\
3! &= 3 * 2 * 1 = 6 \\
4! &= 4 * 3 * 2 * 1 = 24 \\
&\vdots \\
n! &= n * (n-1) * \ldots * 3 * 2 * 1
\end{aligned}$$

Using the formula, we can compute the number of ways to choose $k = 1$ successes in $n = 4$ trials:

$$\binom{4}{1} = \frac{4!}{1!(4-1)!} = \frac{4!}{1!3!} = \frac{4*3*2*1}{(1)(3*2*1)} = 4$$

[25] Other notation for n choose k includes ${}_nC_k$, C_n^k, and $C(n, k)$.

This result is exactly what we found by carefully thinking of each possible scenario in Example 3.37.

Substituting n choose k for the number of scenarios and $p^k(1-p)^{n-k}$ for the single scenario probability in Equation (3.39) yields the general binomial formula.

Binomial distribution

Suppose the probability of a single trial being a success is p. Then the probability of observing exactly k successes in n independent trials is given by

$$\binom{n}{k} p^k (1-p)^{n-k} = \frac{n!}{k!(n-k)!} p^k (1-p)^{n-k} \qquad (3.40)$$

Additionally, the mean, variance, and standard deviation of the number of observed successes are

$$\mu = np\sigma \qquad ^2 = np(1-p) \qquad \sigma = \sqrt{np(1-p)} \qquad (3.41)$$

TIP: Is it Binomial? Four conditions to check.
(1) The trials independent.
(2) The number of trials, n, is fixed.
(3) Each trial outcome can be classified as a *success* or *failure.*
(4) The probability of a success (p) is the same for each trial.

● **Example 3.42** What is the probability 3 of 8 randomly selected students will refuse to administer the worst shock, i.e. 5 of 8 will?

We would like to apply the Binomial model, so we check our conditions. The number of trials is fixed ($n = 3$) (condition 2) and each trial outcome can be classified as a success or failure (condition 3). Because the sample is random, the trials are independent and the probability of a success is the same for each trial (conditions 1 and 4).

In the outcome of interest, there are $k = 3$ successes in $n = 8$ trials, and the probability of a success is $p = 0.35$. So the probability that 3 of 8 will refuse is given by

$$\begin{aligned} \binom{8}{3}(0.35)^3(1-0.35)^{8-3} &= \frac{8!}{3!(8-3)!}(0.35)^3(1-0.35)^{8-3} \\ &= \frac{8!}{3!5!}(0.35)^3(0.65)^5 \end{aligned}$$

Dealing with the factorial part:

$$\frac{8!}{3!5!} = \frac{8*7*6*5*4*3*2*1}{(3*2*1)(5*4*3*2*1)} = \frac{8*7*6}{3*2*1} = 56$$

Using $(0.35)^3(0.65)^5 \approx 0.005$, the final probability is about $56 * 0.005 = 0.28$.

TIP: computing binomial probabilities

The first step in using the Binomial model is to check that the model is appropriate. The second step is to identify n, p, and k. The final step is to apply our formulas and interpret the results.

TIP: computing factorials
In general, it is useful to do some cancelation in the factorials immediately. Alternatively, many computer programs and calculators have built in functions to compute n choose k, factorials, and even entire binomial probabilities.

⊙ **Exercise 3.43** If you ran a study and randomly sampled 40 students, how many would you expect to refuse to administer the worst shock? What is the standard deviation of the number of people who would refuse? Equation (3.41) may be useful. Answers in the footnote[26].

⊙ **Exercise 3.44** The probability a random smoker will develop a severe lung condition in his or her lifetime is about 0.3. If you have 4 friends who smoke, are the conditions for the Binomial model satisfied? One possible answer in the footnote[27].

⊙ **Exercise 3.45** Suppose these four friends do not know each other and we can treat them as if they were a random sample from the population. Is the Binomial model appropriate? What is the probability that (a) none of them will develop a severe lung condition? (b) One will develop a severe lung condition? (c) That no more than one will develop a severe lung condition? Answers in the footnote[28].

⊙ **Exercise 3.46** What is the probability that **at least** 2 of your 4 smoking friends will develop a severe lung condition in their lifetimes?

⊙ **Exercise 3.47** Suppose you have 7 friends who are smokers and they can be treated as a random sample of smokers. (a) How many would you expect to develop a severe lung condition, i.e. what is the mean? (b) What is the probability that at most 2 of your 7 friends will develop a severe lung condition. Hint in the footnote[29].

Below we consider the first term in the Binomial probability, n choose k under some special scenarios.

⊙ **Exercise 3.48** Why is it true that $\binom{n}{0} = 1$ and $\binom{n}{n} = 1$ for any number n? Hint in the footnote[30].

---

[26]We are asked to determine the expected number (the mean) and the standard deviation, both of which can be directly computed from the formulas in Equation (3.41): $\mu = np = 40 * 0.35 = 14$ and $\sigma = \sqrt{np(1-p)} = \sqrt{40 * 0.35 * 0.65} = 3.02$. Because very roughly 95% of observations fall within 2 standard deviations of the mean (see Section 1.3.5), we would probably observe at least 8 but less than 20 individuals in our sample to refuse to administer the shock.

[27]If the friends know each other, then the independence assumption is probably not satisfied.

[28]To check if the Binomial model is appropriate, we must verify the conditions. (i) Since we are supposing we can treat the friends as a random sample, they are independent. (ii) We have a fixed number of trials ($n = 4$). (iii) Each outcome is a success or failure. (iv) The probability of a success is the same for each trials since the individuals are like a random sample ($p = 0.3$ if we say a "success" is someone getting a lung condition, a morbid choice). Compute parts (a) and (b) from the binomial formula in Equation (3.40): $P(0) = \binom{4}{0}(0.3)^0(0.7)^4 = 1 * 1 * 0.7^4 = 0.2401$, $P(1) = \binom{4}{1}(0.3)^1(0.7)^4 = 0.4116$. Note: $0! = 1$, as shown on page 122. Part (c) can be computed as the sum of parts (a) and (b): $P(0)+P(1) = 0.2401+0.4116 = 0.6517$. That is, there is about a 65% chance that no more than one of your four smoking friends will develop a severe lung condition.

[29]First compute the separate probabilities for 0, 1, and 2 friends developing a severe lung condition.

[30]How many different ways are there to arrange 0 successes and n failures in n trials? How many different ways are there to arrange n successes and 0 failures in n trials?

⊙ **Exercise 3.49** How many ways can you arrange one success and $n-1$ failures in n trials? How many ways can you arrange $n-1$ successes and one failure in n trials? Answer in the footnote[31].

3.4.2 Normal approximation to the binomial distribution

The binomial formula is cumbersome when the sample size (n) is large, particularly when we consider a range of observations. In some cases we may use the normal distribution as an easier and faster way to estimate binomial probabilities.

● **Example 3.50** Approximately 20% of the US population smokes cigarettes. A local government believed their community had a lower smoker rate in their community and commissioned a survey of 400 randomly selected individuals. The survey found that only 59 of the 400 participants smoke cigarettes. If the true proportion of smokers in the community was really 20%, what is the probability of observing 59 or fewer smokers in a sample of 400 people?

We leave the usual verification that the four conditions for the binomial model are valid as an exercise.

The question posed is equivalent to asking, what is the probability of observing $k = 0$, 1, ..., 58, or 59 smokers in a sample of $n = 400$ when $p = 0.20$? We can compute these 60 different probabilities and add them together to find the answer:

$$\begin{aligned} &P(k = 0 \text{ or } k = 1 \text{ or } \cdots \text{ or } k = 59) \\ &\quad = P(k = 0) + P(k = 1) + \cdots + P(k = 59) \\ &\quad = 0.0041 \end{aligned}$$

If the true proportion of smokers in the community is $p = 0.20$, then the probability of observing 59 or fewer smokers in a sample of $n = 400$ is less than 0.0041.

The computations in Example 3.50 are tedious and long. In general, we should avoid such work if an alternative method exists that is faster, easier, and still accurate. Recall that calculating probabilities of a range of values is much easier in the normal model. We might wonder, is it possible to use the normal model in place of the binomial distribution? Surprisingly we can, if certain conditions are met.

⊙ **Exercise 3.51** Here we consider the binomial model when the probability of a success is $p = 0.10$. Figure 3.17 shows four hollow histograms for simulated samples from the binomial distribution using four different sample sizes: $n = 10, 30, 100, 300$. What happens to the shape of the distributions as the sample size increases? What distribution does the last hollow histogram resemble?

---

[31] One success and $n-1$ failures: there are exactly n unique places we can put the success, so there are n ways to arrange one success and $n-1$ failures. A similar argument is used for the second question. Mathematically, we show these results by verifying the following two equations:

$$\binom{n}{1} = n, \qquad \binom{n}{n-1} = n$$

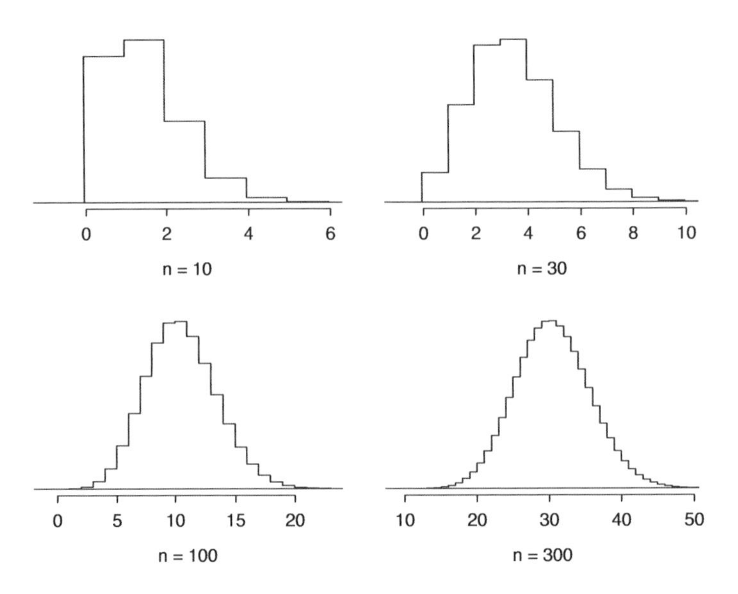

Figure 3.17: Hollow histograms of samples from the binomial model when $p = 0.10$. The sample sizes for the four plots are $n = 10$, 30, 100, and 300, respectively.

Normal approximation of the binomial distribution
The binomial distribution with probability of success p is nearly normal when the sample size n is sufficiently large that np and $n(1-p)$ are both at least 10. The approximate normal distribution has parameters corresponding to the mean and standard deviation of the binomial distribution:

$$\mu = np\sigma \qquad = \sqrt{np(1-p)}$$

The normal approximation may be used when computing the range of many possible successes. For instance, we may apply the normal distribution to the setting of Example 3.50.

● **Example 3.52** How can we use the normal approximation to estimate the probability of observing 59 or fewer smokers in a sample of 400, if the true proportion of smokers is $p = 0.20$?

Showing that the binomial model is reasonable was a suggested exercise in Example 3.50. We also verify that both np and $n(1-p)$ are at least 10:

$$np = 400 * 0.20 = 80 \qquad \text{and} \qquad n(1-p) = 400 * 0.8 = 320$$

With these conditions checked, we may use the normal approximation in place of the binomial distribution using the mean and standard deviation from the binomial model:

$$\mu = np = 80 \qquad \sigma = \sqrt{np(1-p)} = 8$$

We want to find the probability of observing fewer than 59 smokers using this model.

⊙ **Exercise 3.53** Use the normal model $N(\mu = 80, \sigma = 8)$ to estimate the probability of observing fewer than 59 smokers. Your answer should be approximately equal to the solution of Example 3.50: 0.0041.

Caution: The normal approximation may fail on small intervals
The normal approximation to the binomial distribution tends to perform poorly when estimating the probability of a small range of counts, even when the conditions are met.

Suppose we wanted to compute the probability of observing 69, 70, or 71 smokers in 400 when $p = 0.20$. With such a large sample, we might be tempted to apply the normal approximation and use the range 69 to 71. However, we would find that the binomial solution and the normal approximation notably differ:

Binomial: 0.0703 Normal: 0.0476

We can identify the cause of this discrepancy using Figure 3.18, which shows the areas representing the binomial probability (outlined) and normal approximation (shaded). Notice that the width of the area under the normal distribution is 0.5 units too slim on both sides of the interval.

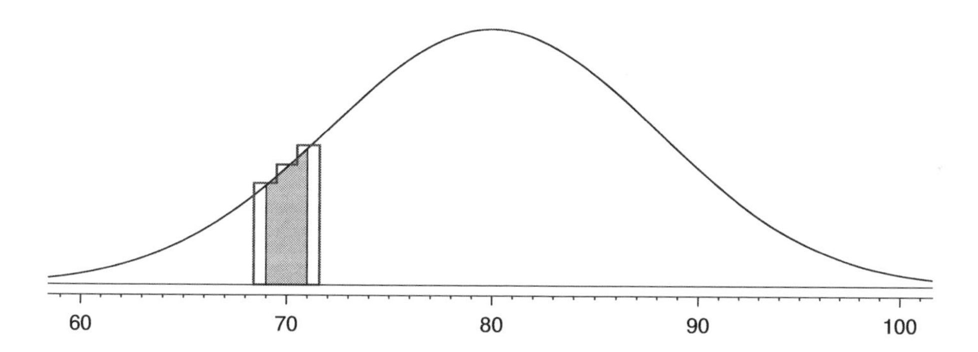

Figure 3.18: A normal curve with the area between 69 and 71 shaded. The outlined area represents the exact binomial probability.

TIP: Improving the accuracy of the normal approximation to the binomial distribution
The normal approximation to the binomial distribution for intervals of values is usually improved if cutoff values are modified slightly. The cutoff values for the lower end of a shaded region should be reduced by 0.5, and the cutoff value for the upper end should be increased by 0.5.

The tip to add extra area when applying the normal approximation is most often useful when examining a range of observations. While it is possible to apply it when computing a tail area, the benefit of the modification usually disappears since the total interval is typically quite wide.

3.5 More discrete distributions (special topic)

3.5.1 Negative binomial distribution

The geometric distribution describes the probability of observing the first success on the nth trial. The **negative binomial distribution** is more general: it describes the probability of observing the kth success on the nth trial.

● **Example 3.54** Each day a high school football coach tells his star kicker, Brian, that he can go home after he successfully kicks four 35 yard field goals. Suppose we say each kick has a probability p of being successful. If p is small – e.g. close to 0.1 – would we expect Brian to need many attempts before he successfully kicks his fourth field goal?

We are waiting for the fourth success ($k = 4$). If the probability of a success (p) is small, then the number of attempts (n) will probably be large. This means that Brian is more likely to need many attempts before he gets $k = 4$ successes. To put this another way, the probability of n being small is low.

To identify a negative binomial case, we check 4 conditions. The first three are common to the binomial distribution[32].

TIP: Is it negative binomial? Four conditions to check.
(1) The trials are independent.
(2) Each trial outcome can be classified as a success or failure.
(3) The probability of a success (p) is the same for each trial.
(4) The last trial must be a success.

⊙ **Exercise 3.55** Suppose Brian is very diligent in his attempts and he makes each 35 yard field goal with probability $p = 0.8$. Take a guess at how many attempts he would need before making his fourth kick. One answer in the footnote[33].

● **Example 3.56** In yesterday's practice, it took Brian only 6 tries to get his fourth field goal. Write out each of the possible sequence of kicks.

Because it took Brian six tries to get the fourth success, we know the last kick must have been a success. That leaves three successful kicks and two unsuccessful kicks (we label these as failures) that make up the first five attempts. There are ten possible sequences of these first five kicks, which are shown in Table 3.19. If Brian achieved his fourth success ($k = 4$) on his sixth attempt ($n = 6$), then his order of successes and failures must be one of these ten possible sequences.

⊙ **Exercise 3.57** Each sequence in Table 3.19 has exactly two failures and four successes with the last attempt always being a success. If the probability of a success is $p = 0.8$, find the probability of the first sequence. Answer in the footnote[34].

---

[32] See a similar guide for the binomial distribution on page 123.

[33] Since he is likely to make each field goal attempt, it will take him at least four but probably not more than 6 or 7.

[34] The first sequence: $0.2 * 0.2 * 0.8 * 0.8 * 0.8 * 0.8 = 0.0164$.

	Kick Attempt					
	1	2	3	4	5	6
1	F	F	$\overset{1}{S}$	$\overset{2}{S}$	$\overset{3}{S}$	$\overset{4}{S}$
2	F	$\overset{1}{S}$	F	$\overset{2}{S}$	$\overset{3}{S}$	$\overset{4}{S}$
3	F	$\overset{1}{S}$	$\overset{2}{S}$	F	$\overset{3}{S}$	$\overset{4}{S}$
4	F	$\overset{1}{S}$	$\overset{2}{S}$	$\overset{3}{S}$	F	$\overset{4}{S}$
5	$\overset{1}{S}$	F	F	$\overset{2}{S}$	$\overset{3}{S}$	$\overset{4}{S}$
6	$\overset{1}{S}$	F	$\overset{2}{S}$	F	$\overset{3}{S}$	$\overset{4}{S}$
7	$\overset{1}{S}$	F	$\overset{2}{S}$	$\overset{3}{S}$	F	$\overset{4}{S}$
8	$\overset{1}{S}$	$\overset{2}{S}$	F	F	$\overset{3}{S}$	$\overset{4}{S}$
9	$\overset{1}{S}$	$\overset{2}{S}$	F	$\overset{3}{S}$	F	$\overset{4}{S}$
10	$\overset{1}{S}$	$\overset{2}{S}$	$\overset{3}{S}$	F	F	$\overset{4}{S}$

Table 3.19: The ten possible sequences when the fourth successful kick is on the sixth attempt.

If the probability Brian kicks a 35 yard field goal is $p = 0.8$, what is the probability it takes Brian exactly six tries to get his fourth successful kick? We can write this probability as

$$
\begin{aligned}
&P(\text{it takes Brian six tries to make four field goals}) \\
&= P(\text{Brian makes three of his first five field goals, and he makes the sixth one}) \\
&= P(1^{st} \text{ sequence from above OR } 2^{nd} \text{ from above OR ... OR } 10^{th} \text{ seq. from above})
\end{aligned}
$$

The second equality holds because the ten sequences from Example 3.56 describe all possible ways it could take Brian six tries to kick four field goals. We can break down this last probability into the sum of ten disjoint possibilities:

$$
\begin{aligned}
&P(1^{st} \text{ sequence from above OR } 2^{nd} \text{ from above OR ... OR } 10^{th} \text{ seq. from above}) \\
&= P(1^{st} \text{ sequence from above}) + P(2^{nd} \text{ from above}) + \cdots + P(10^{th} \text{ seq. from above})
\end{aligned}
$$

The probability of the first sequence was identified in Exercise 3.57 as 0.0164, and each of the other sequences have the same probability. Since each of the ten sequence has the same probability, the total probability is ten times that of any individual sequence.

The way to compute this negative binomial probability is similar to how the binomial problems were solved in Section 3.4. The probability is broken into two pieces:

$$
\begin{aligned}
&P(\text{it takes Brian six tries to make four field goals}) \\
&= [\text{Number of possible sequences}] * P(\text{Single sequence})
\end{aligned}
$$

Each part is examined separately, then we multiply to get the final result.

We first identify the probability of a single sequence. One particular case is to first observe all the failures ($n - k$ of them) followed by the k successes:

$$
\begin{aligned}
&P(\text{Single sequence}) \\
&= P(n - k \text{ failures and then } k \text{ successes}) \\
&= (1 - p)^{n-k} p^k
\end{aligned}
$$

We must also identify the number of sequences for the general case. Above, ten sequences were identified where the fourth success came on the sixth attempt. These sequences were identified by fixing the last observation as a success and looking for all the ways to arrange the other observations. In other words, how many ways could we arrange $k-1$ successes in $n-1$ trials? This can be found using the n choose k coefficient but for $n-1$ and $k-1$ instead:

$$\binom{n-1}{k-1} = \frac{(n-1)!}{(k-1)!\,((n-1)-(k-1))!} = \frac{(n-1)!}{(k-1)!\,(n-k)!}$$

This is the number of different ways we can order $k-1$ successes and $n-k$ failures in $n-1$ trials, and we use it to identify the general probability formula for the kth success coming on the nth trial. If the factorial notation (the exclamation point) is unfamiliar, see page 122.

Negative binomial distribution

The negative binomial distribution describes the probability of observing the kth success on the nth trial:

$$P(\text{the } k^{th} \text{ success on the } n^{th} \text{ trial}) = \binom{n-1}{k-1} p^k (1-p)^{n-k} \qquad (3.58)$$

where p is the probability an individual trial is a success. All trials are assumed to be independent.

● **Example 3.59** Show using Equation (3.58) that the probability Brian kicks his fourth successful field goal on the sixth attempt is 0.164.

The probability of a single success is $p = 0.8$, the number of successes is $k = 4$, and the number of necessary attempts under this scenario is $n = 6$.

$$\binom{n}{k-1} p^k (1-p)^{n-k} = \frac{5!}{3!2!} 0.8^4 * 0.2^2 = 10 * 0.0164 = 0.164$$

⊙ **Exercise 3.60** The negative binomial distribution requires that each kick attempt by Brian is independent. Do you think it is reasonable to suggest that each of Brian's kick attempts are independent? Comment in the footnote[35].

⊙ **Exercise 3.61** Assume Brian's kick attempts are independent. What is the probability that Brian will kick his fourth field goal within 5 attempts? Answer in the footnote[36].

---

[35] We cannot conclusively say they are or are not independent. However, many statistical reviews of athletic performance suggests such attempts are very nearly independent.

[36] If his fourth field goal ($k = 4$) is within five attempts, it either took him four or five tries ($n = 4$ or $n = 5$). We have $p = 0.8$ from earlier. Use Equation (3.58) to compute the probability of $n = 4$ tries and $n = 5$ tries, then add those probabilities together:

$$P(n = 4 \text{ OR } n = 5) = P(n = 4) + P(n = 5)$$
$$= \binom{4-1}{4-1} 0.8^4 + \binom{5-1}{4-1} (0.8)^4 (1-0.8) = 1 * 0.41 + 4 * 0.082 = 0.41 + 0.33 = 0.74$$

TIP: Binomial versus negative binomial
In the binomial case, we typically have a fixed number of trials and instead consider the number of successes. In the negative binomial case, we examine how many trials it takes to observe a fixed number of successes and require that the last observation be a success.

⊙ **Exercise 3.62** On 70% of days, a hospital admits at least one heart attack patient. On 30% of the days, no heart attack patients are admitted. Identify each case below as a binomial or negative binomial case, and compute the probability. Answers are in the footnote[37].

(a) What is the probability the hospital will admit a heart attack patient on exactly three days this week?

(b) What is the probability the second day with a heart attack patient will be the fourth day of the week?

(c) What is the probability the fifth day of next month will be the first day with a heart attack patient?

3.5.2 Poisson distribution

● **Example 3.63** There are about 8 million individuals in New York City. How many individuals might we expect to be hospitalized for acute myocardial infarction (AMI), i.e. a heart attack, each day? According to historical records, the average number is about 4.4 individuals. However, we would also like to know the approximate distribution of counts. What would a histogram of the number of AMI occurrences each day look like if we recorded the daily counts over an entire year?

A histogram of the number of occurrences of AMI on 365 days[38] for NYC is shown in Figure 3.20. The sample mean (4.38) is similar to the historical average of 4.4. The sample standard deviation is about 2, and the histogram indicates that about 70% of the data fall between 2.4 and 6.4. The distribution's shape is unimodal and skewed to the right.

The **Poisson distribution** is often useful for estimating the number of rare events in a large population over a unit of time. For instance, consider each of the following events, which are rare for any given individual:

- having a heart attack,
- getting married, and
- getting struck by lightning.

The Poisson distribution helps us describe the number of such events that will occur in a short unit of time for a fixed population if the individuals within the population are independent.

[37]In each part, $p = 0.7$. (a) The number of days is fixed, so this is binomial. The parameters are $k = 3$ and $n = 7$: 0.097. (b) The last "success" (admitting a heart attack patient) is fixed to the last day, so we should apply the negative binomial distribution. The parameters are $k = 2$, $n = 4$: 0.132. (c) Knowing next month is May doesn't help solve this problem. This problem is negative binomial with $k = 1$ and $n = 5$: 0.006. Note that the negative binomial case when $k = 1$ is the same as using the geometric distribution.

[38]These data are simulated. In practice, we should check for an association between successive days.

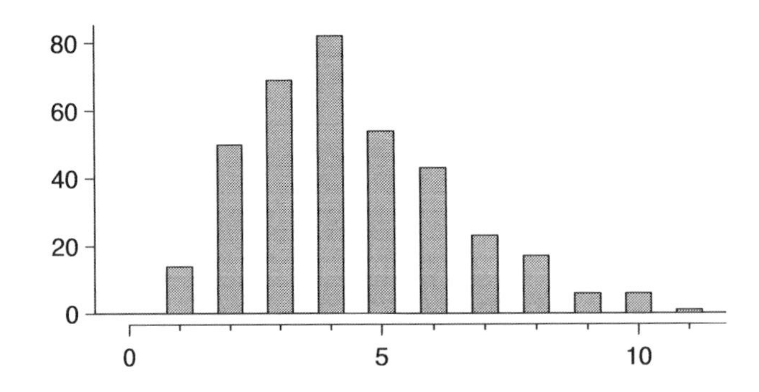

Figure 3.20: A histogram of the number of occurrences of AMI on 365 separate days in NYC.

The histogram in Figure 3.20 approximates a Poisson distribution with rate equal to 4.4. The **rate** for a Poisson distribution is the average number of occurrences in a mostly-fixed population per unit of time. In Example 3.63, the time unit is a day, the population is all New York City residents, and the historical rate is 4.4. The parameter in the Poisson distribution is the rate – or how many rare events we expect to observe – and it is typically denoted by λ (the Greek letter *lambda*) or μ. Using the rate, we can describe the probability of observing exactly k rare events in a single unit of time.

λ
Rate for the Poisson dist.

Poisson distribution
Suppose we are watching for rare events and the number of observed events follows a Poisson distribution with rate λ. Then

$$P(\text{observe } k \text{ rare events}) = \frac{\lambda^k e^{-\lambda}}{k!}$$

where k may take a value 0, 1, 2, ..., and $k!$ represents k-factorial, as described on page 122. The mean and standard deviation of this distribution are λ and $\sqrt{\lambda}$, respectively.

We will leave a rigorous set of conditions for the Poisson distribution to a later course. However, we offer a few simple guidelines that can be used for an initial evaluation of whether the Poisson model would be appropriate.

TIP: Is it Poisson?
A random variable may follow a Poisson distribution if the event being considered is rare, the population is large, and the events occur independently of each other.

Even when rare events are not really independent – for instance, Saturdays and Sundays are especially popular for weddings – a Poisson model may sometimes still be reasonable if we allow it to have a different rate for different times. In the wedding example, the rate would be modeled as higher on weekends than on weekdays. The idea of modeling rates for a Poisson distribution against a second variable such as `dayOfTheWeek` forms the foundation of some more advanced methods that fall in the realm of **generalized linear models**. In Chapters 7 and 8, we will discuss a foundation of linear models, but we leave generalized linear models to a later course.

3.6 Exercises

3.6.1 Normal distribution

3.1 What percent of a standard normal distribution $N(\mu = 0, \sigma = 1)$ is found in each region? Be sure to draw a graph.

(a) $Z < -1.35$ (b) $Z > 1.48$ (c) $-0.4 < Z < 1.5$ (d) $|Z| > 2$

3.2 What percent of a standard normal distribution $N(\mu = 0, \sigma = 1)$ is found in each region? Be sure to draw a graph.

(a) $Z > -1.13$ (b) $Z < 0.18$ (c) $Z > 8$ (d) $|Z| < 0.5$

3.3 A college senior who took the Graduate Record Examination (GRE) exam scored 620 on the Verbal Reasoning section and 670 on the Quantitative Reasoning section. The mean score for Verbal Reasoning section was 462 with a standard deviation of 119, and the mean score for the Quantitative Reasoning was 584 with a standard deviation of 151. Suppose that both distributions are nearly normal.

(a) Write down the short-hand for these two normal distributions.

(b) What is her Z score on the Verbal Reasoning section? On the Quantitative Reasoning section? Draw a standard normal distribution curve and mark these two Z scores.

(c) What do these Z scores tell you?

(d) Find her percentile scores for the two exams.

(e) On which section did she do better compared to the rest of the exam takers?

(f) What percent of the test takers did better than her on the Verbal Reasoning section? On the Quantitative Reasoning section?

(g) Explain why simply comparing her raw scores from the two sections would lead to the incorrect conclusion that she did better on the Quantitative Reasoning section.

3.4 Two friends, Leo (male, age 33) and Mary (female, age 28), both completed the Hermosa Beach Triathlon. In triathlons, it is common for racers to be placed into age and gender groups. Leo competed in the *Men, Ages 30 - 34* group while Mary competed in the *Women, Ages 25 - 29* group. Leo completed the race in 1:22:28 (4948 seconds), while Mary completed the race in 1:31:53 (5513 seconds). Obviously Leo finished faster, but they are curious about how they did within their respective groups. Can you help them? Here is some information on the performance of their groups:

- The finishing times of the *Men, Ages 30 - 34* group has a mean of 4313 seconds with a standard deviation of 583 seconds.
- The finishing times of the *Women, Ages 25 - 29* group has a mean of 5261 seconds with a standard deviation of 807 seconds.
- The distributions of finishing times for both groups are approximately Normal.

Remember, faster finishes are better. So the shorter time it takes to finish, the better the performance.

(a) Write down the short-hand for these two normal distributions.

(b) What are the Z scores for Leo's and Mary's finishing times? What do these Z scores tell you?

(c) What is Leo's percentile?

(d) What is Mary's percentile?

(e) Did Leo or Mary rank better in their respective groups? Explain your reasoning.

3.5 Exercise 3.3 gives the distributions of the scores of the Verbal and Quantitative Reasoning sections of the GRE exam. If the distributions of the scores on these exams are not nearly normal, how would your answers to parts (b)-(e) of Exercise 3.3 change?

3.6 Exercise 3.4 gives the distributions of triathlon finishing times for *Men, Ages 30 - 34* and *Women, Ages 25 - 29* who completed a triathlon. If the distributions of finishing times are not nearly normal, how would your answers to parts (b)-(e) of Exercise 3.4 change?

3.7 Based on the information given in Exercise 3.3, calculate the following:

(a) The score of a student who scored in the 80th percentile on the Quantitative Reasoning section.

(b) The score of a student who scored worse than 70% of the test takers in the Verbal Reasoning section.

3.8 Based on the information given in Exercise 3.4, calculate the following:

(a) The cutoff time for the fastest 5% of athletes in Leo's group, i.e. those who took the shortest 5% of time to finish.

(b) The cutoff time for the slowest 10% of athletes in Mary's group.

3.9 Heights of 10 year olds, regardless of gender, closely follow a normal distribution with mean 55 inches and standard deviation 6 inches.

(a) What is the probability that a randomly chosen 10 year old is shorter than 48 inches?

(b) What is the probability that a randomly chosen 10 year old is between 60 and 65 inches?

(c) If the tallest 10% of the class is considered "very tall", what is the height cutoff for "very tall"?

(d) The height requirement for *Batman the Ride* at Six Flags Magic Mountain is 54 inches. What percent of 10 year olds cannot go on this ride?

3.10 The distribution of speeds of cars traveling on the Interstate 5 Freeway (I-5) in California is nearly normal with a mean of 72.6 miles/hour and a standard deviation of 4.78 miles/hour. [17]

(a) What percent of cars travel slower than 80 miles/hour?

(b) What percent of cars travel between 60 and 80 miles/hour?

(c) How fast to do the fastest 5% of cars travel?

(d) The speed limit on this stretch of the I-5 is 70 miles/hour. Approximate what percentage of the cars travel above the speed limit on this stretch of the I-5.

3.11 The average daily high temperature in June in LA is 77°F with a standard deviation of 5°F. Suppose that the temperatures in June closely follow a normal distribution.

(a) What is the probability of observing an 82.4°F temperature or higher in LA during a randomly chosen day in June?

(b) How cold are the coldest 10% of the days during June in LA?

3.12 The Capital Asset Pricing Model (CAPM) assumes that returns on a portfolio are normally distributed. A portfolio has an average return of 14.7% (i.e. an average gain of 14.7%) with a standard deviation of 33%. A return of 0% means the value of the portfolio doesn't change, a negative return means that the portfolio loses money, and a positive return means that the portfolio gains money.

(a) What percent of the time does this portfolio lose money, i.e. have a return less than 0%?

(b) What is the cutoff for the highest 15% of returns with this portfolio?

3.13 Suppose a newspaper article states that the distribution of auto insurance premiums for residents of California is approximately normal with a mean of $1,650. The article also states that 25% of California residents pay more than $1,800.

(a) What is the Z score that corresponds to the top 25% (or the 75th percentile) of the standard normal distribution?

(b) What is the mean insurance cost? What is the cutoff for the 75th percentile?

(c) Identify the standard deviation of insurance premiums in LA.

3.14 MENSA is an organization whose members have IQs in the top 2% of the population. If IQs are normally distributed with mean 100 and the minimum IQ score required for admission to MENSA is 132, what is the standard deviation of IQ scores?

3.15 The textbook you need to buy for your chemistry class is expensive at the college book store, so you consider buying it on Ebay instead. A look at the past auctions suggest that the prices of that chemistry textbook have an approximately normal distribution with mean $89 and standard deviation $15.

(a) What is the probability that a randomly selected auction for this book closes at more than $100?

(b) Ebay allows you to set your maximum bid price so that if someone outbids you on an auction you can automatically outbid them, up to the maximum bid price you set. If you are only bidding on one auction, what may be the advantages and disadvantages of setting a bid price too high or too low? What if you are bidding on multiple auctions?

(c) If we watched 10 auctions, roughly what percentile might we use for a maximum bid cutoff to be somewhat sure that you will win one of these ten auctions? Is it possible to find a cutoff point that will *ensure* that you win an auction?

(d) If you are patient enough to track ten auctions closely, about what price might you use as your maximum bid price if you want to be somewhat sure that you will buy one of these ten books?

3.16 SAT scores (out of 2400) are distributed normally with a mean of 1500 and a standard deviation of 300. Suppose council awards a certificate of excellence to all students who score at least 1900 on the SAT, and suppose we pick one of the recognized students at random. What is the probability this student's score will be at least 2100? (The material covered in Section 2.3 would be useful for this question.)

3.17 Below are final exam scores of 20 Introductory Statistics students. Also provided are some sample statistics. Use this information to determine if the scores approximately follow the 68-95-99.7% Rule.

79, 83, 57, 82, 94, 83, 72, 74, 73, 71, 66, 89, 78, 81, 78, 81, 88, 69, 77, 79

Mean	77.7
Std. Dev.	8.44

3.18 Below are heights of 25 female college students. Also provided are some sample statistics. Use this information to determine if the heights approximately follow the 68-95-99.7% Rule.

54, 55, 56, 56, 57, 58, 58, 59, 60, 60, 60, 61, 61, 62, 62, 63, 63, 63, 64, 65, 65, 67, 67, 69, 73

Mean	61.52
Std. Dev.	4.58

3.6.2 Evaluating the Normal approximation

3.19 Exercise 3.17 lists the final exam scores of 20 Introductory Statistics students. Do these data appear to follow a normal distribution? Explain your reasoning using the graphs provided below.

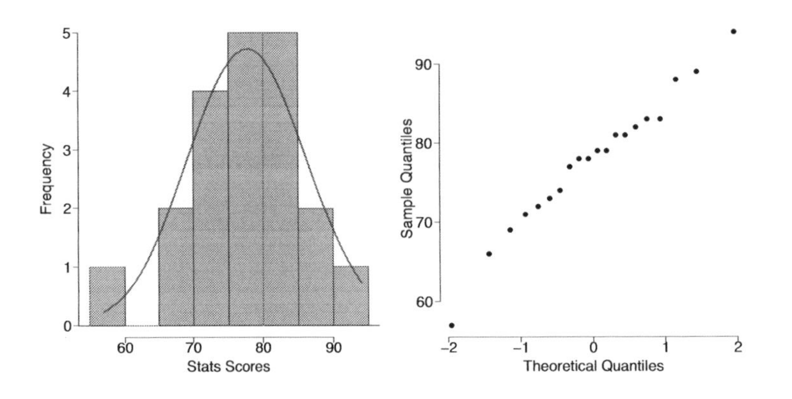

3.20 Exercise 3.18 lists the heights of 25 female college students. Do these data appear to follow a normal distribution? Explain your reasoning using the graphs provided below.

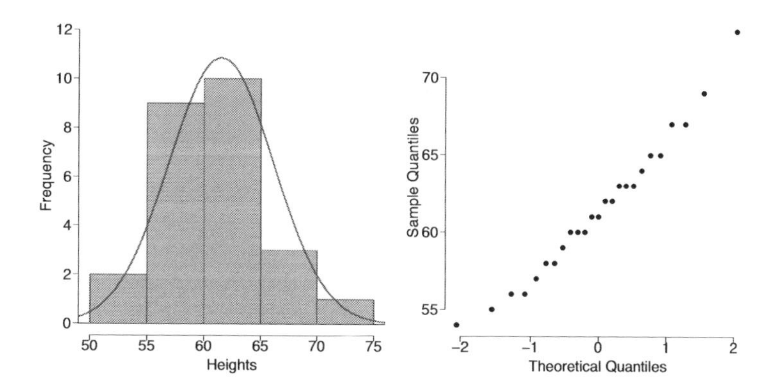

3.6.3 Geometric distribution

3.21 In a hand of poker, can each card dealt be considered an independent Bernoulli trial?

3.22 Can the outcome of each roll of a die be considered an independent Bernouilli trial?

3.23 American Community Surveys conducted between the years 2005 and 2009 indicate that 49% of women ages 15 to 50 are married. [13]

(a) We randomly select three women between these ages. What is the probability that the third woman selected is the only one who is married?

(b) What is the probability that all three randomly selected women are married?

(c) On average, how many women would you expect to sample before selecting a married woman? What is the standard deviation?

(d) If the proportion of married women was actually 30%, how many women would you expect to sample before selecting a married woman? What is the standard deviation?

(e) Based on your answers to parts (c) and (d), how does decreasing the probability of an event affect the mean and standard deviation of the wait time until success?

3.24 A machine that produces a special type of transistors (a component of computers) has a 2% defective rate. The production is considered a random process where each transistor is independent of the others.

(a) What is the probability that the 10th transistor produced is the first with a defect?

(b) What is the probability that the machine produces no defective transistors in a batch of 100?

(c) On average, how many transistors would you expect to be produced before the first with a defect? What is the standard deviation?

(d) Another machine that also produces transistors has a 5% defective rate where each transistor is produced independent of the others. On average how many transistors would you expect to be produced with this machine before the first with a defect? What is the standard deviation?

(e) Based on your answers to parts (c) and (d), how does increasing the probability of an event affect the mean and standard deviation of the wait time until success?

3.25 A husband and wife both have brown eyes but carry genes that make it possible for their children to have brown eyes (probability 0.75), blue eyes (0.125), or green eyes (0.125).

(a) What is the probability the first blue-eyed child they have is their third child? Assume that the eye colors of the children are independent of each other.

(b) On average, how many children would such a pair of parents have before having a blue-eyed child? What is the standard deviation of the number of children they would expect to have?

3.26 Exercise 3.10 states that the distribution of speeds of cars traveling on the Interstate 5 Freeway (I-5) in California is nearly normal with a mean of 72.6 miles/hour and a standard deviation of 4.78 miles/hour. The speed limit on this stretch of the I-5 is 70 miles/hour.

(a) A highway patrol officer is hidden on the side of the freeway. What is the probability that 5 cars pass and none are speeding? Assume that the speeds of the cars are independent of each other.

(b) On average, how many cars would the highway patrol officer expect to watch until the first car that is speeding? What is the standard deviation of the number of cars he would expect to watch?

3.6.4 Binomial distribution

3.27 According to a 2008 study, 69.7% of 18-20 year olds consumed alcoholic beverages in the past year. [18]

(a) Suppose a random sample of the ten 18-20 year olds is taken. Can we use the binomial distribution to calculate the probability that exactly six consumed alcoholic beverages? Explain.

(b) Calculate this probability.

(c) What is the probability that exactly four out of the ten 18-20 year olds have *not* consumed an alcoholic beverage?

(d) What is the probability that at most 2 out of 5 randomly sampled 18-20 year olds have consumed alcoholic beverages?

(e) What is the probability that at least 1 out of 5 randomly sampled 18-20 year olds have consumed alcoholic beverages?

3.28 The Centers for Disease Control estimates that 90% of Americans have had chicken pox by the time they reach adulthood. [19]

(a) Can we use the binomial distribution to calculate the probability of finding exactly 97 people out of a random sample of 100 American adults have had chicken pox in their childhood? Explain.

(b) Calculate this probability.

(c) What is the probability that exactly 3 out of a new sample of 100 American adults have *not* had chicken pox in their childhood?

(d) What is the probability that at least 1 out of 10 randomly sampled American adults have had chicken pox?

(e) What is the probability that at most 3 out of 10 randomly sampled American adults have had chicken pox?

3.29 Exercise 3.27 states that about 69.7% of 18-20 year olds consumed alcoholic beverages in the past year. We consider a sample of fifty 18-20 year olds.

(a) How many people would you expect to have consumed alcoholic beverages? And with what standard deviation?

(b) Would you be surprised if there were 45 or more people who have consumed alcoholic beverages?

(c) What is the probability that 45 or more people in this sample have consumed alcoholic beverages? How does this probability relate to your answer to part (b)?

3.30 Exercise 3.28 states that about 90% of American adults had chicken pox before adulthood. We consider a random sample of 120 American adults.

(a) How many people would you expect to have had chicken pox in their childhood? And with what standard deviation?

(b) Would you be surprised if there were 105 people who have had chicken pox in their childhood?

(c) What is the probability that 105 or fewer people in this sample have had chicken pox in their childhood? How does this probability relate to your answer to part (b)?

3.31 A 2005 Gallup Poll found that that 7% of teenagers (ages 13 to 17) are afraid of spiders. At a summer camp there are 10 teenagers sleeping in each tent. Assume that these 10 teenagers are independent of each other. [20]

(a) Calculate the probability that at least one of them is afraid of spiders.

(b) Calculate the probability that exactly 2 of them are afraid of spiders?

(c) Calculate the probability that at most 1 of them is afraid of spiders?

(d) If the camp counselor wants no make sure no more than 1 teenager in each tent is afraid of spiders, should she randomly assign teenagers to tents?

3.32 A dreidel is a four-sided spinning top with the Hebrew letters *nun*, *gimel*, *hei*, and *shin*, one on each side. Each side is equally likely to come up in a single spin of the dreidel. Suppose you spin a dreidel three times. Calculate the probability of getting

(a) at least one *nun*?

(b) exactly 2 *nuns*?

(c) exactly 1 *hei*?

(d) at most 2 *gimels*?

Photo by Staccabees on Flickr

http://www.flickr.com/photos/44689913@N04/4116667696

3.33 Exercise 3.25 states that a particular husband and wife both have brown but eyes carry genes that make it possible for their children to have brown eyes (probability 0.75), blue eyes (0.125), or green eyes (0.125).

(a) What is the probability that their first child will have green eyes and the second will not?

(b) What is the probability that exactly one of their two children will have green eyes?

(c) If they have six children, what is the probability that exactly two will have green eyes?

(d) If they have six children, what is the probability that at least one will have green eyes?

(e) What is the probability that the first green eyed child will be the 4th child?

(f) Would it be considered unusual if only 2 out of their 6 children had brown eyes?

3.34 Sickle cell anemia is a genetic blood disorder where red blood cells lose their flexibility and assume an abnormal, rigid, "sickle" shape, which results in a risk of various complications. If both parents are carriers of the disease, then a child has a 25% chance of having the disease, 50% chance of being a carrier, and 25% chance of not carrying any sickle cell genes. If two parents who are carriers of the disease have 3 children, what is the probability that

(a) two will have the disease?

(b) none will have the disease?

(c) at least one will not be carrying any sickle cell genes?

(d) the first child with the disease will the be 3rd child?

3.35 In the game of roulette, a wheel is spun and you place bets on where it will stop. One popular bet is that it will stop on a red slot, which has probability 18/38. If it stops on red, you double the money you bet. If not, you lose the money you bet. Suppose you play 3 times, each time with a $1 bet. Let Y represent the total amount won or lost. Write a probability model for Y.

3.36 In a multiple choice exam there are 5 questions and 4 choices for each question (a, b, c, d). Robin has not studied for the exam at all, and decides to randomly guess the answers. What is the probability that

(a) the first question she gets right is the 3rd question?

(b) she gets exactly 3 or exactly 4 questions right?

(c) she gets the majority of the questions right?

3.37 In this exercise, we consider the formula for the number of ways to arrange n unique objects: $n!$. We will derive this formula for a couple of special cases.

Suppose I have five books covering five topics: algebra, biology, English, history, and statistics. Each day, my roommate randomizes the order of my books. For simplicity, we will suppose each possible ordering has an equal chance.

(a) On a given day, what is the probability that the books will be randomly placed into alphabetical order?

(b) If the alphabetical order has an equal chance of occurring relative to all other possible orderings, how many ways must there be to arrange the five books?

(c) How many possible ways could we have ordered 8 unique books?

3.38 While it is often assumed that the probabilities of having a boy or a girl are the same, the actual probability of having a boy is slightly higher at 0.51. Suppose a couple plans to have 3 kids.

(a) Use the binomial model to calculate the probability that two of them will be boys.

(b) Write out all possible orderings of 3 children, 2 of whom are boys. Use these scenarios to calculate the same probability from part (a) but using the addition rule for disjoint outcomes. Confirm that your answers from parts (a) and (b) match.

(c) If we wanted to calculate the probability that a couple who plans to have 8 kids will have 3 boys, briefly describe why the approach from part (b) would be more tedious than the approach from part (a).

3.6.5 More discrete models

3.39 Calculate the following probabilities and indicate which probability distribution model is appropriate in each case. You roll a fair die 5 times. What is the probability of rolling

(a) the first 6 on the fifth roll?

(b) exactly three 6s?

(c) the third 6 on the fifth roll?

3.40 Calculate the following probabilities and indicate which probability distribution model is appropriate in each case. A very good darts player can hit the bull's eye (red circle in the center of the dart board) 65% of the time. What is the probability that he

(a) hits the bullseye for the 10th time on the 15th try?

(b) hits the bullseye 10 times in 15 tries?

(c) hits the first bullseye on the third try?

3.41 For a sociology class project you are asked to conduct a survey on 20 students at your school. You decide to stand outside of your dorm's cafeteria and conduct the survey on the first 20 students leaving the cafeteria after dinner one evening. Your dorm is comprised of 45% males and 55% females.

(a) Which probability model is most appropriate for calculating the probability that the 4th person you survey is the 2nd female? Explain.

(b) Compute the probability from part (a).

(c) The three possible scenarios that lead to 4th person you survey being the 2nd female are

$$\{M, M, F, F\}, \{M, F, M, F\}, \{F, M, M, F\}$$

One common feature among these scenarios is that the last trial is always female. In the first three trials there are 2 males and 1 female. Use the binomial coefficient to confirm that there are 3 ways of ordering 2 males and 1 female.

(d) Use the findings presented in part (c) to explain why the formula for the coefficient for the negative binomial is $\binom{n-1}{k-1}$ while the formula for the binomial coefficient is $\binom{n}{k}$.

3.42 A not-so-skilled volleyball player has a 15% chance of making the serve, which involves hitting the ball so it passes over the net on a trajectory such that it will land in the opposing team's court. Assume that her serves are independent of each other.

(a) What is the probability that on the 10th try she will make her 3rd successful serve?

(b) Suppose she has made two successful serves in nine attempts. What is the probability that her 10th serve will be successful?

(c) Even though parts (a) and (b) discuss the same scenario, the probabilities you calculated should be different. Can you explain the reason for this discrepancy?

3.43 A coffee shop serves an average of 75 customers per hour during the morning rush.

(a) Which distribution we have studied is most appropriate for calculating the probability of a given number of customers arriving within one hour during this time of day?

(b) What are the mean and the standard deviation of the number of customers this coffee shop serves in one hour during this time of day?

(c) Would it be considered unusually low if only 60 customers showed up to this coffee shop in one hour during this time of day?

3.44 A very skilled court stenographer makes one typographical error (typo) per hour on average.

(a) What probability distribution is most appropriate for calculating the probability of a given number of typos this stenographer makes in an hour?

(b) What are the mean and the standard deviation of the number of typos this stenographer makes?

(c) Would it be considered unusual if this stenographer made 4 typos in a given hour?

3.45 Exercise 3.43 gives the average number of customers visiting a particular coffee shop during the morning rush hour as 75. Calculate the probability that this coffee shop serves 70 customers in one hour during this time of day?

3.46 Exercise 3.44 gives the rate at which a very skilled court stenographer makes typos. Calculate the probability that this stenographer makes at most 2 typos in a given hour.

Chapter 4

Foundations for inference

Statistical inference is concerned primarily with understanding the quality of parameter estimates. For example, a classic inferential question is "How sure are we that the estimated mean, $\bar{x}$, is near the true population mean, μ?" While the equations and details change depending on the parameter we are studying[1], the foundations for inference are the same throughout all of statistics. We introduce these common themes in Sections 4.1-4.4 by discussing inference about the population mean, μ, and set the stage for other parameters in Section 4.5. Some advanced considerations are discussed in Section 4.6. Understanding this chapter will make the rest of the book, and indeed the rest of statistics, seem much more familiar.

Throughout the next few sections we consider a data set called `run10`, shown in Table 4.1. The `run10` data set represents all 14,974 runners who finished the 2009 Cherry Blossom 10 mile run in Washington, DC. The variables are described in Table 4.2.

ID	time	age	gender	state
1	94.12	32	M	MD
2	71.53	33	M	MD
⋮	⋮	⋮	⋮	⋮
14974	83.43	29	F	DC

Table 4.1: Three observations from the `run10` data set.

variable	**description**
`time`	ten mile run time, in minutes
`age`	age, in years
`gender`	gender (`M` for male, `F` for female, and `N` for not-listed)
`state`	home state (or country if not from US)

Table 4.2: Variables and their descriptions for the `run10` data set.

These data are special because they include the results for the entire population of runners who finished the 2009 Cherry Blossom Run. We take a simple random sample of this population, `run10Samp`, represented in Table 4.3. Using this sample, we will draw

[1] We have already seen μ, σ, and p, and we will be introduced to others during later chapters.

ID	time	age	gender	state
3976	96.20	46	F	VA
13597	69.43	20	M	MD
⋮	⋮	⋮	⋮	⋮
8999	78.73	61	F	NY

Table 4.3: Three observations for the `run10Samp` data set, which represents a simple random sample of 100 runners from the 2009 Cherry Blossom Race.

conclusions about the entire population. This is the practice of statistical inference in the broadest sense.

4.1 Variability in estimates

We would like to estimate two features of the Cherry Blossom runners using the `run10Samp` sample.

(1) How long does it take a runner, on average, to complete the 10 miles?

(2) What is the average age of the runners?

These questions are important when planning future events[2]. We will use $x_1, ..., x_{100}$ to represent the 10 mile time for each runner in our sample, and $y_1, ..., y_{100}$ will represent the age of each of these participants.

4.1.1 Point estimates

We want to estimate the **population mean** based on the sample. The most intuitive way to go about doing this is to simply take the **sample mean**. That is, to estimate the average 10 mile run time of all participants, take the average time for the sample:

$$\bar{x} = \frac{96.2 + 69.43 + \cdots + 78.73}{100} = 93.65$$

The sample mean $\bar{x} = 93.65$ minutes is called a **point estimate** of the population mean: if we can only choose one value to estimate the population mean, this is our best guess. Will this point estimate be exactly equal to the population mean? No, but it will probably be close.

If we take a new sample of 100 people and recompute the mean, we will probably not get the exact same answer. Our estimate will vary from one sample to another. This **sampling variation** suggests our estimate may be close but not exactly equal to the parameter.

We can also estimate the average age of participants by examining the sample mean of age:

$$\bar{y} = \frac{46 + 20 + \cdots + 61}{100} = 35.22$$

[2]While we focus on the mean in this chapter, questions regarding variation are often just as important in practice. For instance, we would plan an event very differently if the standard deviation of runner age was 2 versus if it was 20.

What about generating point estimates of other **population parameters**, such as the population median or population standard deviation? Here again we might estimate parameters based on sample statistics, as shown in Table 4.4. For example, we estimate the population standard deviation for the running time using the sample standard deviation, 15.66 minutes.

`time`	estimate	parameter
mean	93.65	94.26
median	92.51	93.85
st. dev.	15.66	15.44

Table 4.4: Point estimates and parameter values for the `time` variable.

⊙ **Exercise 4.1** What is the point estimate for the median of the amount of time it takes participants to run 10 miles? What is the population median? Table 4.4 may be helpful.

⊙ **Exercise 4.2** Suppose we want to estimate the difference in run times for men and women. If $\bar{x}_{men} = 90.10$ and $\bar{x}_{women} = 96.92$, then what would be a good point estimate for the population difference? Answer in the footnote[3].

⊙ **Exercise 4.3** If you had to provide a point estimate of the population IQR for the run time of participants, how might you make such an estimate using a sample?

4.1.2 Point estimates are not exact

Estimates are usually not exactly equal to the truth, but they get better as more data become available. We can see this by plotting a running mean from our `run10Samp` sample. A **running mean** is a sequence of means, where each mean uses one more observation in its calculation than the mean directly before it in the sequence. For example, the second mean in the sequence is the average of the first two observations and the third in the sequence is the average of the first three. The running mean for the `time` variable is shown in Figure 4.5, and it approaches the true population average, 94.26 minutes, as more data become available.

Sample point estimates only approximate the population parameter, and they vary from one sample to another. If we took another simple random sample of the Cherry Blossom runners, we would find that the sample mean for `time` would be a little different. It will be useful to quantify how variable an estimate is from one sample to another. If this variability is small (i.e. the sample mean doesn't change much from one sample to another) then that estimate is probably very accurate. If it varies widely from one sample to another, then we should not expect our estimate to be very good.

4.1.3 Standard error of the mean

From the random sample represented in `run10Samp`, we guessed the average time it takes to run 10 miles is 93.65 minutes. Suppose we take another random sample of 100 individuals and take its mean: 95.05 minutes. Suppose we took another (94.35 minutes) and another (93.20 minutes) and so on. If we do this many many times – which we can do only because

[3] We could take the difference of the two sample means: $96.92 - 90.10 = 6.82$. Men appear to have run about 6.82 minutes faster on average in the 2009 Cherry Blossom Run.

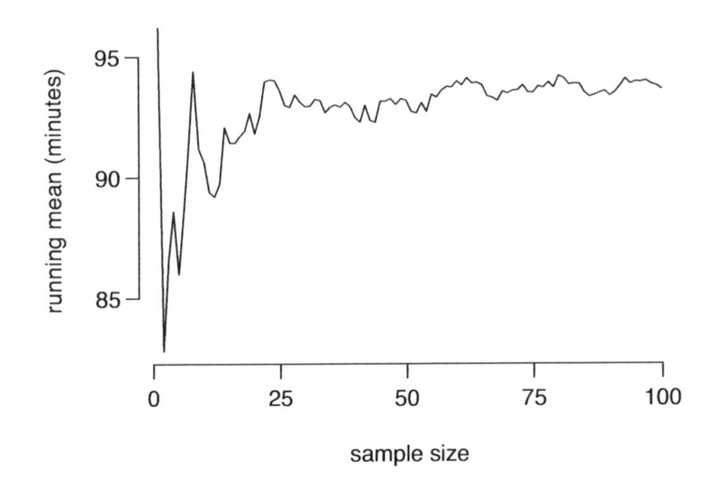

Figure 4.5: The mean computed after adding each individual to the sample. The mean tends to approach the true population average as more data become available.

we have the entire population data set – we can build up a **sampling distribution** for the sample mean when the sample size is 100, shown in Figure 4.6.

> **Sampling distribution**
> The sampling distribution represents the distribution of the point estimate. It is useful to think of the point estimate as being plucked from such a distribution. Understanding the concept of a sampling distribution is central to understanding statistical inference.

The sampling distribution shown in Figure 4.6 can be described as follows. The distribution is approximately symmetric, unimodal, and is centered at the true population mean 94.26 minutes. The variability of the point estimates is described by the standard deviation of this distribution, which is approximately 1.5 minutes.

The center of the distribution, 94.26 minutes, is the actual population mean μ. Intuitively, this makes sense. The sample means should tend to be close to and "fall around" the population mean.

We can see that the sample mean has some variability around the population mean, which we can quantify using the standard deviation of the sample mean: $\sigma_{\bar{x}} = 1.54$. In this case, the standard deviation of the sample mean tells us how far the typical estimate is away from the actual population mean, 94.26 minutes. Thus, this standard deviation describes the typical **error**, and for this reason we usually call this standard deviation the **standard error (SE)** of the estimate.

SE
standard error

> **Standard error of an estimate**
> The standard deviation associated with an estimate is called the *standard error*. It describes the typical error or uncertainty associated with the estimate.

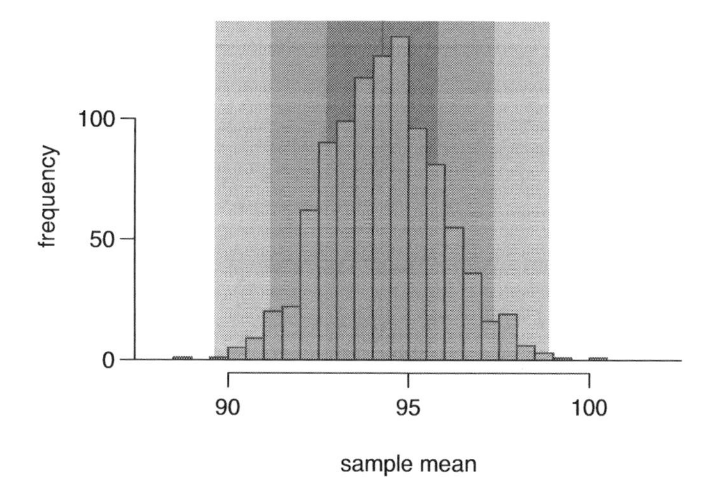

Figure 4.6: A histogram of 1000 sample means for run time, where the samples are of size $n = 100$.

When considering the case of the point estimate $\bar{x}$, there is one problem: there is no obvious way to estimate its standard error from a single sample. However, there is a way around this issue.

⊙ **Exercise 4.4** (a) Would you rather use a small sample or a large sample when estimating a parameter? Why? (b) Using your reasoning from (a), would you expect a point estimate based on a small sample to have smaller or larger standard error than a point estimate based on a larger sample? Answers in the footnote[4].

In the sample of 100 observations, the standard error of the sample mean is equal to one-tenth of the population standard deviation: $1.54 = 15.4/10$. In other words, the standard error of the sample mean based on 100 observations is equal to

$$SE_{\bar{x}} = \sigma_{\bar{x}} = \frac{\sigma_x}{\sqrt{n}} = \frac{15.44}{\sqrt{100}} = 1.54$$

where σ_x is the standard deviation of the individual observations. This is no coincidence. We can show mathematically that this equation is correct when the observations are independent[5]. The most common way to verify that observations in a sample are independent is to collect a simple random sample of less than 10% of the population[6].

[4](a) Consider two random samples: one of size 10 and one of size 1000. Individual observations in the small sample are highly influential on the estimate while in larger samples these individual observations would more often average each other out. The larger sample would tend to provide a more accurate estimate. (b) If we think an estimate is better, we probably mean it typically has less error. Based on (a), our intuition suggests that the larger the sample size the smaller the standard error.

[5]This can be shown based on the probability tools of Section 2.5. For example, see Question 47 in the OpenIntro Online Question Bank.

[6]The choice of 10% is based on the findings of Section 2.4; observations look like they were independently drawn so long as the sample consists of less than 10% of the population.

Computing SE for the sample mean
Given n independent observations from a population with standard deviation σ, the standard error of the sample mean is equal to

$$SE = \frac{\sigma}{\sqrt{n}} \tag{4.5}$$

A reliable method to ensure sample observations are independent is to conduct a simple random sample consisting of less than 10% of the population.

There is one subtle issue of Equation (4.5): the population standard deviation is typically unknown. But you might have already guessed how to resolve this problem: we use the point estimate of the standard deviation from the sample. This estimate tends to be sufficiently good when the sample size is at least 50 and the population distribution is not too strongly skewed[7]. Thus, we often just use the sample standard deviation s instead of σ. When the sample size or skew condition is not met, which is usually checked by examining the sample data, we must use more sophisticated tools to compensate, which are the topic of Chapter 6.

⊙ **Exercise 4.6** In Section 4.1.1, the sample standard deviation of the runners' ages for `run10Samp` was $s_y = 10.93$ based on the sample of size 100. Because the sample is simple random and consists of less than 10% of the population, the observations are independent. (a) Compute the standard error of the sample mean, $\bar{y} = 35.22$ years. (b) Would you be surprised if someone told you the average age of all the runners was actually 35 years? Answers in the footnote[8].

⊙ **Exercise 4.7** (a) Would you be more trusting of a sample that has 100 observations or 400 observations? (b) We want to show mathematically that our estimate tends to be better when the sample size is larger. If the standard deviation of the individual observations is 10, what is our estimate of the standard error when the sample size is 100? What about when it is 400? (c) Explain how your answer to (b) mathematically justifies your intuition in part (a). Answers are in the footnote[9].

4.1.4 Basic properties of point estimates

In this section, we have achieved three goals. First, we have determined that point estimates from a sample may be used to estimate population parameters. We have also determined that these point estimates are not exact: they vary from one sample to another. Lastly, we have quantified the uncertainty of the sample mean using what we call the standard error, mathematically represented in Equation (4.5). While we could also quantify the standard error for other estimates – such as the median, standard deviation, or any other number of statistics – we postpone these extensions until later chapters or courses.

[7]Some books suggest 30 is sufficient; we take a slightly more conservative approach.

[8](a) Because observations are independent, use Equation (4.5) with the *sample* standard deviation to compute the standard error: $SE_{\bar{y}} = 10.93/\sqrt{100} = 1.09$ years. (b) It would not be surprising. Our sample is within about 0.2 standard errors of 35 years. In other words, 35 years old does not seem to be implausible given that our sample was so close to it. (We use our standard error to identify what is close. 0.2 standard errors is very close.)

[9](a) Larger samples tend to have more information, so a point estimate with 400 observations seems more trustworthy. (b) The standard error when the sample size is 100 is given by $SE_{100} = 10/\sqrt{100} = 1$. For 400: $SE_{400} = 10/\sqrt{400} = 0.5$. The larger sample has a smaller standard error. (c) The standard error of the sample with 400 observations is lower than that of the sample with 100 observations. Since the error is lower, this mathematically shows the estimate from the larger sample tends to be better (though it does not guarantee that every large sample will be better than a particular small sample).

4.2 Confidence intervals

A point estimate provides a single plausible value for a parameter. However, a point estimate is rarely perfect; usually there is some error in the estimate. Instead of supplying just a point estimate of a parameter, a next logical step would be to provide a plausible *range of values* for the parameter.

In this section and in Section 4.3, we will emphasize the special case where the point estimate is a sample mean and the parameter is the population mean. In Section 4.5, we will discuss how to generalize these methods to a variety of point estimates and population parameters that we will encounter in Chapter 5.

4.2.1 Capturing the population parameter

A plausible range of values for the population parameter is called a **confidence interval**.

Using only a point estimate is like fishing in a murky lake with a spear, and using a confidence interval is like fishing with a net. We can throw a spear where we saw a fish but we will probably miss. On the other hand if we toss a net in that area, we have a good chance of catching the fish.

If we report a point estimate, we probably will not hit the exact population parameter. On the other hand, if we report a range of plausible values – a confidence interval – we have a good shot at capturing the parameter.

⊙ **Exercise 4.8** If we want to be very certain we capture the population parameter, should we use a wider interval or a smaller interval? Hint in the footnote[10].

4.2.2 An approximate 95% confidence interval

Our point estimate is the most plausible value of the parameter, so it makes sense to cast the confidence interval around the point estimate. The standard error, which is a measure of the uncertainty associated with the point estimate, provides a guide for how large we should make the confidence interval.

The standard error represents the standard deviation associated with the estimate, and roughly 95% of the time the estimate will be within 2 standard errors of the parameter. If the interval spreads out 2 standard errors from the point estimate, we can be roughly 95% **confident** we have captured the true parameter:

$$\text{point estimate} \pm 2 * SE \tag{4.9}$$

But what does "95% confident" mean? Suppose we took many samples and built a confidence interval from each sample using Equation (4.9). Then about 95% of those intervals would contain the actual mean, μ. Figure 4.7 shows this process with 25 samples, where 24 of the resulting confidence intervals contain the average time for all the runners, $\mu = 94.26$ minutes, and one does not.

The rule where about 95% of observations are within 2 standard deviations of the mean is only approximately true. However, it holds very well for the normal distribution. As we will soon see, the mean tends to be normally distributed when the sample size is sufficiently large.

[10] If we want to be more certain we will capture the fish, we might use a wider net.

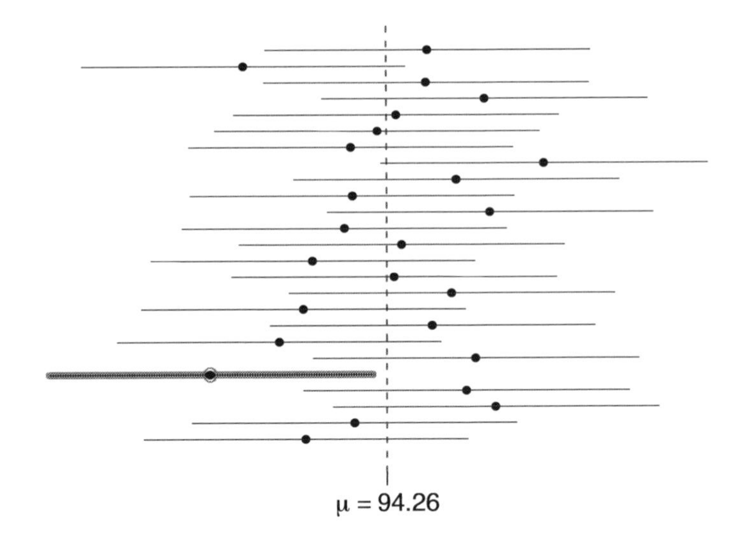

Figure 4.7: Twenty-five samples of size $n = 100$ were taken from the `run10` data set. For each sample, a confidence interval was created to try to capture the average 10 mile time. Only 1 of the 25 intervals does not capture the true mean, $\mu = 94.26$ minutes.

● **Example 4.10** If the sample mean of the 10 mile run times from `run10Samp` is 93.65 minutes and the standard error is 1.57 minutes, what would be an approximate 95% confidence interval for the average 10 mile time for all runners in the race?

We apply Equation (4.9):

$$93.65 \pm 2 * 1.57 \quad \rightarrow \quad (90.51, 96.79)$$

Based on these data, we are about 95% confident that the average 10 mile time for all runners in the race was larger than 90.51 but less than 96.76 minutes. Our interval extends out 2 standard errors from the point estimate, $\bar{x}$.

⊙ **Exercise 4.11** The `run10Samp` sample suggested the average runner age is about 35.22 years with a standard error of 1.09 years. What is a 95% confidence interval for the average age of all of the runners? Answer in the footnote[11].

⊙ **Exercise 4.12** In Figure 4.7, one interval does not contain 94.26 minutes. Does this imply that the mean cannot be 94.26? Answer in the footnote[12].

4.2.3 A sampling distribution for the mean

In Section 4.1.3, we introduced a sampling distribution for $\bar{x}$, the average run time for samples of size 100. Here we examine this distribution more closely by taking samples

[11] Again apply Equation (4.9): $35.22 \pm 2 * 1.09 \rightarrow (33.04, 37.40)$. We interpret this interval as follows: We are about 95% confident the average age of all participants in the 2009 Cherry Blossom Run was between 33.04 and 37.40 years.

[12] Just as some observations occur more than 2 standard deviations from the mean, some point estimates will be more than 2 standard errors from the parameter. A confidence interval only provides a plausible range of values for a parameter. While we might say other values are implausible based on the data, this does not mean they are impossible.

from the `run10` data set, computing the mean of `time` for each sample, and plotting those sample means in a histogram. We did this earlier in Figure 4.6, however, we now take 10,000 samples to get an especially accurate depiction of the sampling distribution. A histogram of the means from the 10,000 samples is shown in the left panel of Figure 4.8.

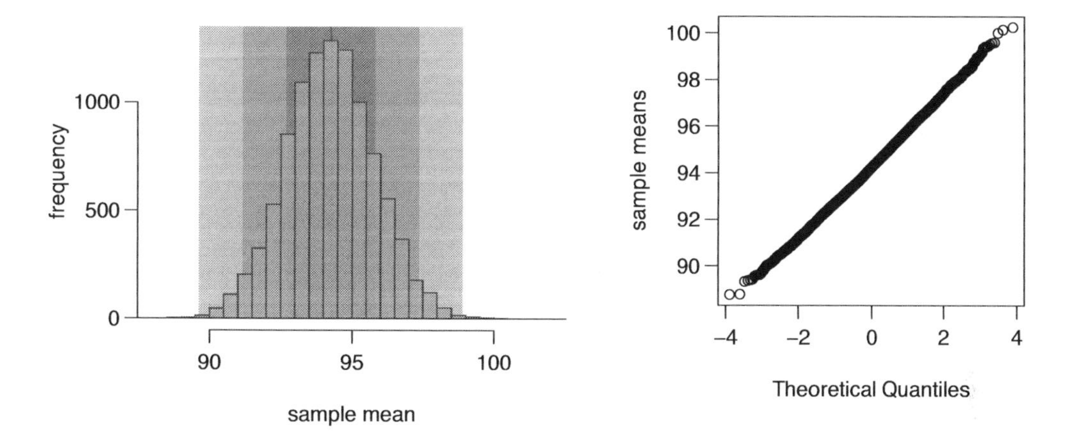

Figure 4.8: The left panel shows a histogram of the sample means for 10,000 different random samples. The right panel shows a normal probability plot of those sample means.

Does this distribution look familiar? Hopefully so! The distribution of sample means closely resembles the normal distribution (see Section 3.1). A normal probability plot of these sample means is shown in the right panel of Figure 4.8. Because all of the points closely fall around a straight line, we can conclude the distribution of sample means is nearly normal (see Section 3.2). This result can be explained by the Central Limit Theorem.

Central Limit Theorem, informal description
If a sample consists of at least 50 independent observations and the data are not extremely skewed, then the distribution of the sample mean is well approximated by a normal model.

We revisit the Central Limit Theorem in more detail in Section 4.4. However, we will apply this informal rule for the time being.

The choice of using 2 standard errors in Equation (4.9) was based on our general guideline that roughly 95% of the time, observations are within two standard deviations of the mean. Under the normal model, we can actually make this more accurate by using 1.96 in place of 2.

$$\text{point estimate} \pm 1.96 * SE \tag{4.13}$$

If a point estimate, such as $\bar{x}$, is associated with a normal model and standard error SE, then we use this more precise 95% confidence interval.

Margin of error
In a 95% confidence interval, $1.96 * SE$ is called the margin of error

4.2.4 Changing the confidence level

Suppose we want to consider confidence intervals where the confidence level is somewhat higher than 95%: perhaps we would like a confidence level of 99%. Think back to the analogy about trying to catch a fish: if we want to be even more sure we will catch the fish, we should use a wider net. To create a 99% confidence level, we must also widen our 95% interval. On the other hand, if we want an interval with lower confidence, such as 90%, we could make our original 95% interval slightly slimmer.

The 95% confidence interval structure provides guidance in how to make intervals with new confidence levels. Below is a general 95% confidence interval:

$$\text{point estimate} \pm 1.96 * SE \tag{4.14}$$

There are three components to this interval: the point estimate, "1.96", and the standard error. The choice of $1.96 * SE$ was based on capturing 95% of the data since the estimate is within *1.96* standard deviations of the parameter about 95% of the time. The choice of 1.96 corresponds to a 95% confidence level.

⊙ **Exercise 4.15** If X is a normally distributed random variable, how often will X be within 2.58 standard deviations of the mean? Answer in the footnote[13].

To create a 99% confidence interval, adjust 1.96 in the 95% confidence interval formula to be 2.58. Exercise 4.15 shows that 99% of the time a normal random variable will be within 2.58 standard deviations of the mean. This approach – using the Z scores in the normal model to compute confidence levels – is appropriate when $\bar{x}$ is associated with a normal distribution with mean μ and standard deviation $SE_{\bar{x}}$. Thus, the formula for a 99% confidence interval for the mean when the Central Limit Theorem applies is

$$\bar{x} \pm 2.58 * SE_{\bar{x}} \tag{4.16}$$

The normal approximation is crucial to the precision of these confidence intervals. Section 4.4 provides a more detailed discussion about when the normal model can safely be applied; there are a number of important considerations when evaluating whether to apply the normal model.

Conditions for $\bar{x}$ being nearly normal and SE being accurate
Important conditions to help ensure the sampling distribution of $\bar{x}$ is nearly normal and the estimate of SE sufficiently accurate:

- The sample observations are independent.
- The sample size is large: $n \geq 50$ is a good rule of thumb.
- The distribution of sample observations is not extremely skewed.

Additionally, the larger the sample size, the more lenient we can be with the sample's skew.

[13] This is equivalent to asking how often the Z score will be larger than -2.58 but less than 2.58. (Draw the picture!) To determine this probability, look up -2.58 and 2.58 in the normal probability table (0.0049 and 0.9951). Thus, there is a $0.9951 - 0.0049 \approx 0.99$ probability that the unobserved random variable X will be within 2.58 standard deviations of μ.

Verifying independence is often the most difficult of the conditions to check, and the way to check for independence varies from one situation to another. However, we can provide simple rules for the most common scenarios.

TIP: How to verify sample observations are independent
Observations in a simple random sample consisting of less than 10% of the population are independent.

Caution: Independence for random processes and experiments
If a sample is from a random process or experiment, it is important to verify the observations from the process or subjects in the experiment are nearly independent and maintain their independence throughout the process or experiment.

⊙ **Exercise 4.17** Create a 99% confidence interval for the average age of all runners in the 2009 Cherry Blossom Run. The point estimate is $\bar{y} = 35.22$ and the standard error is $SE_{\bar{y}} = 1.09$. The interval is in the footnote[14].

Confidence interval for any confidence level
If the point estimate follows the normal model with standard error SE, then a confidence interval for the population parameter is

$$\text{point estimate} \pm z^\star SE$$

where $z^\star$ corresponds to the confidence level selected.

Figure 4.9 provides a picture of how to identify $z^\star$ based on a confidence level. We select $z^\star$ so that the area between -$z^\star$ and $z^\star$ in the normal model corresponds to the confidence level.

⊙ **Exercise 4.18** Use the data in Exercise 4.17 to create a 90% confidence interval for the average age of all runners in the 2009 Cherry Blossom Run. Answer in the footnote[15].

4.2.5 Interpreting confidence intervals

A careful eye might have observed the somewhat awkward language used to describe confidence intervals. Correct interpretation:

We are XX% confident that the population parameter is between...

Incorrect language might try to describe the confidence interval as capturing the population parameter with a certain probability. This is one of the most common errors: while it might

[14] We verified the conditions earlier (though do it again for practice), so the normal approximation and estimate of SE should be reasonable. We can apply the 99% confidence interval formula: $\bar{y} \pm 2.58 * SE_{\bar{y}} \rightarrow (32.4, 38.0)$. We are 99% confident that the average age of all runners is between 32.4 and 38.0 years.

[15] We need to find $z^\star$ such that 90% of the distribution falls between -$z^\star$ and $z^\star$ in the standard normal normal model, $N(\mu = 0, \sigma = 1)$. We can look up -$z^\star$ in the normal probability table by looking for a lower tail of 5% (the other 5% is in the upper tail), thus $z^\star = 1.65$. The 90% confidence interval can then be computed as (33.4, 37.0). (We had already verified conditions for normality and the standard error.) That is, we are 90% confident the average age is larger than 33.4 years but less than 37.0 years.

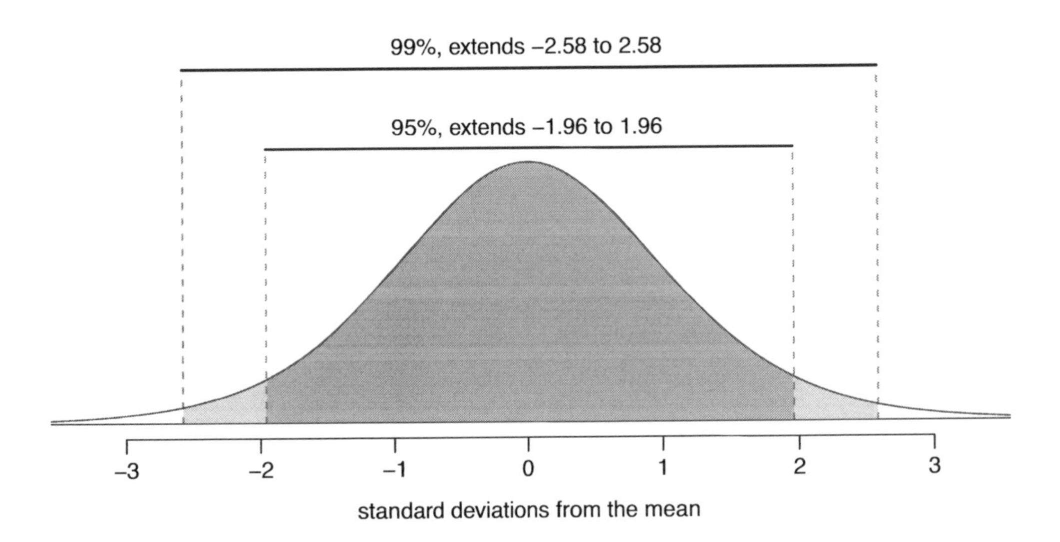

Figure 4.9: The area between -$z^\star$ and $z^\star$ increases as $|z^\star|$ becomes larger. If the confidence level is 99%, we choose $z^\star$ such that 99% of the normal curve is between -$z^\star$ and $z^\star$, which corresponds to 0.5% in the lower tail and 0.5% in the upper tail; for a 99% confidence level, $z^\star = 2.58$.

be useful to *think* of it as a probability, the confidence level only quantifies how plausible it is that the parameter is in the interval.

Another especially important consideration of confidence intervals is that they *only try to capture the population parameter.* Our intervals say nothing about the confidence of capturing individual observations, a proportion of the observations, or about capturing point estimates. Confidence intervals only attempt to capture population parameters.

4.2.6 Nearly normal with known SD (special topic)

In rare circumstances we know important characteristics of a population. For instance, we might know a population is nearly normal and we may also know its parameter values. Even so, we may still like to study characteristics of a random sample from the population. Consider the conditions required for modeling a sample mean using the normal distribution:

(1) The observations are independent.

(2) The sample size n is at least 50.

(3) The data distribution is not extremely skewed.

Conditions (2) and (3) are required for two reasons: so we can adequately estimate the standard deviation and to ensure the sample mean is nearly normal. However, if the population is known to be nearly normal, the sample mean is always nearly normal (this is a special case of the Central Limit Theorem). If the standard deviation is also known, then conditions (2) and (3) are not necessary for those data.

● **Example 4.19** The heights of male seniors in high school closely follow a normal distribution $N(\mu = 70.43, \sigma = 2.73)$, where the units are inches[16]. If we randomly sampled the heights of five male seniors, what distribution should the sample mean follow?

The population is nearly normal, the standard deviation is known, and the heights represent a random sample from a much larger population, satisfying the independence condition. Therefore the sample mean of the heights will follow a nearly normal distribution with mean $\mu = 70.43$ inches and standard error $SE = \sigma/\sqrt{n} = 2.73/\sqrt{5} = 1.22$ inches.

Alternative conditions for using the normal distribution to model the sample mean
If the population of cases is known to be nearly normal and the population standard deviation σ is known, then the sample mean $\bar{x}$ will follow a nearly normal distribution $N(\mu, \sigma/\sqrt{n})$ if the sampled observations are independent.

Sometimes the mean changes over time but the standard deviation remains the same. In such cases, a sample mean of small but nearly normal observations paired with a known standard deviation can be used to produce a confidence interval for the current population mean using the normal distribution.

Even when the population mean and standard deviation are both known, it can be useful to model the sample mean with the normal distribution. Below we consider one such case.

● **Example 4.20** Is there a connection between height and popularity in high school? Many students may suspect as much, but what do the data say? Suppose the top 5 nominees for prom king at a high school have an average height of 71.8 inches. Does this provide strong evidence that these seniors' heights are not representative of all male seniors at their high school?

If these five seniors are height-representative, then their heights should be like a random sample from the distribution given in Example 4.19, $N(\mu = 70.43, \sigma = 2.73)$, and the sample mean will follow $N(\mu = 70.43, \sigma/\sqrt{n} = 1.22)$. Formally we are conducting what is called a *hypothesis test*, which we will discuss in greater detail during the next section. We are weighing two possibilities:

H_0: The prom king nominee heights are representative; $\bar{x}$ will follow a normal distribution with mean 70.43 inches and standard error 1.22 inches.

H_A: The heights are not representative; we suspect the mean height is different from 70.43 inches.

If there is strong evidence that the sample mean is not from the normal distribution provided in the H_0 description, then that suggests the prom king nominees' heights are not a simple random sample (i.e. H_A is true). We can look at how the Z score of the sample mean to tell us how unusual our sample is, if the H_0 description is true:

$$Z = \frac{\bar{x} - \mu}{\sigma/\sqrt{n}} = \frac{71.8 - 70.43}{1.22} = 1.12$$

[16]These values were computed using the USDA Food Commodity Intake Database.

A Z score of just 1.12 is not very unusual (usually we use a cutoff of 2 to decide what is unusual), so there is not strong evidence against the claim that the heights are representative. This does not mean the heights are actually representative, only that this very small sample does not necessarily show otherwise.

TIP: Relaxing the nearly normal condition
As the sample size becomes larger, it is reasonable to *slowly* relax the nearly normal assumption on the data when dealing with small samples.

4.3 Hypothesis testing

Is the typical US runner getting faster or slower over time? We consider this question in the context of the Cherry Blossom Run, comparing runners in 2006 and 2009. Technological advances in shoes, training, and diet might suggest runners would be faster in 2009. An opposing viewpoint might say that with the average body mass index on the rise, people tend to run slower. In fact, all of these components might be influencing run time.

In addition to considering run times in this section, we consider a topic near and dear to most college students: sleep. A recent study found that college students average about 7 hours of sleep per night[17]. However, researchers at a rural college are interested in showing their students sleep longer than seven hours on average. We investigate this matter in Section 4.3.4.

4.3.1 Hypothesis testing framework

In 2006, the average time for all runners who finished the Cherry Blossom Race was 93.29 minutes (93 minutes and about 17 seconds). We want to determine if the `run10Samp` data set provides strong evidence that the participants in 2009 were faster or slower than those runners in 2006, versus the other possibility that there has been no change[18]. We simplify these three options into two competing **hypotheses**:

H_0: The average 10 mile run time was the same for 2006 and 2009.

H_A: The average 10 mile run time for 2009 was *different* than that of 2006.

We call H_0 the null hypothesis and H_A the alternative hypothesis.

H_0
null hypothesis

H_A
alternative hypothesis

Null and alternative hypotheses
The **null hypothesis (H_0)** often represents a skeptical perspective or a claim to be tested. The **alternative hypothesis (H_A)** represents an alternative claim under consideration and is often represented by a range of possible parameter values.

The null hypothesis often represents a skeptical position or a perspective of no difference. (Why should we think run times have actually changed from 2006 to 2009?) The alternative hypothesis often represents a new perspective, such as the possibility that there has been a change.

[17] http://media.www.theloquitur.com/media/storage/paper226/news/2000/10/19/News/Poll-Shows.College.Students.Get.Least.Amount.Of.Sleep-5895.shtml

[18] While we could answer this question by examining the `run10` population data, we only consider the sample data in `run10Samp`.

TIP: Hypothesis testing framework
The hypothesis testing framework is built for a skeptic to consider a new claim. The skeptic will not reject the null hypothesis (H_0), unless the evidence in favor of the alternative hypothesis (H_A) is so strong that she rejects H_0 in favor of H_A.

The hypothesis testing framework is a very general tool, and we often use it without a second thought. If a person makes a somewhat unbelievable claim, we are initially skeptical. However, if there is sufficient evidence that supports the claim, we set aside our skepticism and reject the null hypothesis in favor of the alternative. The hallmarks of hypothesis testing are also found in the US court system.

⊙ **Exercise 4.21** A US court considers two possible claims about a defendant: she is either innocent or guilty. If we set these claims up in a hypothesis framework, which would be the null hypothesis and which the alternative? Answer in the footnote[19].

Jurors examine the evidence to see whether it convincingly shows a defendant is guilty. Even if the jurors leave unconvinced of guilt beyond a reasonable doubt, this does not mean they believe the defendant is innocent. This is also the case with hypothesis testing: *even if we fail to reject the null hypothesis, we typically do not accept the null hypothesis as true.* For this reason, failing to find strong evidence for the alternative hypothesis is not equivalent to accepting the null hypothesis.

In the example with the Cherry Blossom Run, the null hypothesis represents no difference in the average time from 2006 to 2009. The alternative hypothesis represents something new or more interesting: there was a difference, either an increase or a decrease. These hypotheses can be described in mathematical notation using μ_{09} as the average run time for 2009:

$$H_0\text{: } \mu_{09} = 93.29$$

$$H_A\text{: } \mu_{09} \neq 93.29$$

where 93.29 minutes (93 minutes and about 17 seconds) is the average 10 mile time for all runners in 2006 Cherry Blossom Run. In this mathematical notation, the hypotheses can now be evaluated using statistical tools. We call 93.29 the **null value** since it represents the value of the parameter if the null hypothesis is true.

Suppose the `run10Samp` data set is used to evaluate the hypothesis test. If these data provide strong evidence that the null hypothesis is false and the alternative is true, then we reject H_0 in favor of H_A. If we are left unconvinced, then we do not reject the null hypothesis.

4.3.2 Testing hypotheses using confidence intervals

We can start the evaluation of the hypothesis setup by comparing 2006 and 2009 run times using a point estimate: $\bar{x}_{09} = 93.65$ minutes. This estimate suggests the average time is actually longer than the 2006 time, 93.29 minutes. However, to evaluate whether this provides strong evidence that there has been a change, we must consider the uncertainty associated with $\bar{x}_{09}$.

---

[19] The jury considers whether the evidence is so convincing (strong) that innocence is too implausible; in such a case, the jury rejects innocence (the null hypothesis) and concludes the defendant is guilty (alternative hypothesis).

We learned in Section 4.1 that there is fluctuation from one sample to another, and it is very unlikely that the sample mean will be exactly equal to our parameter; we should not expect $\bar{x}_{09}$ to exactly equal μ_{09}. With only knowledge of the `run10Samp`, it might still be possible that the population average in 2009 has remained unchanged from 2006. The difference between $\bar{x}_{09}$ and 93.29 could be due to *sampling variation*, i.e. the variability associated with the point estimate when we take a random sample.

In Section 4.2, confidence intervals were introduced as a way to find a range of plausible values for the population mean. Based on `run10Samp`, a 95% confidence interval for the 2009 population mean, μ_{09}, was found:

$$(90.51, 96.79)$$

Because the 2006 mean (93.29) falls in the range of plausible values, we cannot say the null hypothesis is implausible. That is, we fail to reject the null hypothesis, H_0.

TIP: Double negatives can sometimes be used in statistics
In many statistical explanations, we use double negatives. For instance, we might say that the null hypothesis is *not implausible* or we *failed to reject* the null hypothesis. Double negatives are used to communicate that while we are not rejecting a position, we are also not saying it is correct.

Example 4.22 Next consider whether there is strong evidence that the average age of runners has changed from 2006 to 2009 in the Cherry Blossom Run. In 2006, the average age was 36.13 years, and in the 2009 `run10Samp` data set, the average was 35.22 years with a standard deviation of 10.93 years over 100 runners.

First, set up the hypotheses:

H_0: The average age of runners has not change from 2006 to 2009, $\mu_{age} = 36.13$.
H_A: The average age of runners has changed from 2006 to 2009, $\mu_{age} \neq 36.13$.

Since we have previously verified assumptions on this data set, the normal model may be applied to $\bar{y}$ and the estimate of SE should be very accurate. Using the sample mean and standard error, we can construct a 95% confidence interval for μ_{age} to determine if there is sufficient evidence to reject H_0:

$$\bar{y} \pm 1.96 * \frac{s_{09}}{\sqrt{100}} \quad \rightarrow \quad 35.22 \pm 1.96 * 1.09 \quad \rightarrow \quad (33.08, 37.36)$$

This confidence interval contains the *null parameter*, 36.13. Because 36.13 is not implausible, we cannot reject the null hypothesis. The sample mean of 35.22 years could easily have occurred even if the actual mean was 36.13, so we have not found strong evidence that the average age is different than 36.13 years.

Exercise 4.23 Colleges frequently give estimates of student expenses such as housing. A consultant hired by the school claimed that the average student housing expense was \$650 per month. Set up hypotheses to test whether this claim is accurate. Answer in the footnote[20].

[20] H_0: The average cost is \$650 per month, $\mu = \$650$.
H_A: The average cost is different than \$650 per month, $\mu \neq \$650$.

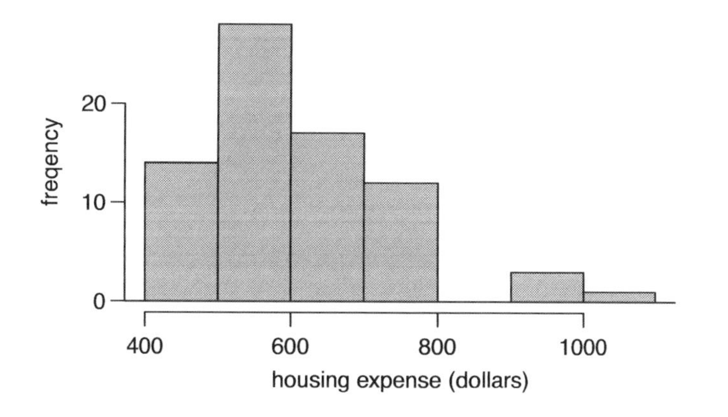

Figure 4.10: Sample distribution of student housing expense.

⊙ **Exercise 4.24** The community college decides to collect data to evaluate the \$650 per month claim. They take a random sample of 75 students at their school and obtain the data represented in Figure 4.10. Can we apply the normal model to the sample mean? Answer in the footnote[21].

● **Example 4.25** The sample mean for student housing is \$611.63 and the sample standard deviation is \$132.85. Construct a 95% confidence interval for the population mean and evaluate the hypotheses of Exercise 4.23.

The standard error associated with the mean may be estimated using the sample standard deviation divided by the square root of the sample size. Recall that $n = 75$ students were sampled.

$$SE = \frac{s}{\sqrt{n}} = \frac{132.85}{\sqrt{75}} = 15.34$$

You showed in Exercise 4.24 that the normal model may be applied to the sample mean. This ensures a 95% confidence interval may be accurately constructed:

$$\bar{x} \pm z^{\star} SE \quad \rightarrow \quad 611.63 \pm 1.96 * (15.34) \quad \rightarrow \quad (581.56, 641.70)$$

Because the null value \$650 is not in the confidence interval, a true mean of \$650 is implausible and we reject the null hypothesis. The data provide statistically significant evidence that the actual housing expense average is less than \$650 per month.

4.3.3 Decision errors

Are hypothesis tests flawless? Absolutely not. Just think of the court system: innocent people are sometimes wrongly convicted and the guilty sometimes walk free. Similarly, we can make a wrong decision in statistical hypothesis tests as well. However, the difference is that we have the tools necessary to quantify how often we make errors in statistics.

---

[21] Applying the normal model requires that certain conditions are met. Because the data are a simple random sample and the sample (presumably) represents no more than 10% of all students at the college, the observations are independent. The sample size is also sufficiently large ($n = 75$) and the data are skewed but not strongly skewed. Thus, the normal model may be applied to the sample mean.

There are two competing hypotheses: the null and the alternative. In a hypothesis test, we make some commitment about which one might be true, but we might choose incorrectly. There are four possible scenarios in a hypothesis test, which are summarized in Table 4.11.

		Test conclusion	
		do not reject H_0	reject H_0 in favor of H_A
Truth	H_0 true	okay	Type 1 Error
	H_A true	Type 2 Error	okay

Table 4.11: Four different scenarios for hypothesis tests.

A **Type 1 Error** is rejecting the null hypothesis when H_0 is actually true. A **Type 2 Error** is failing to reject the null hypothesis when the alternative is actually true.

⊙ **Exercise 4.26** In a US court, the defendant is either innocent (H_0) or guilty (H_A). What does a Type 1 Error represent in this context? What does a Type 2 Error represent? Table 4.11 may be useful. Answers in the footnote[22].

⊙ **Exercise 4.27** How could we reduce the Type 1 Error rate in US courts? What influence would this have on the Type 2 Error rate? Answer in the footnote[23].

⊙ **Exercise 4.28** How could we reduce the Type 2 Error rate in US courts? What influence would this have on the Type 1 Error rate?

Exercises 4.26-4.28 provide an important lesson: if we reduce how often we make one type of error, we generally make more of the other type.

Hypothesis testing is built around rejecting or failing to reject the null hypothesis. That is, we do not reject H_0 unless we have strong evidence. But what precisely does *strong evidence* mean? As a general rule of thumb, for those cases where the null hypothesis is actually true, we do not want to incorrectly reject H_0 more than 5% of those times. This corresponds to a **significance level** of 0.05. We often write the significance level using α (the Greek letter *alpha*): $\alpha = 0.05$. We discuss the appropriateness of different significance levels in Section 4.3.6.

α
significance level of a hypothesis test

If we use a 95% confidence interval to test a hypothesis, then we will make an error whenever, due to sampling variation, the point estimate is at least 1.96 standard errors away from the population parameter. This happens about 5% of the time (2.5% in each tail). Similarly, using a 99% confidence interval to evaluate a hypothesis is equivalent to a significance level of $\alpha = 0.01$.

However, a confidence interval is, in one sense, simplistic in the world of hypothesis tests. Consider the following two scenarios:

- The null value (the parameter value under the null hypothesis) is in the 95% confidence interval but just barely, so we would not reject H_0. However, we might like to somehow say, quantitatively, that it was a close decision.

[22] If the court makes a Type 1 Error, this means the defendant is innocent (H_0 true) but wrongly convicted, i.e. the court wrongly rejects H_0. A Type 2 Error means the court failed to reject H_0 (i.e. fail to convict the person) when she was in fact guilty (H_A true).

[23] To lower the Type 1 Error rate, we might raise our standard for conviction from "beyond a reasonable doubt" to "beyond a conceivable doubt" so fewer people would be wrongly convicted. However, this would also make it more difficult to convict the people who are actually guilty, i.e. we will make more Type 2 Errors.

- The null value is very far outside of the interval, so we reject H_0. However, we want to communicate that, not only did we reject the null hypothesis, but it wasn't even close. Such a case is depicted in Figure 4.12.

In Section 4.3.4, we introduce a tool called the *p-value* that will be helpful in these cases. The p-value method also extends to hypothesis tests where confidence intervals cannot be easily constructed or applied.

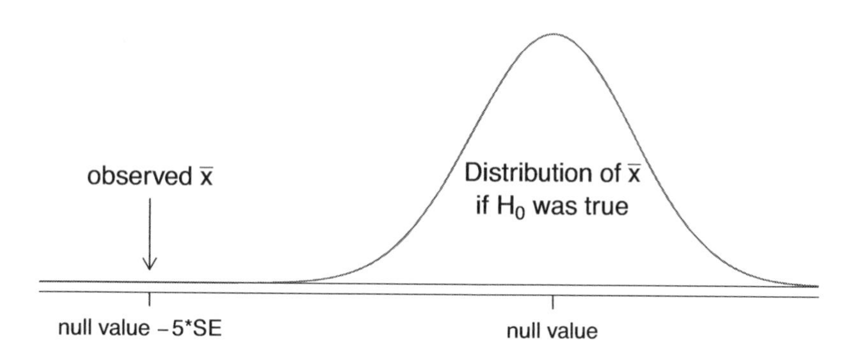

Figure 4.12: It would be helpful to quantify the strength of the evidence against the null hypothesis. In this case, the evidence is extremely strong.

4.3.4 Formal testing using p-values

The p-value is a way to quantify the strength of the evidence against the null hypothesis and in favor of the alternative. Formally the *p-value* is a conditional probability.

> **p-value**
> The **p-value** is the probability of observing data at least as favorable to the alternative hypothesis as our current data set, if the null hypothesis was true. We typically use a summary statistic of the data, in this chapter the sample mean, to help compute the p-value and evaluate the hypotheses.

A poll by the National Sleep Foundation found that college students average about 7 hours of sleep per night. Researchers at a rural school are interested in showing that students at their school sleep longer than seven hours on average, and they would like to demonstrate this using a sample of students. There are three possible cases: the students at the rural school average more than 7 hours of sleep, they average less than 7 hours, or they average exactly 7 hours.

⊙ **Exercise 4.29** What would be a skeptical position in regards to the researchers' claims? Answer in the footnote[24].

We can set up the null hypothesis for this test as a skeptical perspective of the researchers' claim: the students at this school average 7 or fewer hours of sleep per night. The alternative hypothesis takes a new form reflecting the purpose of the study: the students average more than 7 hours of sleep. We can write these hypotheses in statistical language:

[24] A skeptic would have no reason to believe that the researchers are correct: maybe the students at this school average 7 or fewer hours of sleep.

H_0: $\mu \leq 7$.

H_A: $\mu > 7$.

While we might think about the null hypothesis as representing a range of values in this case, it is common to simply write it as an equality:

H_0: $\mu = 7$.

H_A: $\mu > 7$.

Using $\mu > 7$ as the alternative is an example of a **one-sided** hypothesis test. In this investigation we are not concerned with seeing whether the mean is less than 7 hours[25]. Earlier we encountered a **two-sided** hypothesis where we looked for any clear difference, greater than or less than the null value.

Always use a two-sided test unless it was made clear prior to data collection that the test should be one-sided. Switching a two-sided test to a one-sided test after observing the data is dangerous because it can inflate the Type 1 Error rate.

TIP: One-sided and two-sided tests
If the researchers are only interested in showing *either* an increase *or* a decrease, then set up a one-sided test. If the researchers would be interested in any difference from the null value – an increase or decrease – then the test should be two-sided.

TIP: Always write the null hypothesis as an equality
We will find it most useful if we always list the null hypothesis as an equality (e.g. $\mu = 7$) while the alternative always uses an inequality (e.g. $\mu \neq 7$, $\mu > 7$, or $\mu < 7$).

The researchers at the rural school conducted a simple random sample of $n = 110$ students on campus. They found that these students averaged 7.42 hours of sleep and the standard deviation of the amount of sleep for the students was 1.75 hours. A histogram of the sample is shown in Figure 4.13.

Before we can use a normal model for the sample mean or compute the standard error of the sample mean, we must verify conditions. (1) Because the sample is simple random from less than 10% of the student body, the observations are independent. (2) The sample size in the sleep study is sufficiently large (greater than 50). (3) The data are not too strongly skewed in Figure 4.13. With these conditions verified, the normal model can be safely applied to $\bar{x}$ and the standard error estimate will be very accurate.

⊙ **Exercise 4.30** What is the standard deviation associated with $\bar{x}$? That is, estimate the standard error of $\bar{x}$. Answer in the footnote[26].

The hypothesis test will be evaluated using a significance level of $\alpha = 0.05$. We want to consider the data under the scenario that the null hypothesis is true. In this case, the sample mean is from a distribution with mean 7 and standard deviation of about 0.17 that is nearly normal. Such a distribution is shown in Figure 4.14.

[25] This is entirely based on the interests of the researchers. Had they been only interested in the opposite case – showing that their students were actually averaging fewer than seven hours of sleep but not interested in showing more than 7 hours – then our setup would have set the alternative as $\mu < 7$.

[26] The standard error can be estimated from the sample standard deviation and the sample size: $SE_{\bar{x}} = \frac{s_x}{\sqrt{n}} = \frac{1.75}{\sqrt{110}} = 0.17$.

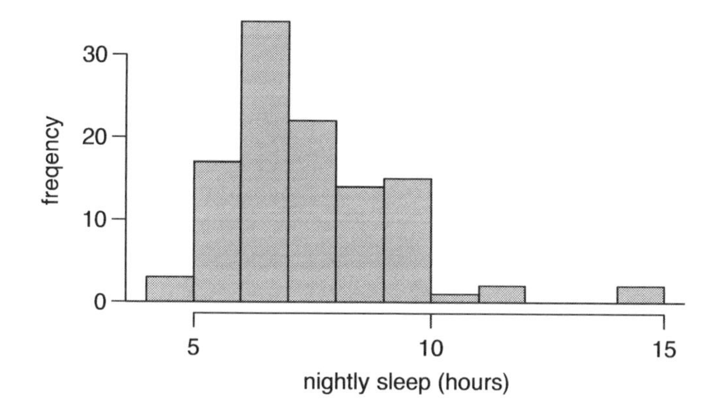

Figure 4.13: Distribution of a night of sleep for 110 college students.

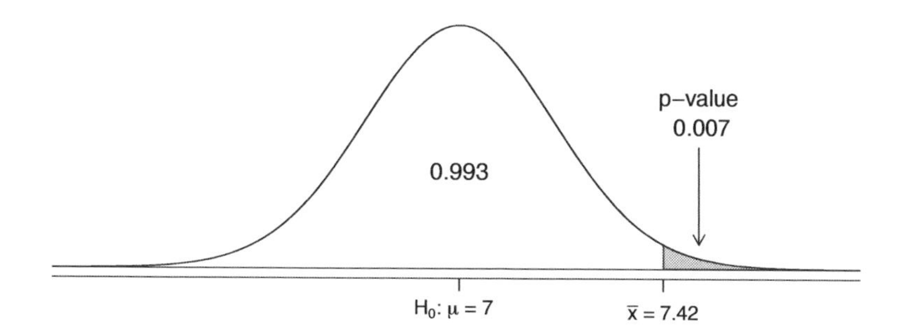

Figure 4.14: If the null hypothesis was true, then the sample mean $\bar{x}$ came from this nearly normal distribution. The right tail describes the probability of observing such a large sample mean if the null hypothesis was true.

The shaded tail in Figure 4.14 represents the chance of observing such an unusually large mean, conditional on the null hypothesis being true. That is, the shaded tail represents the p-value. We shade all means larger than our sample mean, $\bar{x} = 7.42$, because they are more favorable to the alternative hypothesis than the observed mean.

We compute the p-value by finding this tail area of this normal distribution, which we learned to do in Section 3.1. First compute the Z score of the sample mean, $\bar{x} = 7.42$:

$$Z = \frac{\bar{x} - \text{null value}}{SE_{\bar{x}}} = \frac{7.42 - 7}{0.17} = 2.47$$

Using the normal probability table, the lower unshaded area is found to be 0.993. Thus the shaded area is $1 - 0.993 = 0.007$. *If the null hypothesis was true, the probability of observing such a large sample mean is only 0.007.* That is, if the null hypothesis was true, we would not often see such a large mean.

We evaluate the hypotheses by comparing the p-value to the significance level. Because the p-value is less than the significance level (p-value $= 0.007 < 0.05 = \alpha$), we reject the null hypothesis. What we observed is so unusual with respect to the null hypothesis that it casts serious doubt on H_0 and provides strong evidence favoring H_A.

p-value as a tool in hypothesis testing
The p-value quantifies how strongly the data favor H_A over H_0. A small p-value (usually < 0.05) corresponds to sufficient evidence to reject H_0.

TIP: It is useful to first draw a picture to find the p-value
It is useful to draw a picture of the distribution of $\bar{x}$ as though H_0 were true (e.g. μ equals the null value), and shade the region (or regions) of sample means that are even more favorable to the alternative hypothesis. These shaded regions represent the p-value.

The ideas below review the process of evaluating hypothesis tests with p-values:

- The null hypothesis represents a skeptic's position or a position of no difference. We reject this position only if the evidence strongly favors H_A.
- A small p-value means that if the null hypothesis was true, there is a low probability of seeing a point estimate at least as extreme as the one we saw. We interpret this as strong evidence in favor of the alternative.
- We reject the null hypothesis if the p-value is smaller than the significance level, α, which is usually 0.05. Otherwise, we fail to reject H_0.
- We should always state the conclusion of the hypothesis test in plain language so non-statisticians can also understand the results.

The p-value is constructed in such a way that we can directly compare it to the significance level, α, to determine whether or not to reject H_0. This method ensures that the Type 1 Error rate does not exceed the significance level standard.

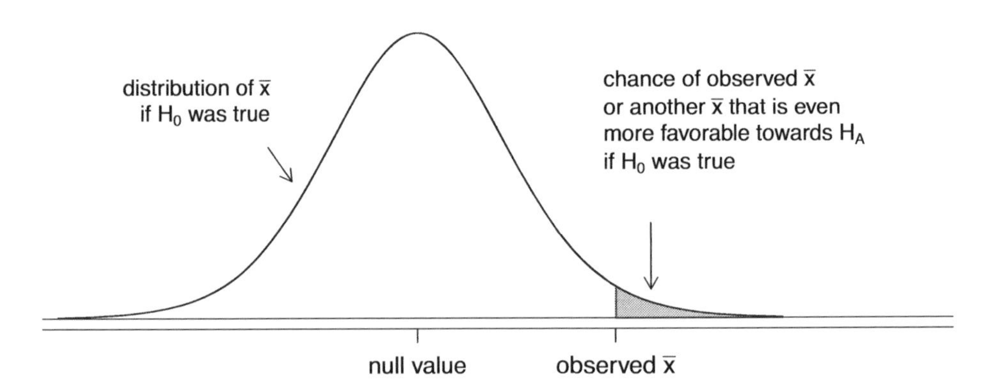

Figure 4.15: To identify the p-value, the distribution of the sample mean is considered if the null hypothesis was true. Then the p-value is defined and computed as the probability of the observed $\bar{x}$ or an $\bar{x}$ even more favorable to H_A under this distribution.

⊙ **Exercise 4.31** If the null hypothesis is true, how often should the test p-value be less than 0.05? Answer in the footnote[27].

[27] About 5% of the time. If the null hypothesis is true, then the data only has a 5% chance of being in the 5% of data most favorable to H_A.

⊙ **Exercise 4.32** Suppose we had used a significance level of 0.01 in the sleep study. Would the evidence have been strong enough to reject the null hypothesis? (The p-value was 0.007.) What if the significance level was $\alpha = 0.001$? Answers in the footnote[28].

⊙ **Exercise 4.33** We would like to evaluate whether Ebay buyers on average pay less than the price on Amazon, and we will evaluate one particular case. During early October 2009, Amazon sold a game called *Mario Kart* for the Nintendo Wii for $46.99. Set up an appropriate (one-sided!) hypothesis test to check the claim that Ebay buyers pay less. Answer in the footnote[29].

⊙ **Exercise 4.34** During early October, 2009, 52 Ebay auctions were recorded for *Wii Mario Kart*. We would like to apply the normal model to the sample mean, however, we must verify three conditions to do so: (1) independence, (2) at least 50 observations, and (3) the data are not extremely skewed. Assume the last condition is satisfied. Do you think the first two conditions would be reasonably met? Answer in the footnote[30].

● **Example 4.35** The average sale price of the 52 Ebay auctions for *Wii Mario Kart* was $44.17 with a standard deviation of $4.15. Does this provide sufficient evidence to reject the null hypothesis in Exercise 4.33? Use a significance level of $\alpha = 0.01$ and assume the conditions for applying the normal model are satisfied.

The hypotheses were set up and the conditions were checked in Exercises 4.33 and 4.34. The next step is to find the standard error of the sample mean and produce a sketch to help find the p-value.

$$SE_{\bar{x}} = s/\sqrt{n} = 4.15/\sqrt{52} = 0.5755$$

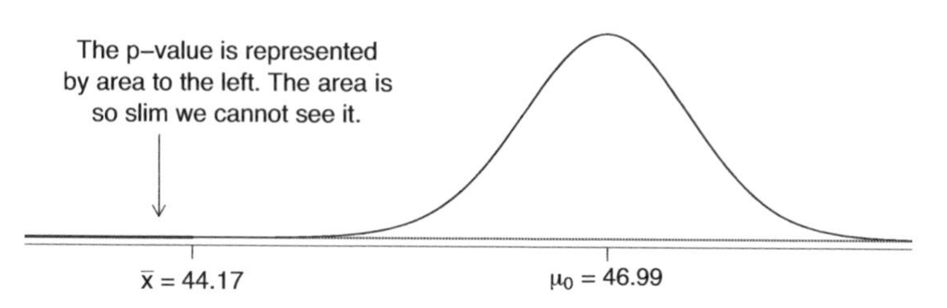

Because the alternative hypothesis says we are looking for a smaller mean, we shaded the lower tail. We find this shaded area by using the Z score and normal probability table: $Z = \frac{44.17-46.99}{0.5755} = -4.90$, which has area less than 0.0002. The area is so small we cannot really see it on the picture. This lower tail area corresponds to the p-value.

[28]We reject the null hypothesis whenever $p\text{-}value < \alpha$. Thus, we would still reject the null hypothesis if $\alpha = 0.01$ but not if the significance level had been $\alpha = 0.001$.

[29]The skeptic would say the average is the same (or more) on Ebay, and we are interested in showing the average price is lower.

H_0: The average auction price on Ebay is equal to (or more than) the price on Amazon. We write only the equality in the statistical notation: $\mu_{ebay} = 46.99$.

H_A: The average price on Ebay is less than the price on Amazon, $\mu_{ebay} < 46.99$.

[30](1) The independence condition is unclear. *We will make the assumption that the observations are independence holds.* (2) The sample size is sufficiently large: $n = 52 \geq 50$. (3) We were told that there is not evidence of extreme skew.

Because the p-value is so small – specifically, smaller than $\alpha = 0.01$ – this provides sufficiently strong evidence to reject the null hypothesis in favor of the alternative. The data provide statistically significant evidence that the average price on Ebay is lower than Amazon's asking price.

4.3.5 Two-sided hypothesis testing with p-values

We now consider how to compute a p-value for a two-sided test. In one-sided tests, we shade the single tail in the direction of the alternative hypothesis. For example, when the alternative had the form $\mu > 7$, then the p-value was represented by the upper tail (Figure 4.15). When the alternative was $\mu < 46.99$, the p-value was the lower tail (Exercise 4.33). In a two-sided test, *we shade two tails* since evidence in either direction is favorable to H_A.

⊙ **Exercise 4.36** Earlier we talked about a research group that examined whether the students at their school slept longer than 7 hours each night. Let's consider a second group of researchers who want to evaluate whether the students at their college differ from the norm of 7 hours. Write out the null and alternative hypotheses for this investigation. Answer in the footnote[31].

● **Example 4.37** The second college randomly samples 72 students and finds a mean of $\bar{x} = 6.83$ hours and a standard deviation of $s = 1.8$ hours. Does this provide strong evidence against H_0 in Exercise 4.36? Use a significance level of $\alpha = 0.05$.

First, we must verify assumptions. (1) A simple random sample of less than 10% of the student body means the observations are independent. (2) The sample size is greater than 50. (3) Based on the earlier distribution and what we already know about college student sleep habits, the distribution is probably not too strongly skewed.

Next we can compute the standard error ($SE_{\bar{x}} = \frac{s}{\sqrt{n}} = 0.21$) of the estimate and create a picture to represent the p-value, shown in Figure 4.16. Both tails are shaded. An estimate of 7.17 or more provides at least as strong of evidence against the null hypothesis and in favor of the alternative as the observed estimate, $\bar{x} = 6.83$.

We can calculate the tail areas by first finding the lower tail corresponding to $\bar{x}$:

$$Z = \frac{6.83 - 7.00}{0.21} = -0.81 \quad \overset{table}{\rightarrow} \quad \text{left tail} = 0.2090$$

Because the normal model is symmetric, the right tail will have the same area as the left tail. The p-value is found as the sum of the two shaded tails:

$$\text{p-value} = \text{left tail} + \text{right tail} = 2 * (\text{left tail}) = 0.4180$$

This p-value is relatively large (larger than $\alpha = 0.05$) so we should not reject H_0. That is, if H_0 were true it would not be very unusual to see a sample mean this far from 7 hours simply due to sampling variation. Thus, we do not have sufficient evidence to conclude that the mean is different from 7 hours.

---

[31] Because the researchers are interested in any difference, they should use a two-sided setup: $H_0 : \mu = 7$, $H_A : \mu \neq 7$.

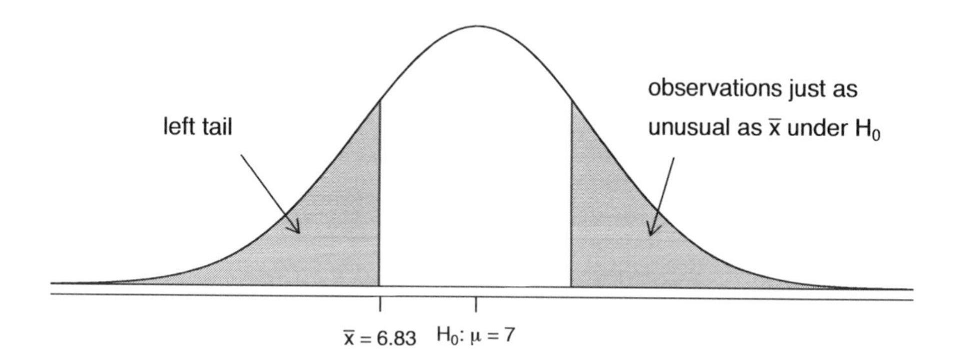

Figure 4.16: H_A is two-sided, so *both* tails must be counted for the p-value.

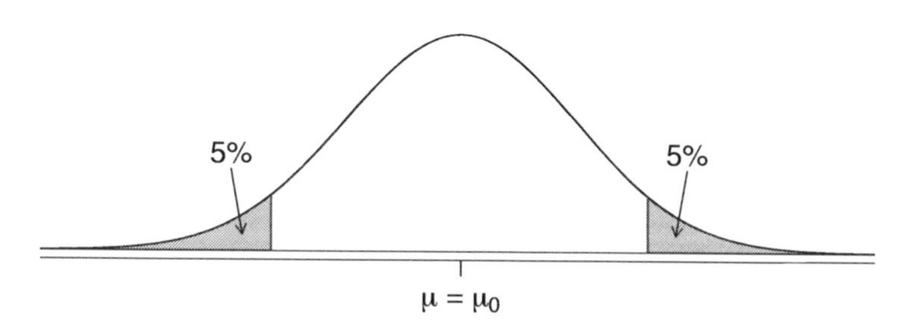

Figure 4.17: The shaded regions represent areas where we would reject H_0 under the bad practices considered in Example 4.38 when $\alpha = 0.05$.

● **Example 4.38** It is never okay to change two-sided tests to one-sided tests after observing the data. In this example we explore the consequences of ignoring this advice. Using $\alpha = 0.05$, we show that freely switching from two-sided tests to one-sided tests will cause us to make twice as many Type 1 Errors as intended.

Suppose the sample mean was larger than the null value, μ_0 (e.g. μ_0 would represent 7 if H_0: $\mu = 7$). Then if we can flip to a one-sided test, we would use H_A: $\mu > \mu_0$. Now if we obtain any observation with a Z score greater than 1.65, we would reject H_0. If the null hypothesis is true, we incorrectly reject the null hypothesis about 5% of the time when the sample mean is above the null value, as shown in Figure 4.17.

Suppose the sample mean was smaller than the null value. Then if we change to a one-sided test, we would use H_A: $\mu < \mu_0$. If $\bar{x}$ had a Z score smaller than -1.65, we would reject H_0. If the null hypothesis was true, then we would observe such a case about 5% of the time.

By examining these two scenarios, we can determine that we will make a Type 1 Error 5% + 5% = 10% of the time if we are allowed to swap to the "best" one-sided test for the data. This is twice the error rate we prescribed with our significance level: $\alpha = 0.05$ (!).

Caution: One-sided hypotheses are allowed only *before* seeing data
After observing data, it is tempting to turn a two-sided test into a one-sided test. Avoid this temptation. Hypotheses must be set up *before* observing the data. If they are not, the test must be two-sided.

4.3.6 Choosing a significance level

Choosing a significance level for a test is important in many contexts, and the traditional level is 0.05. However, it is often helpful to adjust the significance level based on the application. We may select a level that is smaller or larger than 0.05 depending on the consequences of any conclusions reached from the test.

If making a Type 1 Error is dangerous or especially costly, we should choose a small significance level (e.g. 0.01). Under this scenario we want to be very cautious about rejecting the null hypothesis, so we demand very strong evidence favoring H_A before we would reject H_0.

If a Type 2 Error is relatively more dangerous or much more costly than a Type 1 Error, then we should choose a higher significance level (e.g. 0.10). Here we want to be cautious about failing to reject H_0 when the null is actually false. We will discuss this particular case in greater detail in Section 4.6.

Significance levels should reflect consequences of errors
The significance level selected for a test should reflect the consequences associated with Type 1 and Type 2 Errors.

● **Example 4.39** A car manufacturer is considering a higher quality but more expensive supplier for window parts in its vehicles. They sample a number of parts from their current supplier and also parts from the new supplier. They decide that if the high quality parts will last more than 10% longer, it makes financial sense to switch to this more expensive supplier. Is there good reason to modify the significance level in such a hypothesis test?

The null hypothesis is that the more expensive parts last no more than 10% longer while the alternative is that they do last more than 10% longer. This decision is just one of the many regular factors that have a marginal impact on the car and company. A significance level of 0.05 seems reasonable since neither a Type 1 or Type 2 error should be dangerous or (relatively) much more expensive.

● **Example 4.40** The same car manufacturer is considering a slightly more expensive supplier for parts related to safety, not windows. If the durability of these safety components is shown to be better than the current supplier, they will switch manufacturers. Is there good reason to modify the significance level in such an evaluation?

The null hypothesis would be that the suppliers' parts are equally reliable. Because safety is involved, the car company should be eager to switch to the slightly more expensive manufacturer (reject H_0) if the evidence of increased safety is only moderately strong. A slightly larger significance level, such as $\alpha = 0.10$, might be appropriate.

⊙ **Exercise 4.41** A part inside of a machine is very expensive to replace. However, the machine usually functions properly even if this part is broken so we decide to replace the part only if we are extremely certain it is broken based on a series of measurements. Identify appropriate hypotheses for this test (in plain language) and suggest an appropriate significance level. Answer in the footnote[32].

4.4 Examining the Central Limit Theorem

The normal model for the sample mean tends to be very good when the sample consists of at least 50 independent observations and the population data are not strongly skewed. The Central Limit Theorem provides the theory that allows us to make this assumption.

Central Limit Theorem, informal definition
The distribution of $\bar{x}$ is approximately normal. The approximation can be poor if the sample size is small but gets better with larger sample sizes.

The Central Limit Theorem states that when the sample size is small, the normal approximation may not be very good. However, as the sample size becomes large, the normal approximation improves. We will investigate two cases to see roughly when the approximation is reasonable.

We consider two data sets: one from a *uniform* distribution, the other from an *exponential* distribution. These distributions are shown in the top panels of Figure 4.18.

The left panel in the $n = 2$ row represents the sampling distribution of $\bar{x}$ if it is the sample mean of two observations from the uniform distribution shown. The dashed line represents the closest approximation of the normal distribution. Similarly, the right panel of the $n = 2$ row represents the distribution of $\bar{x}$ if its observations came from an exponential distribution, and so on.

⊙ **Exercise 4.42** Examine the distributions in each row of Figure 4.18. What do you notice about the normal approximation for each distribution as the sample size becomes larger?

⊙ **Exercise 4.43** Would the normal approximation be good in all applications where the sample size is at least 50? Answer in the footnote[33].

TIP: With larger n, the sampling distribution of $\bar{x}$ becomes more normal
As the sample size increases, the normal model for $\bar{x}$ becomes more reasonable. We can also relax our condition on skew when the sample size is very large.

We discussed in Section 4.1.3 that the sample standard deviation, s, could be used as a substitute of the population standard deviation, σ, when computing the standard error. This estimate tends to be reasonable when $n \geq 50$. When $n < 50$, we must use small sample inference tools, which we examine in Chapter 6.

[32]Here the null hypothesis is that the part is not broken. If we don't have sufficient evidence to reject H_0, we would not replace the part. It sounds like failing to fix the part if it is broken (H_0 false, H_A true) is not very problematic, and replacing the part is expensive. Thus, we should require very strong evidence against H_0 before we replace the part. Choose a small significance level, such as $\alpha = 0.01$.

[33]Not necessarily. It took the skewed distribution in Figure 4.18 longer for $\bar{x}$ to be nearly normal. If a distribution is even more skewed, then we would need a larger sample to guarantee the distribution of the sample mean is nearly normal.

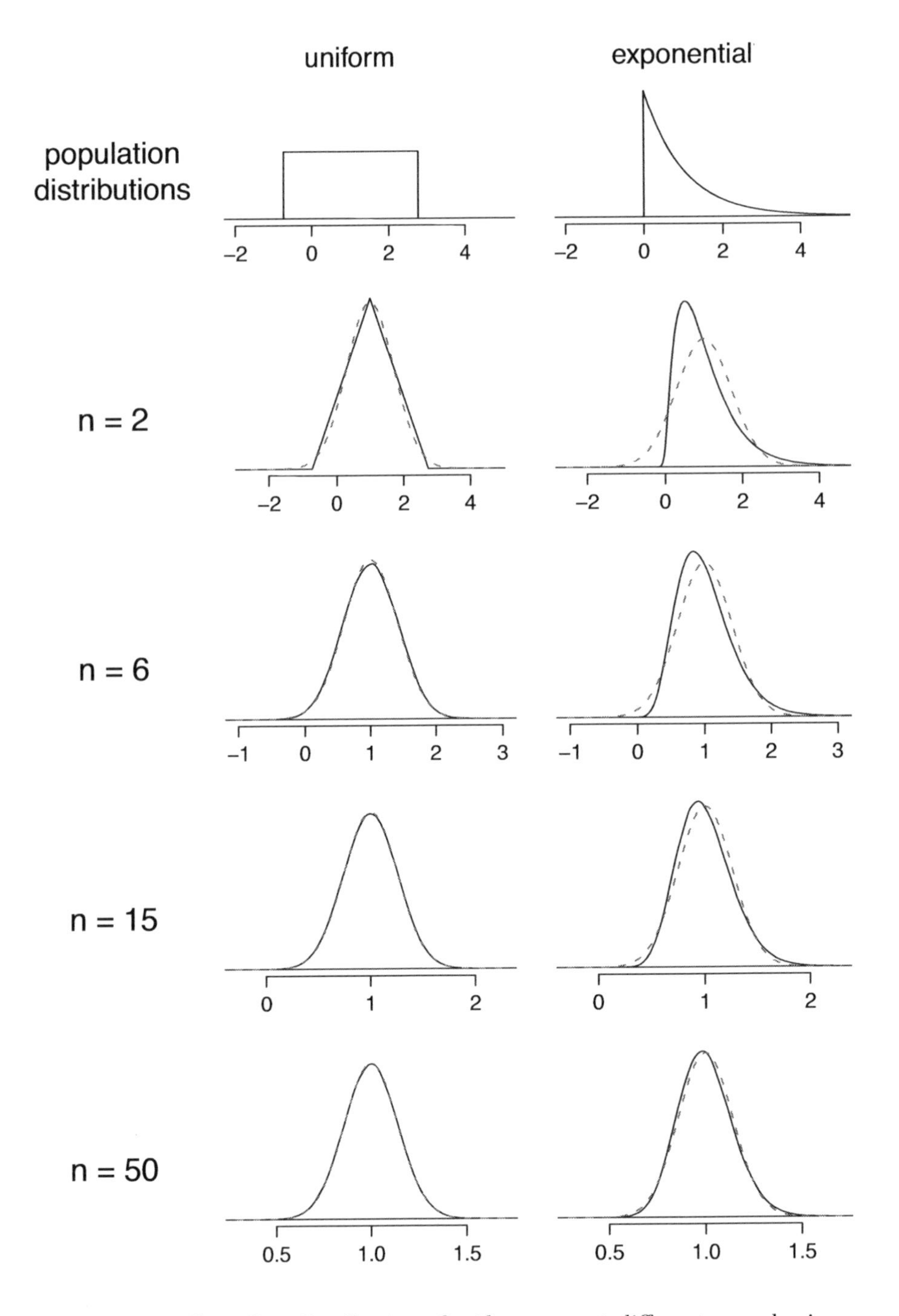

Figure 4.18: Sampling distributions for the mean at different sample sizes and for two data distributions.

 **Example 4.44** Figure 4.19 shows a histogram of 50 observations. These represent winnings and losses from 50 consecutive days of a professional poker player. Can the normal approximation be applied to the sample mean, 90.69?

We should consider each of our conditions.

(1) Because the data arrived in a particular sequence, this is called *time seriesata.* If the player wins on one day, it may influence how she plays the next. To make the assumption of independence we should perform careful checks on such data. While the supporting analysis is not shown, no evidence was found to indicate the observations are not independent.

(2) The sample size is 50; this condition is satisfied.

(3) The data set is very strongly skewed. In strongly skewed data sets, outliers can play an important role and affect the distribution of the sample mean and the estimate of the standard error.

Since we should be skeptical of the independence of observations and the data are very strongly skewed, we should not apply the normal model for the sample mean.

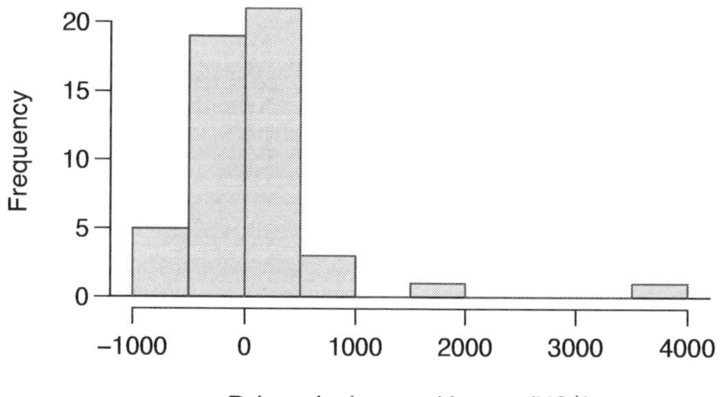

Figure 4.19: Sample distribution of poker winnings.

Caution: Examine data structure when considering independence
Some data sets are collected in such a way that they have a natural underlying structure between observations, most especially consecutive observations. Be especially cautious about independence assumptions regarding such data sets.

Caution: Watch out for strong skew
The normal approximation for the sample mean may not be appropriate when the data are extremely skewed, even when the sample size is 50.

4.5 Inference for other estimators

The sample mean is not the only point estimate for which the sampling distribution is nearly normal. The sampling distribution of sample proportions closely resembles the normal distribution if the sample size is sufficiently large. This is also the case for the difference in two sample means. In this section, we present a general framework for inference that we will use for a wide variety of parameters in Chapter 5. We will also look at point estimates that follow other distributions.

For each point estimate in this section, we make two important assumptions: its associated sampling distribution is nearly normal and the estimate is unbiased. A point estimate is **unbiased** if the sampling distribution of the estimate is centered at the parameter it estimates. That is, an unbiased estimate does not naturally over or underestimate the parameter. Rather, it tends to provide a "good" estimate. The sample mean is an example of an unbiased point estimate.

4.5.1 A general approach to confidence intervals

In Section 4.2, we used the point estimate $\bar{x}$ with a standard error $SE_{\bar{x}}$ to create a 95% confidence interval for the population mean:

$$\bar{x} \pm 1.96 * SE_{\bar{x}} \tag{4.45}$$

We constructed this interval by noting that the sample mean is within 1.96 standard errors of the actual mean about 95% of the time. This same logic generalizes to any unbiased point estimate that is nearly normal.

If a point estimate is nearly normal and unbiased, then a general 95% confidence interval for the estimated parameter is

$$\text{point estimate} \pm 1.96 * SE \tag{4.46}$$

where SE represents the standard error of the estimate.

Recall that the standard error represents the standard deviation associated with a point estimate. Generally the standard error is estimated from the data and computed using a formula. For example, when the point estimate is a sample mean, we compute the standard error using the sample standard deviation and sample size in the following formula:

$$SE_{\bar{x}} = \frac{s}{\sqrt{n}}$$

Similarly, there is a formula for the standard error of many commonly used point estimates, and we will encounter many of these cases in Chapter 5.

● **Example 4.47** In Exercise 4.2 on page 145, we computed a point estimate for the average difference in run times between men and women: $\bar{x}_{women} - \bar{x}_{men} = 6.82$ minutes. Suppose this point estimate is associated with a nearly normal distribution with standard error $SE = 3.05$ minutes. What is a reasonable 95% confidence interval for the difference in average run times?

The normal approximation is said to be valid and the estimate unbiased, so we apply Equation (4.46):

$$6.82 \pm 1.96 * 3.05 \quad \rightarrow \quad (0.84, 12.80)$$

Thus, we are 95% confident that the men were, on average, between 0.84 and 12.80 minutes faster than women in the 2009 Cherry Blossom Run. That is, the actual average difference is plausibly between 0.84 and 12.80 minutes with 95% confidence.

● **Example 4.48** Does Example 4.47 guarantee that if a husband and wife both ran in the race, the husband would run between 0.84 and 12.80 minutes faster than the wife?

Our confidence interval says absolutely nothing about individual observations. It only makes a statement about a plausible range of values for the *average* difference between all men and women who participated in the run.

The 95% confidence interval for the population mean was generalized for other confidence levels in Section 4.2.4. The constant 1.96 was replaced with $z^\star$, where $z^\star$ represented an appropriate Z score for the new confidence level. Similarly, Equation (4.46) can be generalized for a different confidence level.

General confidence interval
A confidence interval based on an unbiased and nearly normal point estimate is

$$\text{point estimate} \pm z^\star SE$$

where SE represents the standard error, and $z^\star$ is selected to correspond to the confidence level. The value $z^\star SE$ is called the **margin of error.**

⊙ **Exercise 4.49** What $z^\star$ would be appropriate for a 99% confidence level? For help, see Figure 4.9 on page 154.

⊙ **Exercise 4.50** The proportion of men in the `run10Samp` sample is $\hat{p} = 0.48$. This sample meets certain conditions that guarantee $\hat{p}$ will be nearly normal, and the standard error of the estimate is $SE_{\hat{p}} = 0.05$. Create a 90% confidence interval for the proportion of participants in the 2009 Cherry Blossom Run who are men. Answer in the footnote[34].

4.5.2 Generalizing hypothesis testing

Just as we generalized our confidence intervals for any unbiased and nearly normal point estimate, we can generalize our hypothesis testing methods. Here we only consider the p-value approach, introduced in Section 4.3.4, since it is the most commonly used technique and also extends to non-normal cases.

⊙ **Exercise 4.51** The Food and Drug Administration would like to evaluate whether a drug (sulphinpyrazone) is effective at reducing the death rate in heart attack patients. Patients were randomly split into two groups: a control group that received a placebo and a treatment group that received the new drug. What would be an appropriate null hypothesis? And the alternative? Answer in the footnote[35].

---

[34]We use $z^\star = 1.65$ (see Exercise 4.18 on page 153), and apply the general confidence interval formula:

$$\hat{p} \pm z^\star SE_{\hat{p}} \quad \rightarrow \quad 0.48 \pm 1.65 * 0.05 \quad \rightarrow \quad (0.3975, 0.5625)$$

Thus, we are 90% confident that between 40% and 56% of the participants were men.

[35]The skeptic's perspective is that the drug does not work (H_0), while the alternative is that the drug does work (H_A).

We can formalize the hypotheses from Exercise 4.51 by letting $p_{control}$ and $p_{treatment}$ represent the proportion of patients who died in the control and treatment groups, respectively. Then the hypotheses can be written as

$$H_0 : p_{control} = p_{treatment} \quad \text{(the drug doesn't work)}$$
$$H_A : p_{control} > p_{treatment} \quad \text{(the drug works)}$$

or equivalently,

$$H_0 : p_{control} - p_{treatment} = 0 \quad \text{(the drug doesn't work)}$$
$$H_A : p_{control} - p_{treatment} > 0 \quad \text{(the drug works)}$$

Strong evidence against the null hypothesis and in favor of the alternative would correspond to an observed difference in death rates,

$$\hat{p}_{control} - \hat{p}_{treatment}$$

being larger than we would expect from chance alone. This difference in sample proportions represents a point estimate that is useful in evaluating the hypotheses.

● **Example 4.52** We want to evaluate the hypothesis setup from Exericse 4.51 using data introduced in Section 1.1. In the control group, 60 of the 742 patients died. In the treatment group, 41 of 733 patients died. The sample difference in death rates can be summarized as

$$\hat{p}_{control} - \hat{p}_{treatment} = \frac{60}{742} - \frac{41}{733} = 0.025$$

As we will find out in Chapter 5, we can safely assume this point estimate is nearly normal and is an unbiased estimate of the actual difference in death rates. The standard error of this sample difference is $SE = 0.013$. We evaluate the hypothesis at a 5% significance level: $\alpha = 0.05$.

We would like to identify the p-value to evaluate the hypotheses. If the null hypothesis was true, then the point estimate would have come from a nearly normal distribution, like the one shown in Figure 4.20. The distribution is centered at zero since $p_{control} - p_{treatment} = 0$ under the null hypothesis. Because a large positive difference provides evidence against the null hypothesis and in favor of the alternative, the upper tail has been shaded to represent the p-value. We need not shade the lower tail since this is a one-sided test: an observation in the lower tail does not support the alternative hypothesis.

The p-value can be computed by using a Z score of our point estimate and the normal probability table.

$$Z = \frac{\text{point estimate} - \text{null value}}{SE_{\text{point estimate}}} = \frac{0.025 - 0}{0.013} = 1.92 \quad (4.53)$$

In our Z score formula, we replaced our sample mean with our general point estimate and included a corresponding standard error. Examining Z in the normal probability table, we find that the lower unshaded tail is about 0.973. Thus, the upper shaded tail representing the p-value is

$$\text{p-value} = 1 - 0.973 = 0.027$$

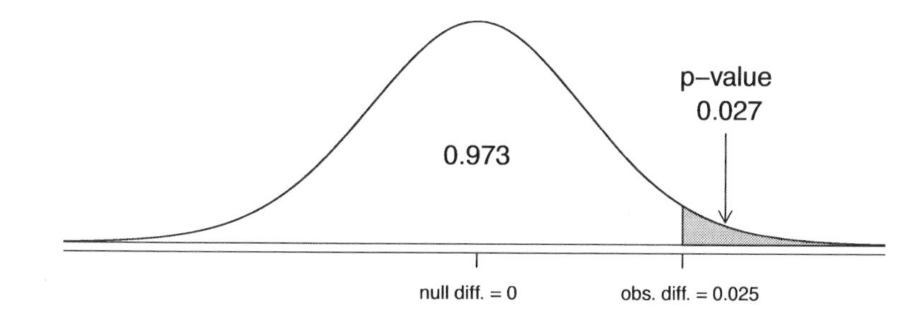

Figure 4.20: The distribution of the sample difference if the null hypothesis was true.

Because the p-value is less than the significance level ($\alpha = 0.05$), we say the null hypothesis is implausible. That is, we reject the null hypothesis in favor of the alternative and conclude that the drug is effective at reducing deaths in heart attack patients.

Provided a point estimate is unbiased and nearly normal, the methods for identifying the p-value and evaluating a hypothesis test change little.

Hypothesis testing using the normal model

1. First set up the hypotheses in plain language, then set them up in mathematical notation.
2. Identify an appropriate point estimate of the parameter in the hypotheses.
3. Verify conditions to ensure the standard error estimate is reasonable and the point estimate is nearly normal and unbiased.
4. Compute the standard error. Draw a picture depicting the distribution of the estimate under the idea that H_0 is true. Shade areas representing the p-value.
5. Using the picture and normal model, compute the *test statistic* (Z score) and identify the p-value to evaluate the hypotheses. Write a conclusion in plain language.

The Z score in Equation (4.53) is called a **test statistic**. In most hypothesis tests, a test statistic is a particular data summary that is especially useful for computing the p-value and evaluating the hypothesis test. In the case of point estimates that are nearly normal, the test statistic is the Z score.

Test statistic
A *test statistic* is a special summary statistic that is particularly useful for evaluating a hypothesis test or identifying the p-value. When a point estimate is nearly normal, we use the Z score of the point estimate as the test statistic. In later chapters we encounter point estimates where other test statistics are helpful.

In Sections 5.1-5.4, we apply the confidence interval and hypothesis testing framework to a variety of point estimates. In these cases, we require the sample size to be sufficiently large for the normal approximation to hold. In Sections 5.6, 5.7, and all of Chapter 6, we apply the ideas of confidence intervals and hypothesis testing to cases where the point estimate and test statistic are not necessarily normal. Such cases occur when the sample size is too small for the normal approximation, the standard error estimate may be poor, or the point estimate tends towards some other distribution.

4.6 Sample size and power (special topic)

The Type 2 Error rate and the magnitude of the error for a point estimate are controlled by the sample size. Real differences from the null value, even large ones, may be difficult to detect with small samples. If we take a very large sample, we might find a statistically significant difference but the magnitude might be so small that it is of no practical value. In this section we describe techniques for selecting an appropriate sample size based on these considerations.

4.6.1 Finding a sample size for a certain margin of error

Many companies are concerned about rising healthcare costs. A company may estimate certain health characteristics of its employees, such as blood pressure, to project its future cost obligations. However, it might be too expensive to measure the blood pressure of every employee at a large company, and the company may choose to take a sample instead.

● **Example 4.54** Blood pressure oscillates with the beating of the heart, and the systolic pressure is defined as the peak pressure when a person is at rest. The average systolic blood pressure for people in the U.S. is about 130 mmHg with a standard deviation of about 25 mmHg. How large of a sample is necessary to estimate the average systolic blood pressure with a margin of error of 4 mmHg using a 95% confidence level?

First, we frame the problem carefully. Recall that the margin of error is the part we add and subtract from the point estimate when computing a confidence interval. The margin of error for a 95% confidence interval estimating a mean can be written as

$$ME_{95\%} = 1.96 * SE = 1.96 * \frac{\sigma_{employee}}{\sqrt{n}}$$

The challenge in this case is to find the sample size n so that this margin of error is less than or equal to 4, which we write as an inequality:

$$1.96 * \frac{\sigma_{employee}}{\sqrt{n}} \le 4$$

In the above equation we wish to solve for the appropriate value of n, but we need a value for $\sigma_{employee}$ before we can proceed. However, we haven't yet collected any data, so we have no direct estimate! Instead we use the best estimate available to us: the approximate standard deviation for the U.S. population, 25. To proceed and solve for n, we substitute 25 for $\sigma_{employee}$:

$$1.96 * \frac{\sigma_{employee}}{\sqrt{n}} \approx 1.96 * \frac{25}{\sqrt{n}} \leq 4$$
$$1.96 * \frac{25}{4} \leq \sqrt{n}$$
$$\left(1.96 * \frac{25}{4}\right)^2 \leq n$$
$$150.06 \leq n$$

The result: we should choose a sample size of at least 151 employees. We round up because the sample size must be *greater than or equal to 150.06*.

The most controversial part of Example 4.54 is the use of the U.S. standard deviation for the employee standard deviation. Usually the standard deviation is not known. In such cases, it is reasonable to review scientific literature or market research to make an educated guess about the standard deviation.

Identify a sample size for a particular margin of error
To estimate the necessary sample size for a maximum margin of error m, we setup an equation to represent this relationship:

$$m \geq ME = z^\star \frac{\sigma}{\sqrt{n}}$$

where $z^\star$ is chosen to correspond to the desired confidence level, and σ is the standard deviation associated with the population. We solve for the sample size, n.

Sample size computations are helpful in planning data collection, and they require careful forethought. Next we consider another topic important in planning data collection and setting a sample size: the Type 2 Error rate.

4.6.2 Power and the Type 2 Error rate

Consider the following two hypotheses:

H_0: The average blood pressure of employees is the same as the national average, $\mu = 130$.

H_A: The average blood pressure of employees is different than the national average, $\mu \neq 130$.

Suppose the alternative hypothesis is actually true. Then we might like to know, what is the chance we make a Type 2 Error, i.e. we fail to reject the null hypothesis even when we should reject it? The answer should not be obvious! If the average blood pressure of the employees is 132 (just 2 mmHg from the null value), it might be very difficult to detect the difference unless we use a large sample size. On the other hand, it would be easier to detect a difference if the real average of employees was 140.

● **Example 4.55** Suppose the actual employee average is 132 and we take a sample of 100 individuals. Then the true sampling distribution of $\bar{x}$ is approximately $N(132, 2.5)$ (since $SE = \frac{25}{\sqrt{100}} = 2.5$). What is the probability of successfully rejecting the null hypothesis?

This problem can be broken into two normal probability questions. Let's first think about what we want to accomplish, which will help us determine the steps to identifying the answer. To know the probability of rejecting the null hypothesis, we must first know which values of $\bar{x}$ represent sufficient evidence to reject H_0. We would reject any $\bar{x}$ that falls in either of the two small tails (2.5% each) of the null distribution. So our first task is to find these two tail cutoff values for rejecting the null hypothesis.

The null distribution could be represented by $N(130, 2.5)$, the same standard deviation as the true distribution but with the null mean as its center. Then we can find the two tail areas by identifying the Z score corresponding to the 2.5% tails (± 1.96), and solving for x in the Z score equation:

$$-1.96 = Z_1 = \frac{x_1 - 130}{2.5} \qquad +1.96 = Z_2 = \frac{x_2 - 130}{2.5}$$
$$x_1 = 125.1 \qquad x_2 = 134.9$$

(An equally valid approach is to recognize that x_1 is $1.96 * SE$ below the mean and x_2 is $1.96 * SE$ above the mean to compute the values.) Figure 4.21 shows the null distribution on the left with these two dotted cutoffs.

Next, we compute the probability that, if $\bar{x}$ did come from $N(132, 2.5)$, we would reject the null hypothesis. This is the same as finding the two shaded tails for the second distribution in Figure 4.21. We use the Z score method:

$$Z_{left} = \frac{125.1 - 132}{2.5} = -2.76 \qquad Z_{right} = \frac{134.9 - 132}{2.5} = 1.16$$
$$area_{left} = 0.003 \qquad area_{right} = 0.123$$

The probability of rejecting the null mean, if the true mean is 132, is the sum of these areas: $0.003 + 0.123 = 0.126$.

The probability of rejecting the null hypothesis is called the **power**. The power varies depending on what we suppose the truth might be. In Example 4.55, the difference between the null value and the supposed true mean was relatively small, so the power was also small: only 0.126. However, when the truth is far from the null value, the power tends to increase.

⊙ **Exercise 4.56** What would the power be if the actual average blood pressure of employees was 140? Suppose the true sampling distribution of $\bar{x}$ under this scenario is $N(140, 2.5)$. It may be helpful to draw and shade an extra distribution on Figure 4.21 and use the cutoff values identified in Example 4.55. Short answer in the footnote[36].

⊙ **Exercise 4.57** If the power of a test is 0.979 for a particular mean, what is the Type 2 Error rate for this mean? Answer in the footnote[37].

---

[36] Find the probability the total area of this distribution below 125.1 (about zero area) and above 134.9 (about 0.979). If the true mean is 140, the power is about 0.979.

[37] The Type 2 Error rate represents the probability of failing to reject the null hypothesis. Since the power is the probability we do reject, the Type 2 Error rate will be $1 - 0.979 = 0.021$.

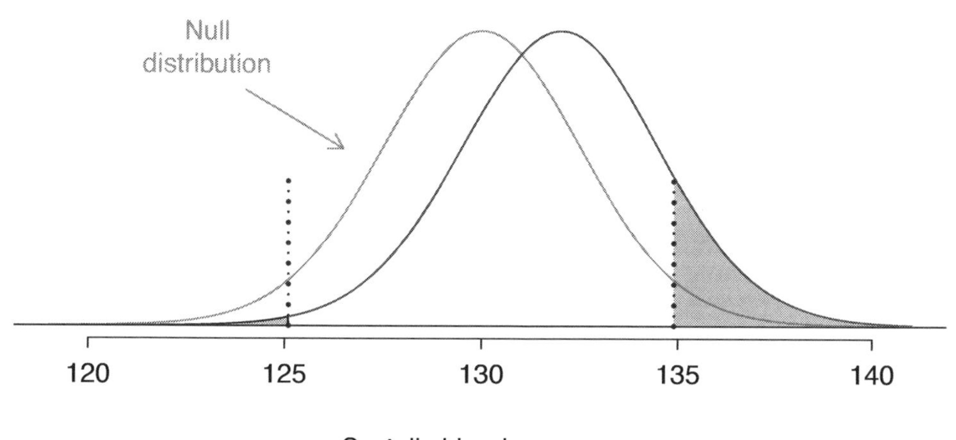

Figure 4.21: The sampling distribution of $\bar{x}$ under three scenarios. Left: $N(125, 2.5)$. Middle: $N(132, 2.5)$, and the shaded areas in this distribution represent the power of the test.

⊙ **Exercise 4.58** Provide an intuitive explanation for why we are more likely to reject H_0 when the true mean is further from the null value. One answer in the footnote[38].

4.6.3 Statistical significance versus practical significance

When the sample size becomes larger, point estimates become more precise and any real differences in the mean and null value become easier to detect and recognize. Even a very small difference would likely be detected if we took a large enough sample. Sometimes researchers will take such large samples that even the slightest difference is detected. While we still say that difference is **statistically significant**, it might not be **practically significant**.

Statistically significant differences are sometimes so minor that they are not practically relevant. This is especially important to research: if we conduct a study, we want to focus on finding a meaningful result. We don't want to spend lots of money finding results that hold no practical value.

The role of a statistician in conducting a study often includes planning the size the study. The statistician might first consult experts or scientific literature to learn what would be the smallest meaningful difference from the null value. She also would obtain some reasonable estimate for the standard deviation. With these important pieces of information, she would choose a sufficiently large sample size so that the power for the meaningful difference is perhaps 80% or 90%. While larger sample sizes may still be used, she might advise against using them in some cases, especially in sensitive areas of research.

[38]When the truth is far from the null value, the point estimate also tends to be far from the null value, making it easier to detect the difference and reject H_0.

4.7 Exercises

4.7.1 Variability in estimates

4.1 For each of the following situations, state whether the parameter of interest is a mean or a proportion. It may be helpful to examine whether individual responses are numerical or categorical.

(a) In a survey, one hundred college students are asked how many hours per week they spend on the Internet.

(b) In a survey, one hundred college students are asked: "What percentage of the time you spend on the Internet is part of your course work?"

(c) In a survey, one hundred college students are asked whether or not they cited information from Wikipedia on their papers.

(d) In a survey, one hundred college students are asked what percentage of their total weekly spending is on alcoholic beverages.

(e) In a sample of one hundred recent college graduates, it is found that 85 percent expect to get a job within one year of their graduation date.

4.2 For each of the following situations, state whether the parameter of interest is a mean or a proportion.

(a) A poll shows that 64% of Americans personally worry a great deal about federal spending and the budget deficit.

(b) A survey reports that local TV news has shown a 17% increase in revenue between 2009 and 2011 while newspaper revenues decreased by 6.4% during this time period.

(c) A survey shows that 52% of Americans aged 18 and older are married.

(d) In a survey, internet users are asked whether or not they purchased any Groupon coupons.

(e) In a survey, internet users are asked how many Groupon coupons they purchased over the last year.

4.3 A college counselor is interested in estimating how many credits a student typically enrolls in each semester. The counselor decides to randomly sample 100 students by using the registrar's database of students. The histogram below shows the distribution of the number of credits taken by these students. Sample statistics for this distribution are also provided.

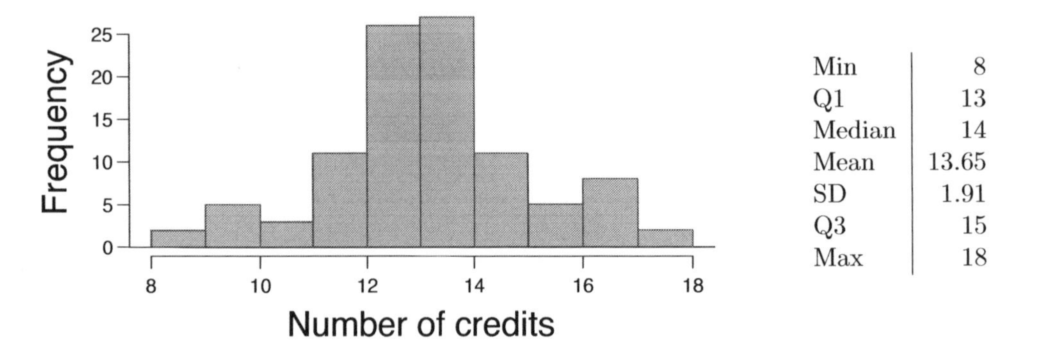

Min	8
Q1	13
Median	14
Mean	13.65
SD	1.91
Q3	15
Max	18

(a) What is the point estimate for the average number of credits taken per semester by students at this college? What about the median?

(b) What is the point estimate for the standard deviation of the number of credits taken per semester by students at this college? What about the IQR?

(c) Is a load of 16 credits unusually high for this college? What about 18 credits? Explain your reasoning.

4.4 Researchers studying anthropometry collected body girth measurements and skeletal diameter measurements, as well as age, weight, height and gender, for 507 physically active individuals. The histogram below shows the sample distribution of heights (in centimeters). [21]

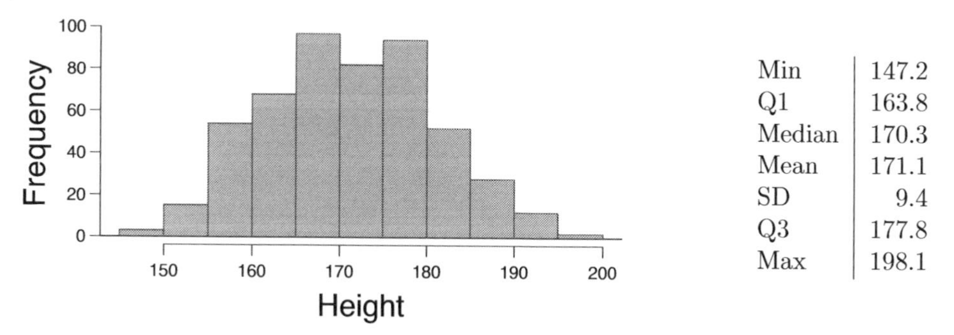

Min	147.2
Q1	163.8
Median	170.3
Mean	171.1
SD	9.4
Q3	177.8
Max	198.1

(a) What is the point estimate for the average height in this sample? What about the median?

(b) What is the point estimate for the standard deviation of the heights in this sample? What about the IQR?

(c) Is a person who is 1m 80cm (180 cm) tall considered unusually tall? What about a person who is 1m 55cm (155cm) tall? Explain your reasoning.

4.5 The college counselor from Exercise 4.3 takes another random sample of 100 students and this time finds a sample mean of 14.02 units and a standard deviation of 1.08 units. Should she be surprised that the sample statistics are slightly different for this sample? Explain your reasoning.

4.6 The researchers from Exercise 4.4 take another random sample of physically active individuals. Would you expect the mean and the standard deviation of this new sample to be the same as those from Exercise 4.4. Explain your reasoning.

4.7 The mean number of credits taken by the random sample of college students in Exercise 4.3 is a point estimate for the mean number of credits taken by all students at that college. In Exercise 4.5 we saw that a new random sample gives a slightly different sample mean. What measures do we use to quantify the variability of such an estimate? Compute this quantity using the data from Exercise 4.3.

4.8 The mean height of the sample of individuals in Exercise 4.4 is a point estimate for the mean height of all active individuals, if we assume the sample of individuals is representative. In Exercise 4.6 we saw that a new random sample gives a slightly different sample mean. What measures do we use to quantify the variability of such an estimate? Compute this quantity this quantity using the data from Exericse 4.4 under the condition that the data are a simple random sample.

4.9 John is shopping for wireless routers and is overwhelmed by the number of available options. In order to get a feel for the average price, he takes a random sample of 75 routers and finds that the average price for this sample is $75 and the standard deviation is $25.

(a) Based on this information, how much variability should he expect to see in the mean prices of repeated samples, each containing 75 randomly selected wireless routers?

(b) A consumer website claims that the average price of routers is $80. Is a true average of $80 consistent with John's sample?

4.10 Students are asked to count the number of chocolate chips in 22 cookies for a class activity. They found that the cookies on average had 14.77 chocolate chips with a standard deviation of 4.37 chocolate chips.

(a) Based on this information, about how much variability should they expect to see in the mean number of chocolate chips in random samples of 22 chocolate chip cookies?

(b) The packaging for these cookies claims that there are at least 20 chocolate chips per cookie. One student thinks this number is unreasonably high since the average they found is much lower. Another student claims the difference might be due to chance. What do you think?

4.7.2 Confidence intervals

4.11 A large sized coffee at Starbucks is called a *venti*, which means *twenty* in Italian, because a venti cup is supposed to hold 20 ounces of coffee. Jonathan believes that his venti always has less than 20 ounces of coffee in it, so he wants to test Starbucks' claim. He randomly chooses 50 Starbucks locations and gets a cup of venti coffee from each one. He then measures the amount of coffee in each cup and finds that the mean amount of coffee is 19.96 ounces and the standard deviation is 0.42 ounces.

(a) Are assumptions and conditions for inference satisfied?

(b) Calculate a 90% confidence interval for the true mean amount of coffee in Starbucks venti cups.

(c) Explain what this interval means in the context of this question.

(d) What does "90% confidence" mean?

(e) Do you think Starbucks' claim that a venti cup has 20 ounces of coffee is reasonable based on the confidence interval you just calculated?

(f) Would you expect a 95% confidence interval to be wider or narrower? Explain your reasoning.

4.12 A factory worker randomly samples 75 ball bearings, and records their lifespans under harsh conditions. He calculates an average lifespan of 6.85 working hours, with a standard deviation of 1.25 working hours. The following histogram shows the distribution of the lifespans of the ball bearings in this sample.

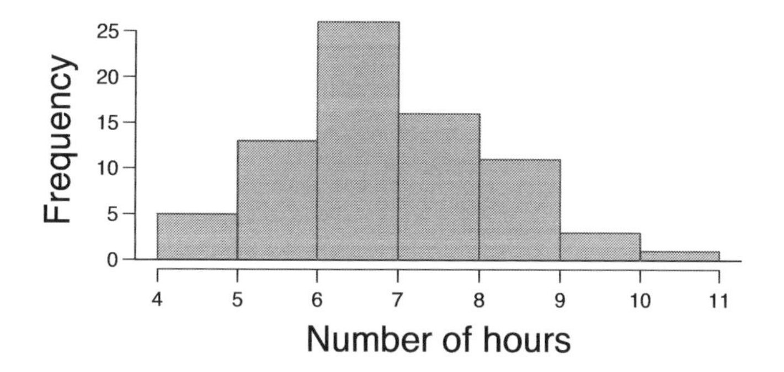

(a) Are assumptions and conditions for inference satisfied?

(b) Calculate a 98% confidence interval for the true average lifespan of ball bearings produced by this machine.

(c) Explain what this interval means in the context of this question.

(d) What does "98% confidence" mean?

(e) The manufacturer of the machine claims that the ball bearings produced by this machine last an average of 7 hours. What do you think about this claim?

(f) Would you expect a 90% confidence interval to be wider or narrower? Explain your reasoning.

4.13 A housing survey was conducted to determine the price of a typical home in Topanga, CA. The mean price of a house was roughly \$1.3 million with a standard deviation of \$300,000. There were no houses listed below \$600,000 but a few houses above \$3 million.

(a) Based on this information, do most houses in Topanga cost more or less than \$1.3 million?

(b) Can we use the normal distribution to calculate the probability that a randomly chosen house in Topanga costs more than \$1.4 million?

(c) What is the probability that the mean of 60 randomly chosen houses in Topanga is more than \$1.4 million?

(d) How would doubling the sample size affect the standard error of the mean?

4.14 Each year about 1500 students take the introductory statistics course at a large university. This year scores on the final exam are distributed with a median of 74 points, a mean of 70 points, and a standard deviation of 8 points. There are no students who scored above 100 (the maximum score attainable on the final) but a few students scored below 20 points.

(a) Based on this information, did most students score above or below 70 points?

(b) Can we calculate the probability that a randomly chosen student scored above 75 using the normal distribution?

(c) What is the probability that the average score for a random sample of 80 students is above 75?

(d) How would cutting the sample size in half affect the standard error of the mean?

4.15 If a higher confidence level means that we are more confident about the number we are reporting, why don't we always report a confidence interval with the highest possible confidence level?

4.16 In Section 4.3.2, a 95% confidence interval for the average age of runners in the 2009 Cherry Blossom Run was calculated to be (33.08, 37.36) based on a sample of 100 runners. How could we decrease the width of this interval without losing confidence?

4.17 A hospital administrator hoping to improve waiting times wishes to estimate the average emergency room waiting time at her hospital. She collects a simple random sample of 64 patients and determines the time (in minutes) between when they checked in to the ER until they were first seen by a doctor. A 95% confidence interval based on this sample is (128 minutes, 147 minutes). Determine whether the following statements are true or false, and explain your reasoning.

(a) This confidence interval is not valid since we do not know if the population distribution of the ER waiting times is nearly Normal.

(b) We are 95% confident that the average waiting time of these 64 emergency room patients is between 128 and 147 minutes.

(c) We are 95% confident that the average waiting time of all patients at this hospital is between 128 and 147 minutes.

(d) 95% of random samples have a sample mean between 128 and 147 minutes.

(e) A 99% confidence interval would be narrower than the 95% confidence interval since we need to be more sure of our estimate.

(f) The margin of error is 9.5 and the sample mean is 137.5.

(g) In order to decrease the margin of error of a 95% confidence interval to half of what it is now, we would need to double the sample size.

4.18 The 2009 holiday retail season, which kicked off on November 27, 2009 (the day after Thanksgiving), had been marked by somewhat lower self-reported consumer spending than was seen during the comparable period in 2008. To get an estimate of consumer spending, 436 randomly sampled American adults were surveyed. Daily consumer spending for the six-day period after Thanksgiving, spanning the Black Friday weekend and Cyber Monday, averaged \$84.71. A 95% confidence interval based on this sample is (\$80.31, \$89.11). Determine whether the following statements are true or false, and explain your reasoning.

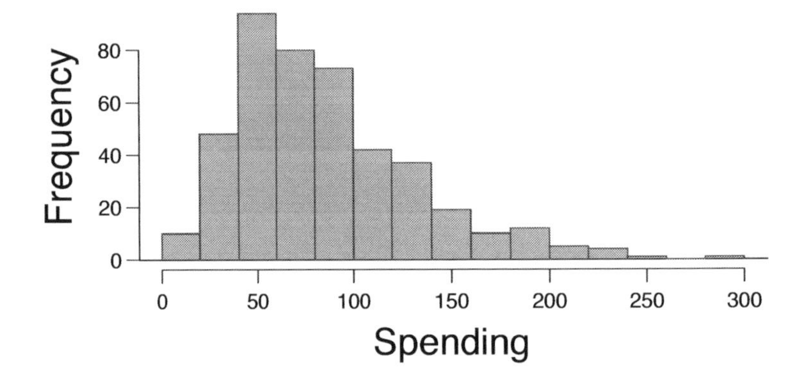

(a) We are 95% confident that the average spending of these 436 American adults is between \$80.31 and \$89.11.

(b) This confidence interval is not valid since the distribution of spending in the sample is right skewed.

(c) 95% of random samples have a sample mean between \$80.31 and \$89.11.

(d) We are 95% confident that the average spending of all American adults is between \$80.31 and \$89.11.

(e) A 99% confidence interval would be narrower than the 95% confidence interval since we need to be more sure of our estimate.

(f) In order to decrease the margin of error of a 95% confidence interval to a third of what it is now, we would need to use a sample 3 times larger.

(g) The margin of error is 4.4.

4.19 The distribution of weights of US pennies is approximately normal with a mean of 2.5 grams and a standard deviation of 0.03 grams.

(a) What is the probability that a randomly chosen penny weighs less than 2.4 grams?

(b) Describe the sampling distribution of the mean weight of 10 randomly chosen pennies.

(c) What is the probability that the mean weight of 10 pennies is less than 2.4 grams?

(d) Sketch the two distributions (population and sampling) on the same scale.

(e) How would your answers in (a) and (b) change if the weights of pennies were not nearly normal?

4.20 A manufacturer of compact fluorescent light bulbs advertises that the distribution of the lifespans of these light bulbs is nearly normal with a mean of 9,000 hours and a standard deviation of 1,000 hours.

(a) What is the probability that a randomly chosen light bulb lasts more than 10,500 hours?

(b) Describe the distribution of the mean lifespan of 60 light bulbs.

(c) What is the probability that the mean lifespan of 60 randomly chosen light bulbs is more than 10,500 hours?

(d) Sketch the two distributions (population and sampling) on the same scale.

(e) How would your answers in (a) and (b) change if the lifespans of light bulbs were not nearly normal?

4.21 Suppose an iPod has 3,000 songs. The histogram below shows the distribution of the lengths of these songs. We also know that, for this iPod, the mean length is 3.45 minutes and the standard deviation is 1.63 minutes.

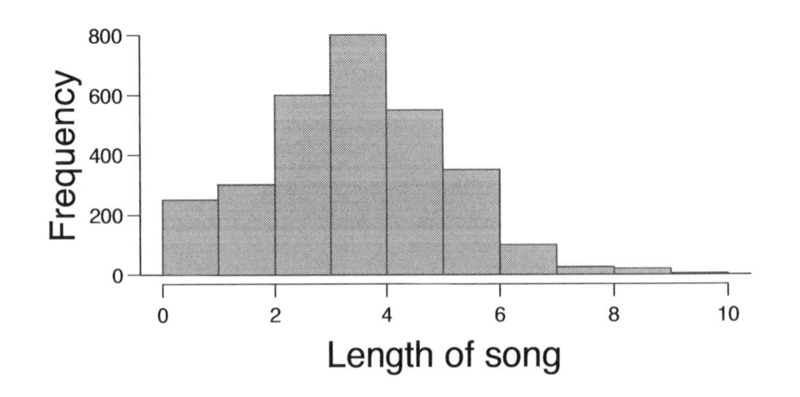

(a) Calculate the probability that a randomly selected song lasts more than 5 minutes.

(b) You are about to go for an hour run and you make a random playlist of 15 songs. What is the probability that your playlist lasts for the entire duration of your run?

(c) You are about to take a trip to visit your parents and the drive is 6 hours. You make a random playlist of 100 songs. What is the probability that your playlist lasts the entire drive?

4.22 Suppose the area that can be painted using a single can of spray paint is slightly variable and follows a normal distribution with a mean of 25 square feet and a standard deviation of 3 square feet.

(a) What is the probability that the area covered by a can of spray paint is more than 27 square feet?

(b) What is the probability that the area covered by 20 cans of spray paint is at least 540 square feet?

(c) How would your answers in (a) and (b) change if the area covered by a can of spray paint is not distributed nearly normally?

4.7.3 Hypothesis testing

4.23 Write the null and alternative hypotheses, in words and then symbols, for each of the following situations.

(a) New York is known as "the city that never sleeps". A random sample of 25 New Yorkers were asked how much sleep they get per night. Is there significant evidence that New Yorkers on average sleep less than 8 hours a night?

(b) Employers at a firm are worried about the effect of March Madness on employee productivity. They estimate that on a regular business day employees spend on average 15 minutes of company time checking personal email, making personal phone calls, etc. They want to test if this type of employee behavior significantly decreases during March Madness.

4.24 Write the null and alternative hypotheses, in words and using symbols, for each of the following situations.

(a) Since 2008, chain restaurants in California have been required to display calorie counts of each menu item. Prior to menus displaying calorie counts, the average calorie intake of diners at a restaurant was 1100 calories. A nutritionist conducts a hypothesis test to see if there has been a significant decrease in the average calorie intake of a diners at this restaurant.

(b) Based on the performance of those who took the GRE exam between July 1, 2004 and June 30, 2007, the average Verbal Reasoning score was calculated to be 462. In 2001 the average verbal score was slightly higher. Is there evidence that the average GRE Verbal Reasoning score has changed since 2004? [22]

4.25 Exercise 4.13 presents the results of a survey showing that the mean price of a house in Topanga, CA is \$1.3 million. A prospective homeowner believes that this figure is an overestimation and decides to collect his own sample for a hypothesis test. Below is how he set up his hypotheses. Indicate any errors you see.

$$H_0 : \bar{x} = \$1.3 \; million$$
$$H_A : \bar{x} > \$1.3 \; million$$

4.26 A study suggests that the average college student spends 2 hours per week communicating with others online. You believe that this is an underestimate and decide to collect your own sample for a hypothesis test. You randomly sample 60 students from your dorm and find that on average they spent 3.5 hours a week communicating with others online. A friend of yours, who offers to help you with the hypothesis test, comes up with the following set of hypotheses. Indicate any errors you see.

$$H_0 : \bar{x} < 2 \; hours$$
$$H_A : \bar{x} > 3.5 \; hours$$

4.27 Exercise 4.17 provides a 95% confidence interval for the mean waiting time at an emergency room (ER) of (128 minutes, 147 minutes).

(a) A local newspaper claims that the average waiting time at this ER exceeds 3 hours. What do you think of their claim?

(b) The Dean of Medicine at this hospital claims the average wait time is 2.2 hours. What do you think of her claim?

(c) Determine if the Dean from part (b) would arrive at the same conclusion using a 99% interval without actually calculating the interval?

4.28 Exercise 4.18 provides a 95% confidence interval for the average spending by American adults during the six-day period after Thanksgiving 2009 of (80.31,89.11).

(a) A local news anchor claims that the average spending during this period in 2009 was approximately \$100. What do you think of her claim?

(b) Can you determine if the news anchor would arrive at the same conclusion using a 90% interval? Why or why not? (Do not actually calculate the interval.)

4.29 A patient is diagnosed with Fibromyalgia, a long-term syndrome of body pain, and is prescribed anti-depressants. Being the skeptic that she is, she doesn't initially believe that anti-depressants will treat her symptoms. However after a couple months of being on the medication she decides that the anti-depressants are working, because she feels like her symptoms are in fact getting better.

(a) Write the hypotheses in words.

(b) What is a Type I error in this context?

(c) What is a Type II error in this context?

(d) How would these errors affect the patient?

4.30 A food safety inspector is called upon to investigate a restaurant with a few customer reports of poor sanitation practices. The food safety inspector uses a hypothesis testing framework to evaluate whether regulations are not being met. If he decides the restaurant is in gross violation, its license to serve food will be revoked.

(a) Write the hypotheses in words.

(b) What is a Type I error in this context?

(c) What is a Type II error in this context?

(d) Which error is more problematic for the restaurant owner?

(e) Which error is more problematic for the diners?

4.31 A 2000 study showed that college students spent an average of 25% of their internet time on coursework. A decade later in 2010 a new survey was given to 238 randomly sampled college students. The responses showed that on average 10% of the time college students spent on the Internet was for coursework with a standard deviation of 30%. Is there evidence that the percentage of time college students spend on the Internet for coursework has changed over the last decade?

(a) Are assumptions and conditions for inference met?

(b) Perform an appropriate hypothesis test and state your conclusion.

(c) Interpret the p-value in context.

(d) What type of an error might you have made?

4.32 The hospital administrator mentioned in Exercise 4.17 randomly selected 64 patients and measures the time (in minutes) between when they checked in to the ER and the time they were first seen by a doctor. The average time is 137.5 minutes and the standard deviation is 39 minutes. She is getting grief from her supervisor on the basis that the wait times in the ER has increased greatly from last year's average of 128 minutes. However, she claims that the increase is probably just due to chance.

(a) Are assumptions and conditions for inference met?

(b) Using a significance level of $\alpha = 0.05$, is the increase in wait times statistically significant?

(c) Would the conclusion of the hypothesis test change if the significance level was changed to $\alpha = 0.01$?

4.33 Exercise 4.12 provides information on the average lifespan and standard deviation of 75 randomly sampled ball bearings produced by a certain machine: $\bar{x} = 6.85, s = 1.25$. We are also told that the manufacturer of the machine claims the ball bearings produced by this machine last 7 hours on average. Conduct a formal hypothesis test of this claim.

4.34 The nutrition label on a bag of potato chips says that a one ounce (28 gram) serving of potato chips has 130 calories and contains ten grams of fat, with three grams of saturated fat. A random sample of 55 bags yielded a sample mean of 138 calories with a standard deviation of 17 calories. Is there evidence that the nutrition label does not provide an accurate measure of calories in the bags of potato chips?

4.35 You are given the following hypotheses:

$$H_0 : \mu = 34$$
$$H_A : \mu > 34$$

We know that the sample standard deviation is 10 and the sample size is 65. For what sample mean would the p-value be equal to 0.05? Assume that all assumptions and conditions necessary for inference are satisfied.

4.36 Suppose regulators monitored 403 drugs last year, each for a particular adverse response. For each drug they conducted a single hypothesis test with a significance level of 5% to determine if the adverse effect was higher in those taking the drug than those who did not take the drug; the regulators ultimately rejected the null hypothesis for 42 drugs.

(a) Describe the error the regulators might have made for a drug where the null hypothesis was rejected.

(b) Describe the error regulators might have made for a drug where the null hypothesis was not rejected.

(c) Suppose the vast majority of the 403 drugs do not have adverse effects. Then, if you picked one of the 42 suspect drugs at random, about how sure would you be that the drug really has an adverse effect?

(d) Can you also say how sure you are that a particular drug from the 361 where the null hypothesis was not rejected does not have the corresponding adverse response?

4.7.4 Examining the Central Limit Theorem

4.37 The histogram below shows the distribution of ages of pennies at a bank.

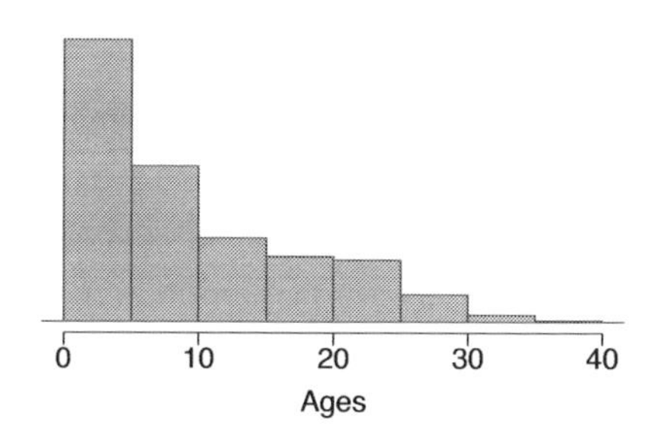

(a) Describe the distribution of the ages of these pennies.

(b) Sampling distributions for means from simple random samples of 5, 50, and 150 pennies is shown in the histograms below. Describe the shapes of these distributions and comment on whether they look like what you would expect to see based on the Central Limit Theorem.

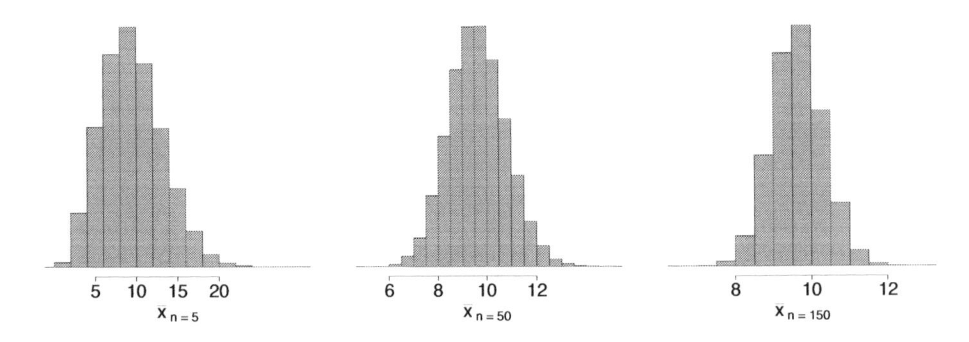

4.38 The mean age of the pennies from Exercise 4.37 is 9.6 years with a standard deviation of 8.2 years. Using the Central Limit Theorem, calculate the means and standard deviations of the distribution of the mean from random samples of size 5, 50, and 150. Comment on whether the sampling distributions shown in Exercise 4.37 agree with the values you computed.

4.7.5 Inference for other estimators

4.39 A car insurance company advertises that customers switching to their insurance save, on average, \$432 on their yearly premiums. A market researcher who thinks this is an overestimate randomly samples 82 customers who recently switched to this insurance. She finds an average savings of \$395, with a standard deviation of \$102.

(a) Are assumptions and conditions for inference satisfied?

(b) Perform a hypothesis test and state your conclusion.

(c) Do you agree with the market researcher that the amount of savings advertised is an overestimate? Explain your reasoning.

(d) Calculate a 90% confidence interval for the average amount of savings of all customers who switch their insurance.

(e) Do your results from the hypothesis test and the confidence interval agree? Explain.

4.40 A restaurant owner is considering extending the happy hour at his restaurant since he believes that an extended happy hour may increase the revenue per customer. He estimates that the current average revenue per customer is \$18 during happy hour. He runs the extended happy hour for a week and finds an average revenue of \$19.25 with standard deviation \$3.02 based on a simple random sample of 70 customers.

(a) Are assumptions and conditions for inference satisfied?

(b) Perform a hypothesis test. Based on the result of the hypothesis test, is there significant evidence that the revenue per customer has increased when happy hour was extended by an hour? Suppose the customers and their buying habits this week were no different than in any other week for this particular bar. (This may not always be a reasonable assumption, in which case, more advanced methods would be necessary than what we have covered.)

(c) Calculate a 90% confidence interval for the average revenue per customer.

(d) Do your results from the hypothesis test and the confidence interval agree? Explain.

(e) If your hypothesis test and confidence interval suggest a significant increase in revenue per customer, why might you still not recommend that the restaurant owner extend the happy hour based on this criterion? What may be a better measure to consider?

4.41 A company offering online speed reading courses claims that students who take their courses show a 5 times (500%) increase in the number of words they can read in a minute without losing comprehension. A random sample of 100 students yielded an average increase of 415% with a standard deviation of 220%. Do these data support the company's claim?

(a) Are assumptions and conditions for inference satisfied?

(b) Perform a hypothesis test and state your conclusion.

(c) Based on the result of the hypothesis test, does the company's claim seem reasonable?

(d) Calculate a 95% confidence interval for the average increase in number of words students can read in a minute without losing comprehension.

(e) Do your results from the hypothesis test and the confidence interval agree? Explain.

4.42 The average price for a slice of pizza in Manhattan was $2.42 in early 2008. A random sample of 50 pizza shops yielded an average price for a slice of pizza of $2.53 with a standard deviation of 48¢ in 2011. Do these data provide strong evidence of an increase in the average price for a slice of pizza in Manhattan?

(a) Are assumptions and conditions for inference satisfied?

(b) Perform a hypothesis test and state your conclusion.

(c) Calculate a 95% confidence interval for the average price for a slice of pizza in Manhattan in 2011.

(d) Do your results from the hypothesis test and the confidence interval agree? Explain.

4.7.6 Sample size and power

4.43 The market researcher at the car insurance company from Exercise 4.39 estimated an annual savings for switching to a particular car insurance company at $395 with a standard deviation of $102. She would like conduct another survey but have a margin of error of approximately $10 at a 99% confidence level. How large of a sample should she collect?

4.44 A random sample of 100 students who took online speed reading courses from the company described in Exercise 4.41 yielded an average increase in reading speed of 415% and a standard deviation of 220%. We would like to calculate a 95% confidence interval for the average increase in reading speed with a margin of error of at most 15%. How many students should we sample?

Chapter 5

Large sample inference

Chapter 4 introduced a framework of statistical inference focused on estimation of the population mean μ. Chapter 5 explores statistical inference for six slightly more interesting cases: differences of two population means (5.1, 5.2); a single population proportion (5.3); differences of two population proportions (5.4); and introduce inference tools for categorical variables with many levels (5.6, 5.7). For each application included in Sections 5.1-5.4, we verify conditions that ensure the sampling distribution of the point estimates are nearly normal and then apply the general framework from Section 4.5. Sections 5.6 and 5.7 introduce a new distribution called the chi-square distribution, which is useful for evaluating independence among categorical variables. In Chapter 6 we will extend our reach to smaller samples.

5.1 Paired data

Are textbooks actually cheaper online? Here we compare the price of textbooks at UCLA's bookstore and prices at Amazon.com. Seventy-three UCLA courses were randomly sampled in Spring 2010, representing less than 10% of all UCLA courses[1]. A portion of this `textbooks` data set is shown in Table 5.1.

	deptAbbr	course	uclaNew	amazNew	diff
1	Am Ind	C170	27.67	27.95	-0.28
2	Anthro	9	40.59	31.14	9.45
⋮	⋮	⋮	⋮	⋮	⋮
73	Wom Std	285	27.70	18.22	9.48

Table 5.1: Three cases of the `textbooks` data set.

5.1.1 Paired observations and samples

Each textbook has two corresponding prices in the data set: one for the UCLA bookstore and one for Amazon. In this way, each textbook price from the UCLA bookstore has a natural correspondence with a textbook price from Amazon. When two sets of observations have this special correspondence, they are said to be **paired**.

[1]When a class had multiple books, only the most expensive text was considered.

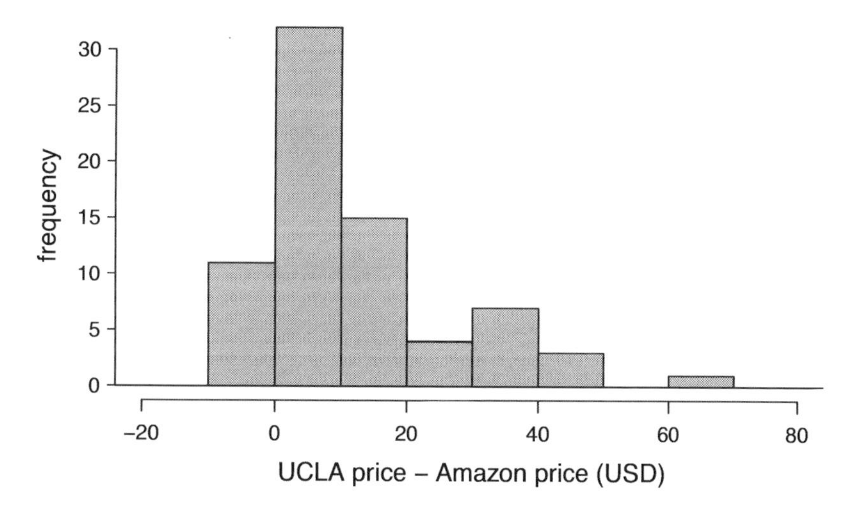

Figure 5.2: Histogram of the difference in price for each book sampled.

Paired data
Two sets of observations are *paired* if each observation in one set has a special correspondence or connection with exactly one observation in the other data set.

To analyze paired data, it is often useful to look at the difference in outcomes of each pair of observations. In the `textbook` data set, we look at the difference in prices, which is represented as the `diff` variable in the `textbooks` data. Here the differences are taken as

$$\text{UCLA price} - \text{Amazon price}$$

for each book. It is important that we always subtract using a consistent order; here Amazon prices are always subtracted from UCLA prices. A histogram of these differences is shown in Figure 5.2. Using differences between paired observations is a common and useful way to analyze paired data.

⊙ **Exercise 5.1** The first difference shown in Table 5.1 is computed as $27.67-27.95 = -0.28$. Verify the other two differences in the table are computed correctly.

5.1.2 Inference for paired data

To analyze a paired data set, we use the exact same tools we developed in Chapter 4 and we simply apply them to the differences in the paired observations.

n_{diff}	$\bar{x}_{diff}$	s_{diff}
73	12.76	14.26

Table 5.3: Summary statistics for the `diff` variable. There were 73 books, so there are 73 differences.

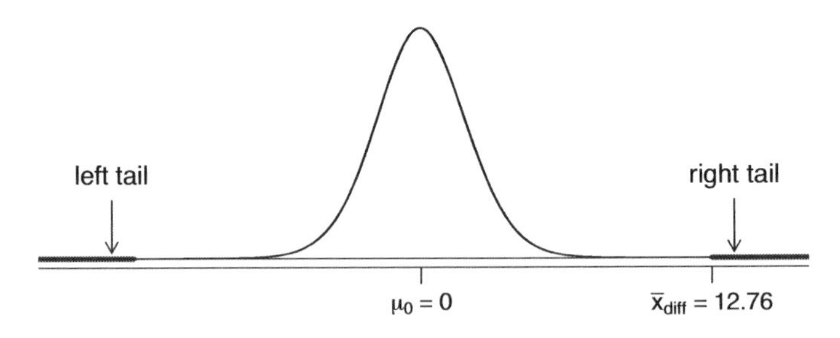

Figure 5.4: Sampling distribution for the mean difference in book prices, if the true average difference is zero.

Example 5.2 Set up and implement a hypothesis test to determine whether, on average, there is a difference between Amazon's price for a book and the UCLA bookstore's price.

There are two scenarios: there is no difference or there is a difference in average prices. The *no difference* scenario is always the null hypothesis:

H_0: $\mu_{diff} = 0$. There is no difference in the average textbook price. The notation μ_{diff} is used as a notational reminder that we should only work with the difference in prices.

H_A: $\mu_{diff} \neq 0$. There is a difference in average prices.

Can the normal model be used to describe the sampling distribution of $\bar{x}_{diff}$? We must check that the `diff` data meet the conditions established in Chapter 4. The observations are based on a simple random sample from less than 10% of all books sold at the bookstore, so independence is reasonable. There is skew in the differences (Figure 5.2), however, it is not extreme. There are also more than 50 differences. Thus, we can conclude the sampling distribution of $\bar{x}_{diff}$ is nearly normal.

We compute the standard error associated with $\bar{x}_{diff}$ using the standard deviation of the differences and the number of differences:

$$SE_{\bar{x}_{diff}} = \frac{s_{diff}}{\sqrt{n_{diff}}} = \frac{14.26}{\sqrt{73}} = 1.67$$

To visualize the p-value, the sampling distribution of $\bar{x}_{diff}$ is drawn under the condition as though H_0 was true, which is shown in Figure 5.4. The p-value is represented by the two (very) small tails.

To find the tail areas, we compute the test statistic, which is the Z score of $\bar{x}_{diff}$ under the condition $\mu_{diff} = 0$:

$$Z = \frac{\bar{x}_{diff} - 0}{SE_{x_{diff}}} = \frac{12.76 - 0}{1.67} = 7.59$$

This Z score is so large it isn't even in the table, which ensures the single tail area will be 0.0002 or smaller. Since the p-value is both tails and the normal distribution is symmetric, the p-value can be estimated as twice the one-tail area:

$$\text{p-value} = 2 * (\text{one tail area}) \approx 2 * 0.0002 = 0.0004$$

Because the p-value is less than 0.05, we reject the null hypothesis. We have found convincing evidence that Amazon is, on average, cheaper than the UCLA bookstore for UCLA course textbooks.

⊙ **Exercise 5.3** Create a 95% confidence interval for the average price difference between books at the UCLA bookstore and books on Amazon. Answer in the footnote[2].

5.2 Difference of two means

In this section we consider a difference in two population means, $\mu_1 - \mu_2$, under the condition that the data are not paired. The methods are similar in theory but different in the details. Just as with a single sample, we identify conditions to ensure a point estimate of the difference, $\bar{x}_1 - \bar{x}_2$, is nearly normal. Next we introduce a formula for the standard error, which allows us to apply our general tools from Section 4.5.

We apply these methods to two examples: participants in the 2009 Cherry Blossom Run and newborn infants. This section is motivated by questions like "Is there convincing evidence that newborns from mothers who smoke have a different average birth weight than newborns from mothers who don't smoke?"

5.2.1 Point estimates and standard errors for differences of means

We would like to estimate the average difference in run times for men and women using a random sample of 100 men and 80 women from the run10 population. Table 5.5 presents relevant summary statistics, and box plots of each sample are shown in Figure 5.6.

	men	women
$\bar{x}$	88.08	96.28
s	15.74	13.66
n	100	80

Table 5.5: Summary statistics for the run time of 180 participants in the 2009 Cherry Blossom Run.

The two samples are independent of one-another, so the data are not paired. Instead a point estimate of the difference in average 10 mile times for men and women, $\mu_w - \mu_m$, can be found using the two sample means:

$$\bar{x}_w - \bar{x}_m = 8.20$$

Because we are examining two simple random samples from less than 10% of the population, each sample contains at least 50 observations, and neither distribution is strongly skewed, we can safely conclude the sampling distribution of each sample means is nearly normal. Finally, because each sample is independent of the other (e.g. the data are not

[2] Conditions have already verified and the standard error computed in Example 5.2. To find the interval, identify $z^\star$ (1.96 for 95% confidence) and plug it, the point estimate, and the standard error into the general confidence interval formula:

$$\text{point estimate} \pm z^\star SE \quad \rightarrow \quad 12.76 \pm (1.96)(1.67) \quad \rightarrow \quad (9.49, 16.03)$$

We are 95% confident that Amazon is, on average, between \$9.49 and \$16.03 cheaper than the UCLA bookstore for UCLA course books.

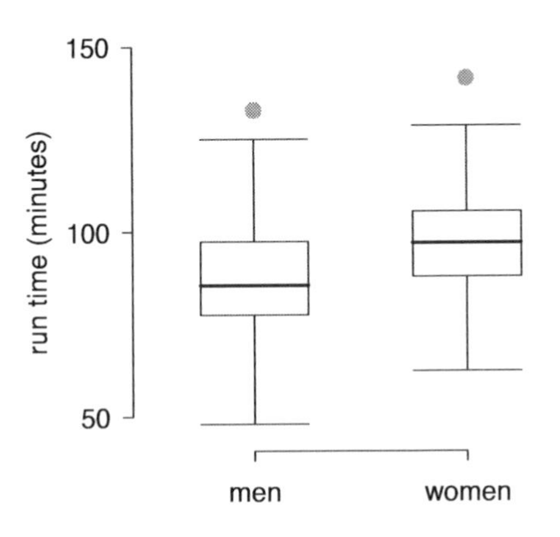

Figure 5.6: Side-by-side box plots for the sample of 2009 Cherry Blossom Run participants.

paired), we can conclude that the difference in sample means can be modeled using a normal distribution[3].

> **Conditions for normality of $\bar{x}_1 - \bar{x}_2$**
> If the sample means, $\bar{x}_1$ and $\bar{x}_2$, each meet the criteria for having nearly normal sampling distributions and the observations in the two samples are independent, then the difference in sample means, $\bar{x}_1 - \bar{x}_2$, will have a sampling distribution that is nearly normal.

We can quantify the variability in the point estimate, $\bar{x}_w - \bar{x}_m$, using the following formula for its standard error:

$$SE_{\bar{x}_w - \bar{x}_m} = \sqrt{\frac{\sigma_w^2}{n_w} + \frac{\sigma_m^2}{n_m}}$$

We usually estimate this standard error using standard deviation estimates based on the samples:

$$\begin{aligned} SE_{\bar{x}_w - \bar{x}_m} &= \sqrt{\frac{\sigma_w^2}{n_w} + \frac{\sigma_m^2}{n_m}} \\ &\approx \sqrt{\frac{s_w^2}{n_w} + \frac{s_m^2}{n_m}} = \sqrt{\frac{13.7^2}{80} + \frac{15.7^2}{100}} = 2.19 \end{aligned}$$

Because each sample has at least 50 observations ($n_w = 80$ and $n_m = 100$), this substitution using the sample standard deviation tends to be very good.

[3]Probability theory guarantees that the difference of two independent normal random variables is also normal. Because each sample mean is nearly normal and observations in the samples are independent, we are assured the difference is also nearly normal.

Distribution of a difference of sample means
The sample difference of two means, $\bar{x}_1 - \bar{x}_2$, is nearly normal with mean $\mu_1 - \mu_2$ and estimated standard error

$$SE_{\bar{x}_1 - \bar{x}_2} = \sqrt{\frac{s_1^2}{n_1} + \frac{s_2^2}{n_2}} \qquad (5.4)$$

when each sample mean is nearly normal and all observations are independent.

5.2.2 Confidence interval for the difference

When the conditions are met for the sampling distribution of $\bar{x}_1 - \bar{x}_2$ to be nearly normal, we can construct a 95% confidence interval for the difference in two means from the framework built in Chapter 4. Here a point estimate, $\bar{x}_w - \bar{x}_m = 8.20$, is associated with a normal model with standard error $SE = 2.19$. Using this information, the general confidence interval formula may be applied in an attempt to capture the true difference in means:

$$\text{point estimate} \pm z^{\star} SE \quad \rightarrow \quad 8.20 \pm 1.96 * 2.19 \quad \rightarrow \quad (3.91, 12.49)$$

Based on the samples, we are 95% confident that men ran, on average, between 3.91 and 12.49 minutes faster than women in the 2009 Cherry Blossom Run.

⊙ **Exercise 5.5** What does 95% confidence mean? Answer in the footnote[4].

⊙ **Exercise 5.6** We may be interested in a different confidence level. Construct the 99% confidence interval for the population difference in average run times based on the sample data. Hint in the footnote[5].

5.2.3 Hypothesis tests based on a difference in means

A data set called `babySmoke` represents a random sample of 150 cases of mothers and their newborns in North Carolina over a year. Four cases from this data set are represented in Table 5.7. We are particularly interested in two variables: `weight` and `smoke`. The `weight` variable represents the weights of the newborns and the `smoke` variable describes which mothers smoked during pregnancy. We would like to know if there is convincing evidence that newborns from mothers who smoke have a different average birth weight than newborns from mothers who don't smoke? We will answer this question using a hypothesis test. The smoking group includes 50 cases and the nonsmoking group contains 100 cases, represented in Figure 5.8.

---

[4] If we were to collected many such samples and create 95% confidence intervals for each, then about 95% of these intervals would contain the population difference, $\mu_w - \mu_m$.

[5] The only thing that changes is $z^{\star}$: we use $z^{\star} = 2.58$ for a 99% confidence level. If the selection of $z^{\star}$ is confusing, see Section 4.2.4 for an explanation.

	fAge	mAge	weeks	weight	sexBaby	smoke
1	NA	13	37	5.00	female	nonsmoker
2	NA	14	36	5.88	female	nonsmoker
3	19	15	41	8.13	male	smoker
⋮	⋮	⋮	⋮	⋮	⋮	
150	45	50	36	9.25	female	nonsmoker

Table 5.7: Four cases from the `babySmoke` data set. An observation listed as "NA" means that particular piece of data is missing.

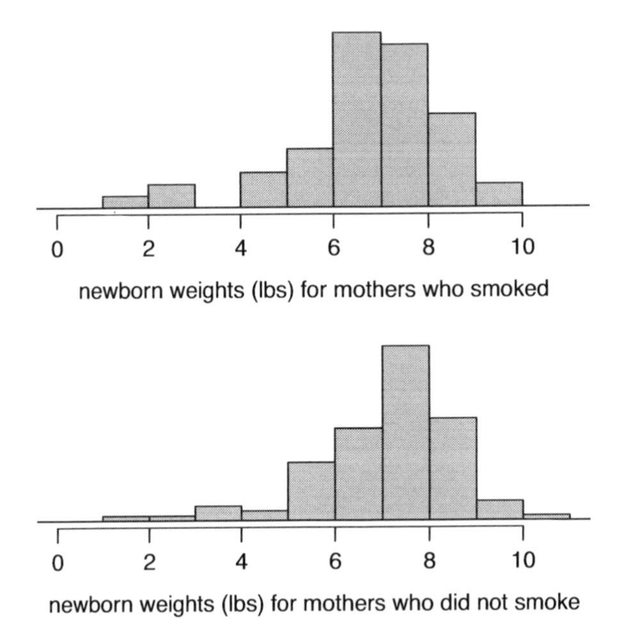

Figure 5.8: The top panel represents birth weights for infants whose mothers smoked. The bottom panel represents the birth weights for infants whose mothers who did not smoke.

● **Example 5.7** Set up appropriate hypotheses to evaluate whether there is a relationship between a mother smoking and average birth weight.

The null hypothesis represents the case of no difference between the groups.

H_0: There is no difference in average birth weight for newborns from mothers who did and did not smoke. In statistical notation: $\mu_n - \mu_s = 0$, where μ_n represents non-smoking mothers and μ_s represents mothers who smoked.

H_A: there is some difference in average newborn weights from mothers who did and did not smoke ($\mu_n - \mu_s \neq 0$).

Summary statistics are shown for each sample in Table 5.9. Because each sample is simple random and consists of less than 10% of all such cases, the observations are independent. Additionally, each sample size is at least 50 and neither sample distribution is strongly skewed (see Figure 5.8), so both sample means are associated with a nearly normal distribution.

	smoker	nonsmoker
mean	6.78	7.18
st. dev.	1.43	1.60
samp. size	50	100

Table 5.9: Summary statistics for the `babySmoke` data set.

⊙ **Exercise 5.8** (a) What is the point estimate of the population difference, $\mu_n - \mu_s$? (b) Can we use a normal distribution to model this difference? (c) Compute the standard error of the point estimate from part (a). Answers in the footnote[6].

● **Example 5.9** If the null hypothesis was true, what would be the expected value of the point estimate from Exercise 5.8? And the standard deviation of this estimate? Draw a picture to represent the p-value.

If the null hypothesis was true, then we expect to see a difference near 0. The standard error corresponds to the standard deviation of the point estimate: 0.26. To depict the p-value, we draw the distribution of the point estimate as though H_0 was true and shade areas representing at least as much evidence against H_0 as what was observed. Both tails are shaded because it is a two-sided test.

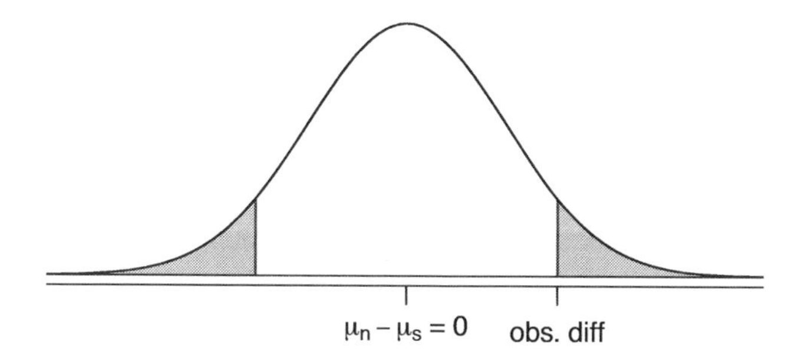

● **Example 5.10** Compute the p-value of the hypothesis test using the figure in Example 5.9 and evaluate the hypotheses using a significance level of $\alpha = 0.05$.

Since the point estimate is nearly normal, we can find the upper tail using the Z score and normal probability table:

$$Z = \frac{0.40 - 0}{0.26} = 1.54 \quad \rightarrow \quad \text{upper tail} = 1 - 0.938 = 0.062$$

[6](a) The difference in sample means is an appropriate point estimate: $\bar{x}_n - \bar{x}_s = 0.40$. (b) Because the samples are independent and each sample mean is nearly normal, their difference is also nearly normal. (c) The standard error of the estimate can be estimated using Equation (5.4):

$$SE = \sqrt{\frac{\sigma_n^2}{n_n} + \frac{\sigma_s^2}{n_s}} \approx \sqrt{\frac{s_n^2}{n_n} + \frac{s_s^2}{n_s}} = \sqrt{\frac{1.60^2}{100} + \frac{1.43^2}{50}} = 0.26$$

The sample standard deviations can be used because we have large sample sizes and neither distribution is too strongly skewed.

Because this is a two-sided test and we want the area of both tails, we double this single tail to get the p-value: 0.124. This p-value is larger than the significance value, 0.05, so we fail to reject the null hypothesis. There is insufficient evidence to say there is a difference in average birth weight of newborns from mothers who did smoke during pregnancy and newborns from mothers who did not smoke during pregnancy.

⊙ **Exercise 5.11** Does the conclusion to Example 5.10 mean that smoking and average birth weight are unrelated? Answer in the footnote[7].

⊙ **Exercise 5.12** If we actually did make a Type 2 Error and there is a difference, what could we have done differently in data collection to be more likely to detect such a difference? Answer in the footnote[8].

5.2.4 Summary for inference of the difference of two means

When considering the difference of two means, there are two common cases: the data observed may be paired or the samples may independent. The first case was treated in Section 5.1, where the one-sample methods were applied to the differences from the paired observations. We examined the second and more complex scenario in this section.

When applying the normal model to the point estimate $\bar{x}_1 - \bar{x}_2$ (corresponding to unpaired data), it is important to verify conditions before applying the inference framework using the normal model. First, each sample mean must meet the conditions for normality; these conditions are described in Chapter 4 on page 152. Secondly, the samples must be collected independently (e.g. not paired data). When these conditions are satisfied, the general inference tools of Chapter 4 may be applied.

For example, a general confidence interval takes the following form:

$$\text{point estimate} \pm z^{\star} SE$$

When estimating $\mu_1 - \mu_2$, the point estimate is the difference in sample means, the value $z^{\star}$ corresponds to the confidence level, and the standard error is computed from Equation (5.4) on page 197. While the point estimate and standard error formulas change a little, the general framework for a confidence interval stays the same. This is also true in hypothesis tests for differences of means.

In a hypothesis test, we apply the standard framework and use the specific formulas for the point estimate and standard error of a difference in two means. The test statistic represented by the Z score may be computed as

$$Z = \frac{\text{point estimate} - \text{null value}}{SE}$$

When assessing the difference in two means, the point estimate takes the form $\bar{x}_1 - \bar{x}_2$, and the standard error again takes the form of Equation (5.4) on page 197. Finally, the null value is the difference in sample means under the null hypothesis. Just as in Chapter 4, the test statistic Z is used to identify the p-value.

---

[7] Absolutely not. It is possible that there is some difference but we did not detect it. If this is the case, we made a Type 2 Error.

[8] We could have collected larger samples. If the sample size is larger, we tend to have a better shot at finding a difference if one exists.

5.2.5 Examining the standard error formula

The formula for the standard error of the difference in two means is similar to the formula for other standard errors. Recall that the standard error of a single mean, $\bar{x}_1$, can be approximated by

$$SE_{\bar{x}_1} = \frac{s_1}{\sqrt{n_1}}$$

where s_1 and n_1 represent the sample standard deviation and sample size.

The standard error of the difference of two sample means can be constructed from the standard errors of the separate sample means:

$$SE_{\bar{x}_1 - \bar{x}_2} = \sqrt{SE_{\bar{x}_1}^2 + SE_{\bar{x}_2}^2} = \sqrt{\frac{s_1^2}{n_1} + \frac{s_2^2}{n_2}} \qquad (5.13)$$

This special relationship follows from probability theory.

⊙ **Exercise 5.14** Prerequisite: Section 2.5. We can rewrite Equation (5.13) in a different way:

$$SE_{\bar{x}_1 - \bar{x}_2}^2 = SE_{\bar{x}_1}^2 + SE_{\bar{x}_2}^2$$

Explain where this formula comes from using the ideas of probability theory. Hint in the footnote[9].

5.3 Single population proportion

According to a poll taken by CNN/Opinion Research Corporation in February 2010, only about 26% of the American public trust the federal government most of the time[10]. This poll included responses of 1,023 Americans.

We will find that the sampling distribution of sample proportions, like sample means, tends to be nearly normal when the sample size is sufficiently large.

5.3.1 Identifying when a sample proportion is nearly normal

A sample proportion can be described as a sample mean. If we represent each "success" as a 1 and each "failure" as a 0, then the sample proportion is the mean of these numerical outcomes:

$$\hat{p} = \frac{0 + 0 + 1 + \cdots + 0}{1023} = 0.26$$

The distribution of $\hat{p}$ is nearly normal when the distribution of 0's and 1's is not too strongly skewed (in the case of a proportion, we mean there are almost no 0's or almost no 1's in the sample) or the sample size is sufficiently large. The most common guideline for sample size and skew when working with proportions is to ensure that we expect to observe a minimum number of successes and failures, typically at least 10 of each.

[9] The standard error squared represents the variance of the estimate. If X and Y are two random variables with variances σ_x^2 and σ_y^2, what is the variance of $X - Y$?

[10] http://www.cnn.com/2010/POLITICS/02/23/poll.government.trust/index.html

$\hat{p}$
sample proportion

p
population proportion

Conditions for the sampling distribution of $\hat{p}$ being nearly normal
The sampling distribution for $\hat{p}$, taken from a sample of size n from a population with a true proportion p, is nearly normal when

1. the sample observations are independent and
2. we expected to see at least 10 successes and 10 failures in our sample, i.e. $np \geq 10$ and $n(1-p) \geq 10$. This is called the **success-failure condition**.

If these conditions are met, then the sampling distribution of $\hat{p}$ is nearly normal with mean p and standard error

$$SE_{\hat{p}} = \sqrt{\frac{p(1-p)}{n}} \qquad (5.15)$$

Typically we do not know the true proportion, p, so must substitute some value to check conditions and to estimate the standard error. For confidence intervals, usually $\hat{p}$ is used to check the success-failure condition and compute the standard error. For hypothesis tests, typically the null value – that is, the proportion claimed in the null hypothesis – is used in place of p. Examples are presented for each of these cases in Sections 5.3.2 and 5.3.3.

TIP: Reminder on checking independence of observations
If our data come from a simple random sample and consist of less than 10% of the population, then the independence assumption is reasonable. Alternatively, if the data come from a random process, we must evaluate the independence condition more carefully.

5.3.2 Confidence intervals for a proportion

We may want a confidence interval for the proportion of Americans who trust federal officials most of the time. Our estimate, based on a sample of size $n = 1023$ from the CNN poll, is $\hat{p} = 0.26$. To use the general confidence interval formula from Section 4.5, we must check the conditions to ensure that the sampling distribution of $\hat{p}$ is nearly normal, and we must determine the standard error of the estimate.

The data are based on a simple random sample and consist of far fewer than 10% of the U.S. population, so independence is confirmed. The sample size must also be sufficiently large, which is checked via the success-failure condition: there were approximately $1023 * \hat{p} = 266$ "successes" and $1023 * (1-\hat{p}) = 757$ "failures" in the sample, both easily greater than 10.

With the conditions met, we are assured that the sampling distribution of $\hat{p}$ is nearly normal. Next, a standard error for $\hat{p}$ is needed, and then we can employ the usual method to construct a confidence interval.

⊙ **Exercise 5.16** Estimate the standard error of $\hat{p} = 0.26$ using Equation (5.15). Because p is unknown and the standard error is for a confidence interval, use $\hat{p}$ in place of p. Answer in the footnote[11].

[11] $SE = \sqrt{\frac{p(1-p)}{n}} \approx \sqrt{\frac{0.26(1-0.26)}{1023}} = 0.014$

 **Example 5.17** Construct a 95% confidence interval for p, the proportion of Americans who trust federal officials most of the time.

Using the standard error estimate from Exercise 5.16, the point estimate 0.26, and $z^\star = 1.96$ for a 95% confidence interval, the general confidence interval formula from Section 4.5 may be used:

$$\text{point estimate } \pm z^\star SE \quad \rightarrow \quad 0.26 \pm 1.96 * 0.014 \quad \rightarrow \quad (0.233, 0.287)$$

We are 95% confident that the true proportion of Americans who trusted federal officials most of the time (in February 2010) is between 0.233 and 0.287. If the proportion has not changed since this poll, not many Americans are very trusting of their federal government.

Constructing a confidence interval for a proportion

- Verify the observations are independent and also verify the success-failure condition using $\hat{p}$.
- If the conditions are met, the sampling distribution of $\hat{p}$ may be well-approximated by the normal model.
- Construct the standard error using $\hat{p}$ in place of p and apply the general confidence interval formula.

5.3.3 Hypothesis testing for a proportion

To apply the normal distribution framework to the context of a hypothesis test for a proportion, the independence and success-failure conditions must be satisfied. In a hypothesis test, the success-failure condition is checked using the null proportion: we verify np_0 and $n(1 - p_0)$ are at least 10, where p_0 is the null value.

⊙ **Exercise 5.18** Deborah Toohey is running for Congress, and her campaign manager claims she has more than 50% support from the district's electorate. Set up a hypothesis test to evaluate this (one-sided) claim. Answer in the footnote[12].

● **Example 5.19** A newspaper collects a simple random sample of 500 likely voters in the district and finds Toohey's support at 52%. Does this provide convincing evidence for the claim of Toohey's manager at the 5% significance level?

Because this is a simple random sample that includes fewer than 10% of the population, the observations are independent. In a one-proportion hypothesis test, the success-failure condition is checked using the null proportion, $p_0 = 0.5$: $np_0 = n(1 - p_0) = 500 * 0.5 = 250 > 10$. With these conditions verified, the normal model may be applied to $\hat{p}$.

Next the standard error can be computed. Again the null value is used because this is a hypothesis test for a single proportion.

$$SE = \sqrt{\frac{p_0 * (1 - p_0)}{n}} = \sqrt{\frac{0.5 * (1 - 0.5)}{500}} = 0.022$$

[12] Is there convincing evidence that the campaign manager is correct? $H_0 : p = 0.50$, $H_A : p > 0.50$.

A picture of the normal model is shown in Figure 5.10 with the p-value represented. Based on the normal model, the test statistic can be computed as the Z score of the point estimate:

$$Z = \frac{\text{point estimate} - \text{null value}}{SE} = \frac{0.52 - 0.50}{0.022} = 0.89$$

The upper tail area, representing the p-value, is 0.186. Because the p-value is larger than 0.05, we failed to reject the null hypothesis. We did not find convincing evidence to support the campaign manager's claim.

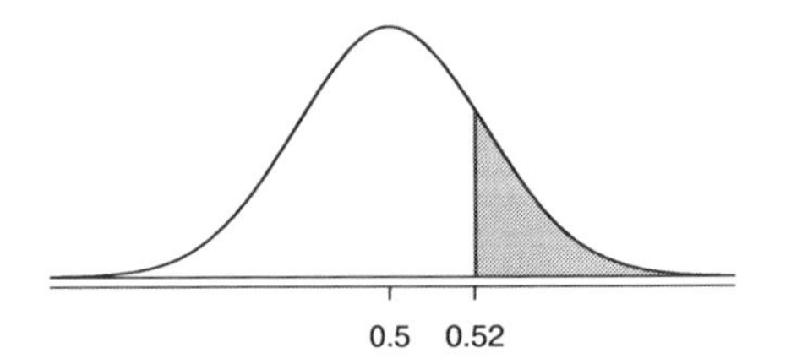

Figure 5.10: Sampling distribution of the sample proportion if the null hypothesis were true. The p-value for the test is shaded.

Hypothesis test for a proportion
Set up hypotheses and verify the conditions using the null value, p_0, to ensure $\hat{p}$ is nearly normal under H_0. If the conditions hold, construct the standard error, again using p_0, and show the p-value in a drawing. Lastly, compute the p-value and evaluate the hypotheses.

5.3.4 Choosing a sample size when estimating a proportion

We first encountered sample size computations in Section 4.6, which considered the case of estimating a single mean. We found that these computations were helpful in planning a study to control the uncertainty of a point estimate. The task was to find a sample size n so that the sample mean would be within some margin of error m of the true value with a certain level of confidence. For example, the margin of error for a point estimate using 95% confidence is $1.96 * SE$, so we had set up an equation to represent the problem:

$$ME = z^{\star} SE \le m$$

where ME represented the margin of error and $z^{\star}$ was chosen to correspond to the confidence level. This problem also arises when the point estimate is a proportion instead of a mean. For instance, if we are conducting a political survey to find the proportion of folks who approve of the job the president is doing, how big of a sample do we need in order to be sure the margin of error is less than 0.04 using a 95% confidence level?

● **Example 5.20** Find the smallest sample size n so that the margin of error of the point estimate $\hat{p}$ will be no larger than $m = 0.04$.

This expression can be written out mathematically as follows:

$$ME = z^{\star} SE = 1.96 * \sqrt{\frac{p(1-p)}{n}} \le 0.04$$

There are two unknowns: p and n. If we have an estimate of p – perhaps from a recent poll – we could use that value. If we have no such estimate, we must use some other value for p. It turns out that the margin of error is largest when p is 0.5, so we typically use this *worst case estimate* if no other estimate is available:

$$1.96 * \sqrt{\frac{0.5(1-0.5)}{n}} \leq 0.04$$
$$1.96^2 * \frac{0.5(1-0.5)}{n} \leq 0.04^2$$
$$600.25 = 1.96^2 * \frac{0.5(1-0.5)}{0.04^2} \leq n$$

We would need at least 600.25 participants – i.e. 601 participants – to ensure the sample proportion is within 0.04 of the true proportion with 95% confidence.

No estimate of the true proportion is required in sample size computations for a proportion, whereas an estimate of the standard deviation is always needed when computing a sample size for an error of a mean. However, if we have an estimate of the proportion, we should use it in place of the worst case estimate of the proportion, 0.5.

⊙ **Exercise 5.21** A recent estimate of the President's approval rating was 47%[13]. What sample size does this estimate suggest we should use for a margin of error of 0.04 with 95% confidence? Answer in the footnote[14].

⊙ **Exercise 5.22** A manager is about to oversee the mass production of a new tire model in her factory, and she would like to estimate what proportion of these tires will be rejected through quality control. The quality control team has monitored the last three tire models produced by the factory, failing 1.7% of tires in the first model, 6.2% of the second model, and 1.3% of the third model. The manager would like to examine enough tires to estimate the failure rate of the new tire model to within about 2% with a 90% confidence level. Answers for parts (a) and (b) below are in the footnote[15].

(a) This manufacturer wants to use 90% confidence. Compared to a 95% confidence level how will this change any sample size computations we complete?
(b) There are three different failure rates to choose from. Perform the sample size computation for each separately, and identify three sample sizes to consider.
(c) The sample sizes in (b) vary widely. Which of the three would you suggest using? What would influence your choice?

---

[13] Gallup Poll for President Obama's approval rating, June 13-19, 2011.

[14] We complete the same computations as before, except now we use 0.47 instead of 0.5 for p:

$$1.96 * \sqrt{\frac{p(1-p)}{n}} \approx 1.96 * \sqrt{\frac{0.47(1-0.47)}{n}} \leq 0.04 \quad \rightarrow \quad n \geq 598.1$$

A sample size of 599 or more would be reasonable.

[15] (a) We choose $z^\star$ to correspond to 90% confidence instead of 95% confidence. That is, we use 1.65 in place of 1.96.

(b) For the 1.7% estimate of p, we estimate the appropriate sample size as follows:

$$1.65 * \sqrt{\frac{p(1-p)}{n}} \approx 1.65 * \sqrt{\frac{0.017(1-0.017)}{n}} \leq 0.02 \quad \rightarrow \quad n \geq 113.7$$

Using the estimate from the first model, we would suggest examining 114 tires (round up!). A similar computation can be accomplished using 0.062 and 0.013 for p: 396 and 88.

5.4 Difference of two proportions

We would like to make conclusions about the difference in two population proportions: $p_1 - p_2$. We consider three examples. In the first, we construct a confidence interval for the difference in support for healthcare reform between Democrats and Republicans. In the second application, a company weighs whether they should switch to a higher quality parts manufacturer. In the last example, we examine the cancer risk to dogs from the use of yard herbicides.

In our investigations, we first identify a reasonable point estimate of $p_1 - p_2$ based on the sample. You may have already guessed its form: $\hat{p}_1 - \hat{p}_2$. Next, we verify that the point estimate follows the normal model by checking conditions. Finally, we compute the estimate's standard error and apply our inferential framework, just as we have done in many cases before.

5.4.1 Sampling distribution of the difference of two proportions

We check two conditions before applying the normal model to $\hat{p}_1 - \hat{p}_2$. First, the sampling distribution for each sample proportion must be nearly normal, and secondly, the samples must be independent. Under these two conditions, the sampling distribution of $\hat{p}_1 - \hat{p}_2$ may be well approximated using the normal model.

Conditions for the sampling distribution of $\hat{p}_1 - \hat{p}_2$ to be normal
The difference in two sample proportions, $\hat{p}_1 - \hat{p}_2$, tends to follow a normal model when

- the sampling distribution for each proportion separately follows a normal model and
- the samples are independent.

The standard error of the difference in sample proportions is

$$SE_{\hat{p}_1 - \hat{p}_2} = \sqrt{SE_{\hat{p}_1}^2 + SE_{\hat{p}_2}^2} = \sqrt{\frac{p_1(1-p_1)}{n_1} + \frac{p_2(1-p_2)}{n_2}} \tag{5.23}$$

where p_1 and p_2 represent the population proportions, and n_1 and n_2 represent the sample sizes.

For the difference in two means, the standard error formula took the following form:

$$SE_{\bar{x}_1 - \bar{x}_2} = \sqrt{SE_{\bar{x}_1}^2 + SE_{\bar{x}_2}^2}$$

The standard error for the difference in two proportions takes a similar form. The reasons behind this similarity are rooted in the probability theory of Section 2.5.

5.4.2 Intervals and tests for $p_1 - p_2$

Just as with the case of a single proportion, the sample proportions are used to verify the success-failure condition and also compute standard error when constructing confidence intervals.

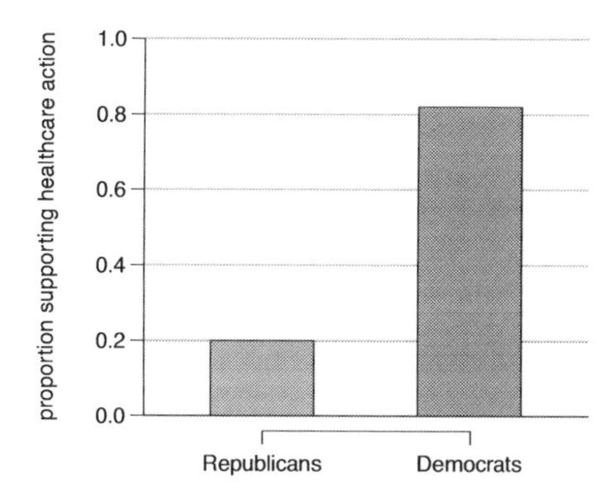

Figure 5.11: Support for Congressional action on Healthcare, by political party (Gallup poll, January 2010).

Example 5.24 One way to measure how bipartisan an issue is would be to compare the issue's support by Democrats and Republicans. In a January 2010 Gallup poll, 82% of Democrats supported a vote on healthcare in 2010 while only 20% of Republicans support such action[16]. This support is summarized in Figure 5.11. The sample sizes were 325 and 172 for Democrats and Republicans, respectively. Create and interpret a 90% confidence interval of the difference in support between the two parties.

First the conditions must be verified. Because this is a random sample from less than 10% of the population, the observations are independent, both within the samples and between the samples. The success-failure condition also holds using the sample proportions (for each sample)[17]. Because our conditions are met, the normal model can be used for the point estimate of the difference in support:

$$\hat{p}_D - \hat{p}_R = 0.82 - 0.20 = 0.62$$

The standard error may be computed using Equation (5.23) with each sample proportion:

$$SE \approx \sqrt{\frac{0.82(1-0.82)}{325} + \frac{0.20(1-0.20)}{172}} = 0.037$$

For a 90% confidence interval, we use $z^\star = 1.65$:

$$\text{point estimate} \pm z^\star SE \quad \rightarrow \quad 0.62 \pm 1.65 * 0.037 \quad \rightarrow \quad (0.56, 0.68)$$

We are 90% confident that the difference in support for healthcare action between the two parties is between 56% and 68%. Healthcare is a very partisan issue, which may not be a surprise to anyone who follows the ongoing health care debates.

[16] http://www.gallup.com/poll/125030/Healthcare-Bill-Support-Ticks-Up-Public-Divided.aspx

[17] Sometimes for the two proportion case, the success-failure threshold is lowered to 5. In this book, we will still use 10.

⊙ **Exercise 5.25** A remote control car company is considering a new manufacturer for wheel gears. The new manufacturer would be more expensive but their higher quality gears are more reliable, resulting in happier customers and fewer warranty claims. However, management must be convinced that the more expensive gears are worth the conversion before they approve the switch. If there is strong evidence that more than a 3% improvement in the percent of gears that pass inspection, management says they will switch suppliers, otherwise they will maintain the current supplier. Setup appropriate hypotheses for the test. Answer in the footnote[18].

● **Example 5.26** The quality control engineer from Exercise 5.25 collects a sample of gears, examining 1000 gears from each company and finds that 899 gears pass inspection from the current supplier and 958 pass inspection from the prospective supplier. Using these data, evaluate the hypothesis setup of Exercise 5.25 using a significance level of 5%.

First, we check the conditions. The sample is not necessarily random, so to proceed we must assume the gears are all independent; for this sample we will suppose this assumption is reasonable, but the engineer would be more knowledgeable as to whether this assumption is appropriate. The success-failure condition also holds for each sample. Thus, the difference in sample proportions, $0.958 - 0.899 = 0.059$, can be said to come from a nearly normal distribution.

The standard error can be found using Equation (5.23):

$$SE = \sqrt{\frac{0.958(1-0.958)}{1000} + \frac{0.899(1-0.899)}{1000}} = 0.0114$$

In this hypothesis test, the sample proportions were used. We will discuss this choice more in Section 5.4.3.

Next, we compute the test statistic and use it to find the p-value, which is depicted in Figure 5.12.

$$Z = \frac{\text{point estimate} - \text{null value}}{SE} = \frac{0.059 - 0.03}{0.0114} = 2.54$$

Using the normal model for this test statistic, we identify the right tail area as 0.006. Since this is a one-sided test, this single tail area is also the p-value, and we reject the null hypothesis because 0.006 is less than 0.05. That is, we have statistically significant evidence that the higher quality gears actually do pass inspection more than 3% as often as the currently used gears. Based on these results, management will approve the switch to the new supplier.

5.4.3 Hypothesis testing when $H_0: p_1 = p_2$

Here we use a new example to examine a special estimate of standard error when $H_0: p_1 = p_2$. We investigate whether there is an increased risk of cancer in dogs that are exposed to the herbicide 2,4-dichlorophenoxyacetic acid (2,4-D). A study in 1994 examined 491

---

[18] H_0: The higher quality gears will pass inspection no more than 3% more frequently than the standard quality gears. $p_{highQ} - p_{standard} = 0.03$. H_A: The higher quality gears will pass inspection more than 3% more often than the standard quality gears. $p_{highQ} - p_{standard} > 0.03$.

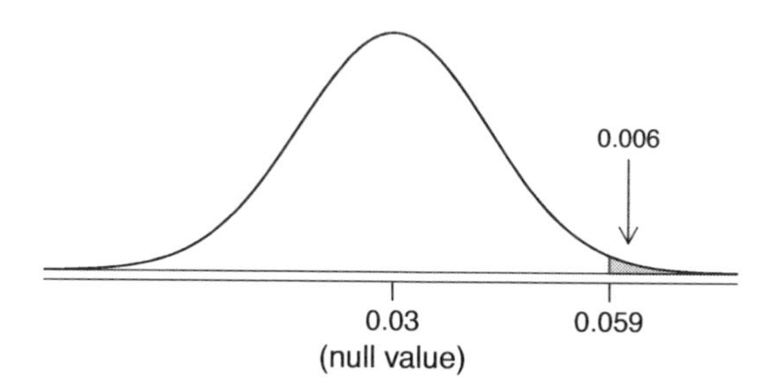

Figure 5.12: Distribution of the test statistic if the null hypothesis was true. The p-value is represented by the shaded area.

dogs that had developed cancer and 945 dogs as a control group[19]. Of these two groups, researchers identified which dogs had been exposed to 2,4-D in their owner's yard. The results are shown in Table 5.13.

	cancer	noCancer
2,4-D	191	304
no 2,4-D	300	641

Table 5.13: Summary results for cancer in dogs and the use of 2,4-D by the dog's owner.

⊙ **Exercise 5.27** Is this study an experiment or an observational study? Answer in the footnote[20].

⊙ **Exercise 5.28** Set up hypotheses to test whether 2,4-D and the occurrence of cancer in dogs are related. Use a one-sided test and compare across the cancer and no cancer groups. Comment and answer in the footnote[21].

● **Example 5.29** Are the conditions met to use the normal model and make inference on the results?

(1) It is unclear whether this is a random sample. However, if we believe the dogs in both the cancer and no cancer groups are representative of each respective population

[19] Hayes HM, Tarone RE, Cantor KP, Jessen CR, McCurnin DM, and Richardson RC. 1991. Case-Control Study of Canine Malignant Lymphoma: Positive Association With Dog Owner's Use of 2, 4-Dichlorophenoxyacetic Acid Herbicides. Journal of the National Cancer Institute 83(17):1226-1231.

[20] The owners were not instructed to apply or not apply the herbicide, so this is an observational study. This question was especially tricky because one group was called the *control group*, which is a term usually seen in experiments.

[21] Using the proportions within the `cancer` and `noCancer` groups rather than examining the rates for cancer in the `2,4-D` and `no 2,4-D` groups may seem odd; we might prefer to condition on the use of the herbicide, which is an explanatory variable in this case. However, the cancer rates in each group do not necessarily reflect the cancer rates in reality due to the way the data were collected. For this reason, computing cancer rates may greatly alarm dog owners.
H_0: the proportion of dogs with exposure to 2,4-D is the same in the `cancer` and `noCancer` groups ($p_c - p_n = 0$).
H_A: the dogs with cancer are more likely to have been exposed to 2,4-D than dogs without cancer ($p_c - p_n > 0$).

and that the dogs in the study do not interact in any way, then we may find it reasonable to assume independence between observations holds. (2) The success-failure condition holds for each sample.

Under the assumption of independence, we can use the normal model and make statements regarding the canine population based on the data.

In your hypotheses for Exercise 5.28, the null is that the proportion of dogs with exposure to 2,4-D is the same in each group. The point estimate of the difference in sample proportions is $\hat{p}_c - \hat{p}_n = 0.067$. To identify the p-value for this test, we first check conditions (Example 5.29) and compute the standard error of the difference:

$$SE = \sqrt{\frac{p_c(1-p_c)}{n_c} + \frac{p_n(1-p_n)}{n_n}}$$

In a hypothesis test, the distribution of the test statistic is always examined as though the null hypothesis was true, i.e. in this case, $p_c = p_n$. The standard error formula should reflect this equality in the null hypothesis. We will use p to represent the common rate of dogs that are exposed to 2,4-D in the two groups:

$$SE = \sqrt{\frac{p(1-p)}{n_c} + \frac{p(1-p)}{n_n}}$$

We don't know the exposure rate, p, but we can obtain a good estimate of it by *pooling* the results of both samples:

$$\hat{p} = \frac{\text{\# of ``successes''}}{\text{\# of cases}} = \frac{191 + 304}{191 + 300 + 304 + 641} = 0.345$$

This is called the **pooled estimate** of the sample proportion, and we use it to compute the standard error when the null hypothesis is that $p_1 = p_2$ (e.g. $p_c = p_n$ or $p_c - p_n = 0$). We also typically use it to verify the success-failure condition.

Pooled estimate of a proportion
When the null hypothesis is $p_1 = p_2$, it is useful to find the pooled estimate of the shared proportion:

$$\hat{p} = \frac{\text{number of ``successes''}}{\text{number of cases}} = \frac{\hat{p}_1 n_1 + \hat{p}_2 n_2}{n_1 + n_2}$$

Here $\hat{p}_1 n_1$ represents the number of successes in sample 1 since

$$\hat{p}_1 = \frac{\text{number of successes in sample 1}}{n_1}$$

Similarly, $\hat{p}_2 n_2$ represents the number of successes in sample 2.

TIP: Using the pooled estimate of a proportion when $H_0 : p_1 = p_2$
When the null hypothesis suggests the proportions are equal, we use the pooled proportion estimate ($\hat{p}$) to verify the success-failure condition and also to estimate the standard error:

$$SE = \sqrt{\frac{\hat{p}(1-\hat{p})}{n_c} + \frac{\hat{p}(1-\hat{p})}{n_n}} \qquad (5.30)$$

⊙ **Exercise 5.31** Using Equation (5.30), $\hat{p} = 0.345$, $n_1 = 491$, and $n_2 = 945$, verify the estimate for the standard error is $SE = 0.026$. Next, complete the hypothesis test using a significance level of 0.05. Be certain to draw a picture, compute the p-value, and state your conclusion in both statistical language and plain language. A short answer is provided in the footnote[22].

5.5 When to retreat

The conditions described for each statistical method ensure each test statistic follows the prescribed distribution when the null hypothesis is true. When the conditions are not met, these methods are not reliable and drawing conclusions from them is treacherous. The conditions for each test typically come in two forms.

- The individual observations must be independent. A random sample from less than 10% of the population ensures the observations are independent (not to be confused with a minimum of 10 successes and 10 failures). In experiments, other considerations must be made to ensure observations are independent. If independence fails, then advanced techniques must be used, and in some such cases, inference regarding the target population may not be possible.
- Other conditions focus on sample size and skew. For example, if the sample size is too small or the skew too strong, then the normal model for the sample mean will fail and the methods described in this chapter are not reliable.

For analyzing smaller samples, we refer to Chapter 6.

Verification of conditions for statistical tools is always necessary. Whenever conditions are not satisfied for a statistical technique, there are three options. The first is to learn new methods that are appropriate for the data. The second route is to hire a statistician[23]. The third route is to ignore the failure of conditions. This last option effectively invalidates any analysis and may discredit novel and interesting findings.

Finally, we caution that there may be no inference tools helpful when considering data that include unknown biases, such as convenience samples. For this reason, there are books, courses, and researchers devoted to the techniques of sample and experimental design. See Sections 1.5-1.7 for basic principles of sampling and data collection.

---

[22] Compute the test statistic:

$$Z = \frac{\text{point estimate} - \text{null value}}{SE} = \frac{0.067 - 0}{0.026} = 2.58$$

Looking this value up in the normal probability table: 0.9951. However this is the lower tail, and the upper tail represents the p-value: $1 - 0.9951 = 0.0049$. We reject the null hypothesis and conclude that dogs getting cancer and owners using 2,4-D are associated.

[23] If you work at a university, then there may be campus consulting services to assist you. Alternatively, there are many private consulting firms that are also available for hire.

5.6 Testing for goodness of fit using chi-square (special topic)

In this section, we develop a method for assessing a null model when the data are binned. This technique is commonly used in two circumstances:

- Given a sample of cases that can be classified into several groups, we would like to determine if the sample is representative of the general population.
- Evaluate whether data resemble a particular distribution, such as the normal distribution or a geometric distribution.

Each of these scenarios can be addressed using the same statistical test: a chi-square test.

In the first case, we consider data from a random sample of 275 jurors in a small county. Jurors identified their racial group, as shown in Table 5.14, and we would like to determine if these jurors are racially representative of the population. If the jury is representative of the population, then the proportions in the sample should roughly reflect the population of eligible jurors, i.e. registered voters.

Race	White	Black	Hispanic	Other	Total
Representation in juries	205	26	25	19	275
Registered voters	0.72	0.07	0.12	0.09	1.00

Table 5.14: Representation by race in a city's juries and population.

While the proportions in the juries do not precisely represent the population proportions, it is unclear whether these data provide convincing evidence that the sample is not representative. If the jurors really were randomly sampled from the registered voters, we might expect small differences due to chance. However, unusually large differences may provide convincing evidence that the juries were not representative.

A second application, assessing the fit of a distribution, is presented at the end of this section. Daily stock returns from the S&P500 for 1990-2009 are used to assess whether stock activity each day is independent of the stock's behavior on previous days.

In these problems, the strategy is not to assess or compare only one or two bins at a time. We would like to examine all bins simultaneously, which will require us to develop a new test statistic. While the details of the test will be new, the general ideas of hypothesis testing will be familiar.

5.6.1 Creating a test statistic for one-way tables

● **Example 5.32** Of the folks in the city, 275 served on a jury. If the individuals are randomly selected to serve on a jury, about how many of the 275 people would we expect to be white? How many would we expect to be black?

About 72% of the population is white, so we would expect about 72% of the jurors to be white: $0.72 * 275 = 198$.

Similarly, we would expect about 7% of the jurors to be black, which would correspond to about $0.07 * 275 = 19.25$ black jurors.

⊙ **Exercise 5.33** Twelve percent of the population is Hispanic and 9% represent other races. How many of the 275 jurors would we expect to be Hispanic or from another race? Answers can be found in Table 5.15.

Race	White	Black	Hispanic	Other	Total
Observed data	205	26	25	19	275
Expected counts	198	19.25	33	24.75	275

Table 5.15: Actual and expected make-up of the jurors.

The sample proportion represented from each race among the 275 jurors was not a precise match for any ethnic group. While some sampling variation is expected, we would expect the sample proportions to be fairly similar to the population proportions if there is no bias on juries. We need to test whether the differences are strong enough to provide convincing evidence that the jurors are not a random sample. These ideas can be organized into hypotheses:

H_0: The jurors are a random sample. That is, there is no racial bias in who serves on a jury, and the observed counts reflect natural sampling fluctuation.

H_A: The jurors are not randomly sampled, i.e. there is racial bias in juror selection.

To assess these hypotheses, we quantify how different the observed counts are from the expected counts. Strong evidence for the alternative hypothesis would come in the form of unusually large deviations in the groups from what would be expected based on sampling variation alone.

5.6.2 The chi-square test statistic

In previous hypothesis tests, we constructed a test statistic of the following form:

$$\frac{\text{point estimate} - \text{null value}}{\text{SE of point estimate}}$$

This construction was based on (1) identifying the difference between a point estimate and an expected value if the null hypothesis was true, and (2) standardizing that difference using the standard error of the point estimate. These two ideas will help in the construction of an appropriate test statistic for count data.

Our strategy will be to first compute the difference between the observed counts and the counts we would expect if the null hypothesis was true, then we will standardize the difference:

$$Z_1 = \frac{\text{observed white count} - \text{null white count}}{\text{SE of observed white count}}$$

The standard error for the point estimate of the count in binned data is the square root of the count under the null[24]. Therefore:

$$Z_1 = \frac{205 - 198}{\sqrt{198}} = 0.50$$

[24]Using some of the rules learned in earlier chapters, we might think that the standard error would be $np(1-p)$, where n is the sample size and p is the proportion in the population. This would be correct if we were looking only at one count. However, we are computing many standardized differences and adding them together. It can be shown – though not here – that the square root of the count is a better way to standardize the count differences.

The fraction is very similar to previous test statistics: first compute a difference, then standardize it. These computations should also be completed for the black, Hispanic, and other groups:

Black

$$Z_2 = \frac{26 - 19.25}{\sqrt{19.25}} = 1.54$$

Hispanic

$$Z_3 = \frac{25 - 33}{\sqrt{33}} = -1.39$$

Other

$$Z_4 = \frac{19 - 24.75}{\sqrt{24.75}} = -1.16$$

We would like to determine if these four standardized differences are irregularly far from zero using a single test statistic. That is, Z_1, Z_2, Z_3, and Z_4 must be combined somehow to help determine if they – as a group – tend to be unusually far from zero. A first thought might be to take the absolute value of these four standardized differences and add them up:

$$|Z_1| + |Z_2| + |Z_3| + |Z_4| = 4.58$$

Indeed, this does give one number summarizing how far the actual counts are from what was expected. However, it is more common to add the squared values:

$$Z_1^2 + Z_2^2 + Z_3^2 + Z_4^2 = 5.89$$

Squaring each standardized difference before adding them together does two things:

- Any standardized difference that is squared will now be positive.
- Differences that already looked unusual – e.g. a standardized difference of 2.5 – would become much larger after being squared.

We commonly use the test statistic X^2, which is the sum of the Z^2 values:

X^2
chi-square test statistic

$$X^2 = Z_1^2 + Z_2^2 + Z_3^2 + Z_4^2 = 5.89$$

This expression can also be written using the observed counts and null counts:

$$X^2 = \frac{(\text{observed count}_1 - \text{null count}_1)^2}{\text{null count}_1} + \cdots + \frac{(\text{observed count}_4 - \text{null count}_4)^2}{\text{null count}_4}$$

The final number X^2 summarizes how strongly the observed counts tend to deviate from the null counts. In Section 5.6.4, we will see that if the null hypothesis is true, then X^2 follows a new distribution called a *chi-square distribution*. Using this distribution, we will be able to obtain a p-value to evaluate the hypotheses.

5.6.3 The chi-square distribution and finding areas

The **chi-square distribution** is sometimes used to characterize data sets and statistics that are always positive and typically right skewed. Recall the normal distribution had two parameters – mean and standard deviation – that could be used to describe its exact characteristics. The chi-square distribution has just one parameter called **degrees of freedom (df)**, which influences the shape, center, and spread of the distribution.

⊙ **Exercise 5.34** Figure 5.16 shows three chi-square distributions. (a) How does the center of the distribution change when the degrees of freedom is larger? (b) What about the variability (spread)? (c) How does the shape change?

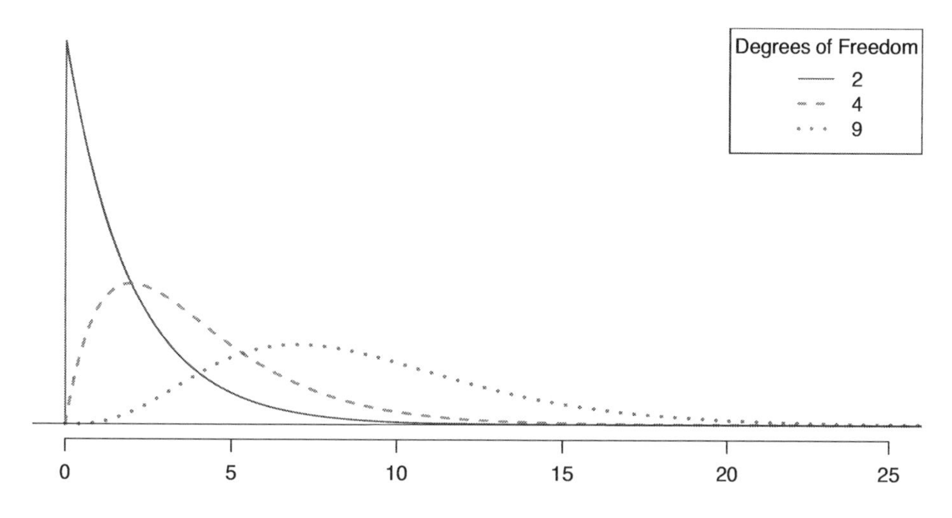

Figure 5.16: Four chi-square distributions with varying degrees of freedom.

Figure 5.16 and Exercise 5.34 demonstrate three general properties of chi-square distributions as the degrees of freedom increases: the distribution becomes more symmetric, the center moves to the right, and the variability inflates.

Our principal interest in the chi-square distribution is the calculation of p-values, which (as we have seen before) is related to finding the relevant area in the tail of a distribution. To do so, a new table is needed: the **chi-square probability table**, partially shown in Table 5.17. A more complete table is presented in Appendix C.3 on page 366. This table differs a bit from the normal probability table, in that we typically do not find the tail area very precisely. Instead, we identify a range for the area. Additionally, the chi-square probability table only provides upper tail values as opposed to the normal probability table, which shows lower tail areas.

Upper tail		0.3	0.2	0.1	0.05	0.02	0.01	0.005	0.001
df	1	1.07	1.64	2.71	3.84	5.41	6.63	7.88	10.83
	2	2.41	3.22	4.61	5.99	7.82	9.21	10.60	13.82
	3	*3.66*	*4.64*	*6.25*	*7.81*	*9.84*	*11.34*	*12.84*	*16.27*
	4	4.88	5.99	7.78	9.49	11.67	13.28	14.86	18.47
	5	6.06	7.29	9.24	11.07	13.39	15.09	16.75	20.52
	6	7.23	8.56	10.64	12.59	15.03	16.81	18.55	22.46
	7	8.38	9.80	12.02	14.07	16.62	18.48	20.28	24.32

Table 5.17: A section of the chi-square probability table. A complete table is in Appendix C.3 on page 366.

● **Example 5.35** Figure 5.18(a) shows a chi-square distribution with 3 degrees of freedom and an upper shaded tail starting at 6.25. Use Table 5.17 to estimate the shaded area.

This distribution has three degrees of freedom, so only the row with 3 degrees of freedom (df) is relevant. This row has been italicized in the table. Next, we see that the value – 6.25 – falls in the column with upper tail area 0.1. That is, the shaded upper tail of Figure 5.18(a) has area 0.1.

● **Example 5.36** We rarely observe the *exact* value in the table. For instance, Figure 5.18(b) shows the upper tail of a chi-square distribution with 2 degrees of freedom. The bound for this upper tail is at 4.3, which does not fall in Table 5.17.

The cutoff of interest – 4.3 – falls between the second and third columns in the 2 degrees of freedom row. Because these columns correspond to tail areas of 0.2 and 0.1, we can be certain that the area shaded in Figure 5.18(b) is between 0.1 and 0.2.

● **Example 5.37** Figure 5.18(c) shows an upper tail area for a chi-square distribution with 5 degrees of freedom and a cutoff of 5.1. Find the tail area.

Looking in the row with 5 df, 5.1 falls below the smallest cutoff for this row (6.06). That means we can only say that the area is *greater than 0.3*.

⊙ **Exercise 5.38** Figure 5.18(d) shows a cutoff of 11.7 on a chi-square distribution with 7 degrees of freedom. Find the area of the upper tail. Answer in the footnote[25].

⊙ **Exercise 5.39** Figure 5.18(e) shows a cutoff of 10 on a chi-square distribution with 4 degrees of freedom. Find the area of the upper tail. Short answer in the footnote[26].

⊙ **Exercise 5.40** Figure 5.18(f) shows a cutoff of 9.21 with a chi-square distribution with 3 df. Find the area of the upper tail. Short answer in the footnote[27].

5.6.4 Finding a p-value for a chi-square test

In Section 5.6.2, we identified a new test statistic (X^2) within the context of assessing whether there was evidence of racial bias in how jurors were sampled. The null hypothesis represented the claim that jurors were randomly sampled and there was no racial bias. The alternative hypothesis was that there was racial bias in how the jurors were sampled.

We determined that a large X^2 value would suggest strong evidence favoring the alternative hypothesis: that there was racial bias. However, we could not quantify what the chance was of observing such a large test statistic ($X^2 = 5.89$) if the null hypothesis actually was true. This is where the chi-square distribution becomes useful. If the null hypothesis was true and there was no racial bias, then X^2 would follow a chi-square distribution, in this case with three degrees of freedom. In general, the statistic X^2 follows a chi-square distribution with $k - 1$ degrees of freedom, where k is the number of bins.

● **Example 5.41** How many categories were there in the juror example? How many degrees of freedom should be associated with the chi-square distribution used for X^2?

In the jurors example, there were $k = 4$ categories: white, black, Hispanic, and other. According to the rule above, the test statistic X^2 should then follow a chi-square distribution with $k - 1 = 3$ degrees of freedom if H_0 is true.

Just like we checked sample size conditions to use the normal model in earlier sections, we must also check a sample size condition to safely apply the chi-square distribution for X^2. Each expected count must be at least[28] 10. In the juror example, the expected counts were 198, 19.25, 33, and 24.75, all easily above 10, so we can apply the chi-square model to the test statistic, $X^2 = 5.89$.

---

[25]The value 11.7 falls between 9.80 and 12.02 in the 7 df row. Thus, the area is between 0.1 and 0.2.

[26]The area is between 0.02 and 0.05.

[27]Between 0.02 and 0.05.

[28]Some books recommend a threshold of 5.

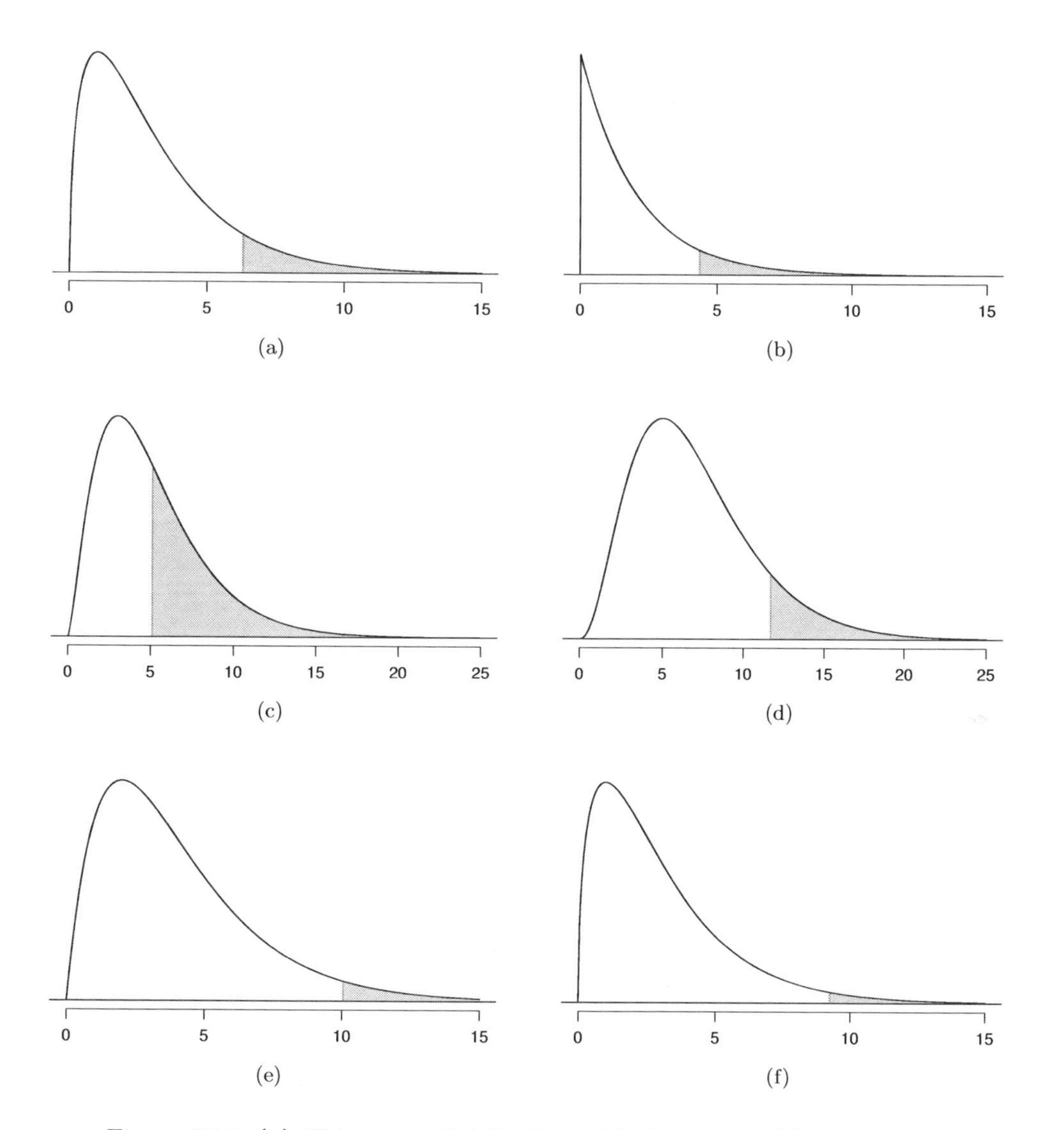

Figure 5.18: **(a)** Chi-square distribution with 3 degrees of freedom, area above 6.25 shaded. **(b)** 2 degrees of freedom, area above 4.3 shaded. **(c)** 5 degrees of freedom, area above 5.1 shaded. **(d)** 7 degrees of freedom, area above 11.7 shaded. **(e)** 4 degrees of freedom, area above 10 shaded. **(f)** 3 degrees of freedom, area above 9.21 shaded.

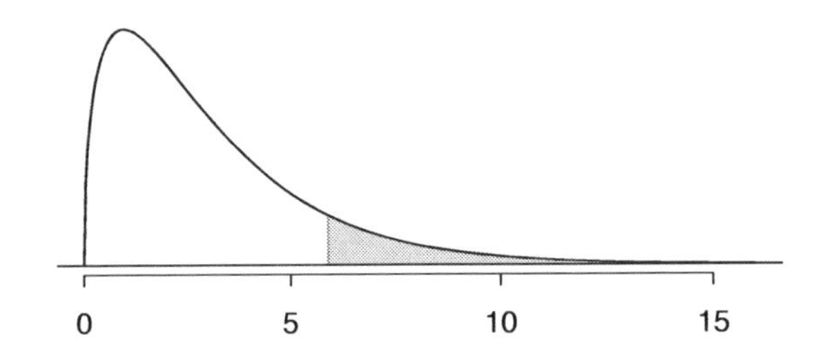

Figure 5.19: The p-value for the juror hypothesis test is shaded in the chi-square distribution with $df = 3$.

● **Example 5.42** If the null hypothesis were true, the test statistic $X^2 = 5.89$ would be closely associated with a chi-square distribution with three degrees of freedom. Using this distribution and test statistic, identify the p-value.

The chi-square distribution and p-value are shown in Figure 5.19. Because larger chi-square values correspond to stronger evidence against the null hypothesis – i.e. larger deviations from what we expect – we shade the upper tail to represent the p-value. Using the chi-square probability table in Appendix C.3 or the short table on page 215, we can determine that the area is between 0.1 and 0.2. That is, the p-value is larger than 0.1 but smaller than 0.2. Generally we do not reject the null hypothesis with such a large p-value. In other words, the data do not provide convincing evidence of racial bias in the juror selection.

Chi-square test for one-way table
Suppose we are to evaluate whether there is convincing evidence that a set of observed counts O_1, O_2, ..., O_k in k categories are unusually different from what might be expected under a null hypothesis. Call the *expected counts* that are based on the null hypothesis E_1, E_2, ..., E_k. If each expected count is at least 10 and the null hypothesis is true, then the following test statistic follows a chi-square distribution with $k - 1$ degrees of freedom:

$$X^2 = \frac{(O_1 - E_1)^2}{E_1} + \frac{(O_2 - E_2)^2}{E_2} + \cdots + \frac{(O_k - E_k)^2}{E_k}$$

The p-value for this test statistic is found by looking at the upper tail of this chi-square distribution. We consider the upper tail because larger values of X^2 would provide greater evidence against the null hypothesis.

TIP: Conditions for the chi-square test
There are two conditions that must be checked before performing a chi-square test:

Independence. Each case that contributes a count to the table must be independent of all the other cases in the table.

Sample size / distribution. Just like for proportions, each particular scenario (i.e. cell count) must have at least 10 cases.

Failing to check conditions may unintentionally affect the test's error rates.

5.6.5 Evaluating goodness of fit for a distribution

Section 3.3 would be useful background reading for this example, but it is not a prerequisite.

We can apply our new chi-square testing framework to the second problem in this section: evaluating whether a certain statistical model fits a data set. Daily stock returns from the S&P500 for 1990-2009 can be used to assess whether stock activity each day is independent of the stock's behavior on previous days. This sounds like a very complex question – and it is – but a chi-square test can be used to study the problem. We will label each day as `Up` or `Down` (`D`) depending on whether the market was up or down that day. For example, consider the following changes in price, their new labels of up and down, and then the number of days that must be observed before each `Up` day:

Change in price	2.52	-1.46	0.51	-4.07	3.36	1.10	-5.46	-1.03	-2.99	1.71
Outcome	Up	D	Up	D	Up	Up	D	D	D	Up
Days to `Up`	1	-	2	-	2	1	-	-	-	4

If the days really are independent, then the number of days until a positive trading day should follow a geometric distribution. The geometric distribution describes the probability of waiting for the kth trial to observe the first success. Here each `Up` day represents a success, and down (`D`) days represent failures. In the data above, it took only one day until the market was up, so the first wait time was 1 day. It took two more days before we observed our next `Up` trading day, and two more for the third `Up` day. We would like to determine if these counts (1, 2, 2, 1, 4, and so on) follow the geometric distribution. Table 5.20 shows the number of waiting days for a positive trading day during 1990-2009 for the S&P500.

Days	1	2	3	4	5	6	7+	Total
Observed	1298	685	367	157	77	33	20	2587

Table 5.20: Distribution of the waiting time until a positive trading day.

We consider how many days one must wait until observing an `Up` day on the S&P500 stock exchange. If the stock activity was independent from one day to the next and the probability of a positive trading day was constant, then we would expect this waiting time to follow a *geometric distribution*. We can organize this into a hypothesis framework:

H_0: Whether the stock market is up or down on a given day is independent from all other days. We will consider the number of days that pass until an `Up` day is observed. Under this hypothesis, the number of days until an `Up` day should follow a geometric distribution.

H_A: The days are not independent. Since we know the number of days until an `Up` day would follow a geometric distribution under the null, we look for deviations from the geometric distribution, which would support the alternative hypothesis.

There are important implications in our result for stock traders: if information from past trading days is useful in telling what will happen today, that may provide an edge over other traders.

We consider data for the S&P500 from 1990 to 2009 and summarize the waiting times in Table 5.21 and Figure 5.22. The S&P500 was positive on 52.9% of those days.

Because applying the chi-square framework requires expected counts to be at least 10, we have *binned* together all the cases where the waiting time was at least 7 days. The actual data, shown in the *Observed* row in Table 5.21, can be compared to the expected

Days	1	2	3	4	5	6	7+	Total
Observed	1298	685	367	157	77	33	20	2587
Geom. Model	1368	644	304	143	67	32	28	2587

Table 5.21: Distribution of the waiting time until a positive trading day. The expected counts based on the geometric model are shown in the last row. To find each expected count, we identify the probability of waiting D days based on the geometric model ($P(D) = (1 - 0.529)^{D-1}(0.529)$) and multiply by the total number of streaks, 2587. For example, waiting for three days occurs under the geometric model about $0.471^2 * 0.529 = 11.7\%$ of the time, which corresponds to $0.117 * 2587 = 304$ streaks.

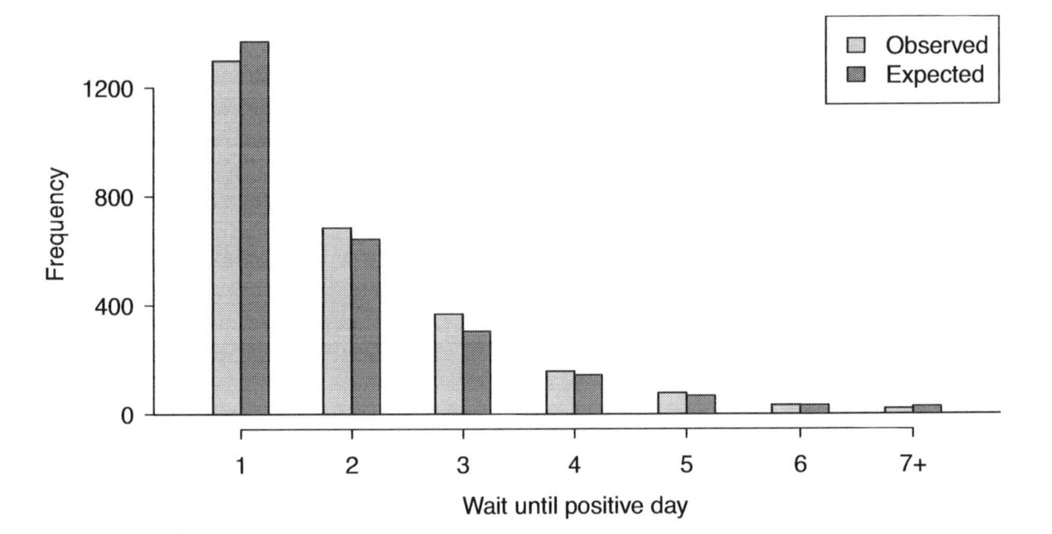

Figure 5.22: Side-by-side bar plot of the observed and expected counts for each waiting time.

counts from the *Geom. Model* row. The method for computing expected counts is discussed in Table 5.21. In general, the expected counts are determined by (1) identifying the null proportion associated with each bin, then (2) multiplying each null proportion by the total count to obtain the expected counts. That is, this strategy identifies what proportion of the total count we would expect to be in each bin.

⊙ **Exercise 5.43** Do you notice any unusually large deviations in the graph? Can you tell if these deviations are due to chance just by looking?

It is not obvious whether differences in the observed counts and the expected counts from the geometric distribution are significantly different. That is, it is not clear whether these deviations might be due to chance or whether they are so strong that the data provide convincing evidence against the null hypothesis. However, we can perform a chi-square test using the counts in Table 5.21.

⊙ **Exercise 5.44** Table 5.21 provides a set of count data for waiting times ($O_1 = 1298$, $O_2 = 685$, ...) and expected counts under the geometric distribution ($E_1 = 1368$, $E_2 = 644$, ...). Compute the chi-square test statistic, X^2. Answer in the footnote[29].

⊙ **Exercise 5.45** Because the expected counts are all at least 10, we can safely apply the chi-square distribution to X^2. However, how many degrees of freedom should we use? Hint: How many groups are there? Solution in the footnote[30].

● **Example 5.46** If the observed counts follow the geometric model, then the chi-square test statistic $X^2 = 24.43$ would closely follow a chi-square distribution with $df = 6$. Using this information, compute a p-value.

Figure 5.23 shows the chi-square distribution, cutoff, and the shaded p-value. If we look up the statistic $X^2 = 24.43$ in Appendix C.3, we find that the p-value is less than 0.001. In other words, we have very strong evidence to reject the notion that the wait times follow a geometric distribution, i.e. trading days are not independent and past days may help predict what the stock market will do today.

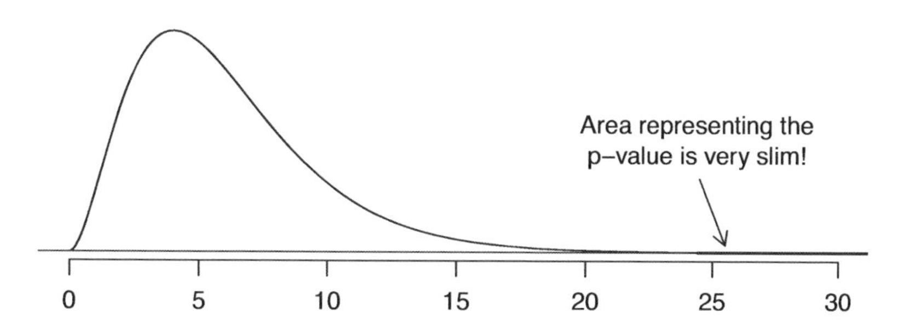

Figure 5.23: Chi-square distribution with 6 degrees of freedom. The p-value for the stock analysis is shaded.

[29] $X^2 = \frac{(1298-1368)^2}{1368} + \frac{(685-644)^2}{644} + \cdots + \frac{(20-28)^2}{28} = 24.43$

[30] There are $k = 7$ groups, so we use $df = k - 1 = 6$.

● **Example 5.47** In Example 5.46, we rejected the null hypothesis that the trading days are independent. Why is this so important?

Because the data provided strong evidence that the geometric distribution is not appropriate, we reject the claim that trading days are independent. While it is not obvious how to exploit this information, it suggests there are some hidden patterns in the data that could be interesting and possibly useful to a stock trader.

5.7 Testing for independence in two-way tables (special topic)

Google is constantly running experiments to test new search algorithms. For example, Google might test three algorithms using a sample of 10,000 google.com search queries. Table 5.24 shows the 10,000 queries split into three algorithm groups[31]. The group sizes were specified before the start of the experiment to be 5000 for the current algorithm and 2500 for each test algorithm.

Search algorithm	current	test 1	test 2	Total
Counts	5000	2500	2500	10000

Table 5.24: Google experiment breakdown of test subjects into three search groups.

● **Example 5.48** What is the ultimate goal of the Google experiment? What are the null and alternative hypotheses, in regular words?

The ultimate goal is to see whether there is a difference in the performance of the algorithms. The hypotheses can be described as the following:

H_0: The algorithms each perform equally well.

H_A: The algorithms do not perform equally well.

In this experiment, the explanatory variable is the search algorithm. However, an outcome variable is also needed. This outcome variable should somehow reflect whether the search results align with the user's interests. One possible way to quantify this is to determine whether (1) the user clicked one of the links provided and did not try a new search, or (2) the user performed a related search. Under scenario (1), we might think that the user was satisfied with the search results. Under scenario (2), the search results probably were not relevant, so the user tried a second search.

Table 5.25 provides the results from the experiment. These data are very similar to the count data in Section 5.6. However, now the different combinations of two variables are binned in a *two-way* table. In examining these data, we want to evaluate whether there is strong evidence that at least one algorithm is performing better than the others. To do so, we apply the chi-square test to this two-way table. The ideas of this test are similar to those ideas in the one-way table case. However, degrees of freedom and expected counts are computed a little differently than before.

---

[31]Google regularly runs experiments in this manner to help improve their search engine. It is entirely possible that if you perform a search and so does your friend, that you will have different search results. While the data presented in this section resemble what might be encountered in a real experiment, these data are simulated.

Search algorithm	current	test 1	test 2	Total
No new search	3511	1749	1818	7078
New search	1489	751	682	2922
Total	5000	2500	2500	10000

Table 5.25: Results of the Google search algorithm experiment.

What is so different about one-way tables and two-way tables?
A one-way table describes counts for each outcome in a single variable. A two-way table describes counts for *combinations* of outcomes for two variables. When we consider a two-way table, we often would like to know, are these variables related in any way? That is, are they dependent (versus independent)?

The hypothesis test for this Google experiment is really about assessing whether there is statistically significant evidence that the choice of the algorithm affects whether a user performs a second search. In other words, the goal is to check whether the `search` variable is independent of the `algorithm` variable.

5.7.1 Expected counts in two-way tables

● **Example 5.49** From the experiment, we estimate the proportion of users who were satisfied with their initial search (no new search) as $7078/10000 = 0.7078$. If there really is no difference among the algorithms and 70.78% of people are satisfied with the search results, how many of the 5000 folks in the current algorithm group would be expected to not perform a new search?

About 70.78% of the 5000 would be satisfied with the initial search:

$$0.7078 * 5000 = 3539 \text{ users}$$

That is, if there was no difference between the three groups, then we would expect 3539 of the current algorithm users not to perform a new search.

⊙ **Exercise 5.50** Using the same rationale described in Example 5.49, about how many users in each of the test groups would not perform a new search if the algorithms were equally helpful? Short answer in the footnote[32].

● **Example 5.51** If 3539 of the 5000 users who are given the current algorithm are not expected to perform a new search, how many would we expect to perform a new search?

We would expect the other 1461 users in the "current" group to perform a new search.

⊙ **Exercise 5.52** If 1769.5 of the 2500 users in each test group are not expected to perform a new search, how many would be expected to perform a new search in each of these groups? Answer in the footnote[33].

---

[32] About 1769.5. It is okay that this is a fraction.
[33] $2500 - 1769.5 = 730.5$.

The expected counts from Examples 5.49 and 5.51 and Exercises 5.50 and 5.52 were used to construct Table 5.26. This is the same as Table 5.25, except now the expected counts have been added in parentheses.

Search algorithm	current		test 1		test 2		Total
No new search	3511	(3539)	1749	(1769.5)	1818	(1769.5)	7078
New search	1489	(1461)	751	(730.5)	682	(730.5)	2922
Total	5000		2500		2500		10000

Table 5.26: The observed counts and the (expected counts).

The examples and exercises above provided some help in computing expected counts. In general, expected counts for a two-way table may be computed using only the row totals, column totals, and the table total. For instance, if there was no difference between the groups, then about 70.78% of each column should be in the first row:

$$\begin{aligned} 0.7078 * (\text{column 1 total}) &= 3539 \\ 0.7078 * (\text{column 2 total}) &= 1769.5 \\ 0.7078 * (\text{column 3 total}) &= 1769.5 \end{aligned}$$

Looking back to how the fraction 0.7078 was computed – as the fraction of users who did not perform a new search (7078/10000) – these three expected counts could have been computed as

$$\begin{aligned} \left(\frac{\text{row 1 total}}{\text{table total}}\right)(\text{column 1 total}) &= 3539 \\ \left(\frac{\text{row 1 total}}{\text{table total}}\right)(\text{column 2 total}) &= 1769.5 \\ \left(\frac{\text{row 1 total}}{\text{table total}}\right)(\text{column 3 total}) &= 1769.5 \end{aligned}$$

This leads us to a general formula for computing expected counts in a two-way table when we would like to test whether there is strong evidence of an association between the column variable and row variable.

Computing expected counts in a two-way table
To identify the expected count for the ith row and j^{th} column, compute

$$\text{Expected Count}_{\text{row } i,\text{ col } j} = \frac{(\text{row } i \text{ total}) * (\text{column } j \text{ total})}{\text{table total}}$$

5.7.2 The chi-square test statistic for two-way tables

The chi-square test statistic for a two-way table is found the same way it is found for a one-way table. For each table count, compute

General formula	$\dfrac{(\text{observed count} - \text{expected count})^2}{\text{expected count}}$
Row 1, Col 1	$\dfrac{(3511 - 3539)^2}{3539} = 0.222$
Row 1, Col 2	$\dfrac{(1749 - 1769.5)^2}{1769.5} = 0.237$
$\vdots$	$\vdots$
Row 2, Col 3	$\dfrac{(682 - 730.5)^2}{730.5} = 3.220$

Adding the computed value for each cell gives the chi-square test statistic X^2:

$$X^2 = 0.222 + 0.237 + \cdots + 3.220 = 6.120$$

Just like before, this test statistic follows a chi-square distribution. However, the degrees of freedom are computed a little differently for a two-way table[34]. For two way tables, the degrees of freedom is equal to

$$df = (\text{number of rows minus 1}) * (\text{number of columns minus 1})$$

In our example, the degrees of freedom parameter is

$$df = (2 - 1) * (3 - 1) = 2$$

If the null hypothesis is true (i.e. the algorithms are equally useful), then the test statistic $X^2 = 6.12$ closely follows a chi-square distribution with 2 degrees of freedom. Using this information, we can compute the p-value for the test, which is depicted in Figure 5.27.

Computing degrees of freedom for a two-way table
When applying the chi-square test to a two-way table, we use

$$df = (R - 1) * (C - 1)$$

where R is the number of rows in the table and C is the number of columns.

● **Example 5.53** Compute the p-value and draw a conclusion about whether the search algorithms have different performances.

Looking in Appendix C.3 on page 366, we examine the row corresponding to 2 degrees of freedom. The test statistic, $X^2 = 6.120$, falls between the fourth and fifth columns, which means the p-value is between 0.02 and 0.05. Because we typically test at a significance level of $\alpha = 0.05$ and the p-value is less than 0.05, the null hypothesis is rejected. That is, the data provide convincing evidence that there is some difference in performance among the algorithms.

[34] Recall: in the one-way table, the degrees of freedom was the number of cells minus 1.

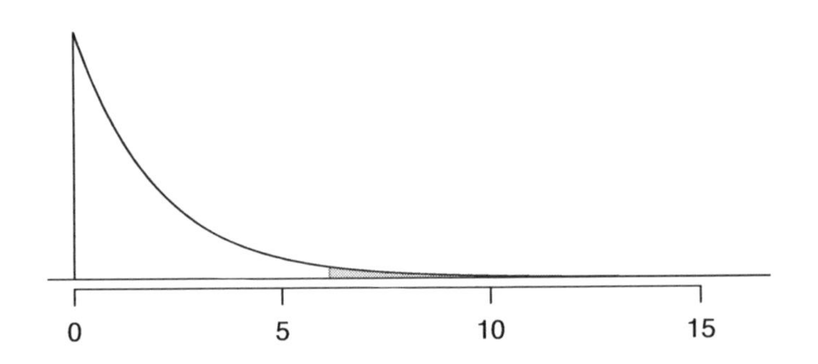

Figure 5.27: Computing the p-value for the Google hypothesis test.

	Obama	Dem. leaders	Rep. leaders	Total
Approve	683	465	375	1523
Disapprove	639	855	682	2379
Total	1322	1320	1260	3902

Table 5.28: Pew Research poll results of a March 2010 poll.

● **Example 5.54** Table 5.28 summarizes the results of a Pew Research poll[35]. We would like to determine if there are actually differences in the approval ratings of Barack Obama, the Democratic leaders, and the Republican leaders. What are appropriate hypotheses for such a test?

H_0: There is no difference in approval ratings between the three groups.

H_A: There is some difference in approval ratings between the three groups, e.g. perhaps Obama's approval differs from other Democratic leaders.

⊙ **Exercise 5.55** A chi-square test for a two-way table may be used to test the hypotheses in Example 5.54. As a first step, compute the expected values for each of the six table cells. The computations for the expected counts in the cells of the first column are shown in the footnote[36].

⊙ **Exercise 5.56** Compute the chi-square test statistic. Solution in the footnote[37].

⊙ **Exercise 5.57** Because there are 2 rows and 3 columns, the degrees of freedom for the test is $df = (2-1)*(3-1) = 2$. Use $X^2 = 142.2$, $df = 2$, and the chi-square table (page 366) to evaluate whether to reject the null hypothesis. Answer in the footnote[38].

[35] See the Pew Research website: http://people-press.org/report/598/healthcare-reform

[36] The expected count for row one / column one is found by multiplying the row one total (1523) and column one total (1322), then dividing by the table total (3902): $\frac{1523*1322}{3902} = 516.0$. Similarly for the first column and the second row: $\frac{2379*1322}{3902} = 806.0$.

[37] For each cell, compute $\frac{(\text{obs}-\text{exp})^2}{exp}$. For instance, the first row and first column: $\frac{(683-516)^2}{516} = 54.0$. Adding the results of each cell gives the chi-square test statistic: $X^2 = 54.0 + \cdots + 17.8 = 142.2$.

[38] The test statistic is larger than the right-most column of the $df = 2$ row of the chi-square probability table, meaning the p-value is less than 0.001. That is, we reject the null hypothesis because the p-value is less than 0.05, and we conclude that Americans' approval has differences among the party leaders and the president.

5.8 Exercises

5.8.1 Paired data

5.1 Is there strong evidence that the continental U.S. is warming? We might take a simple approach to this problem and compare how temperatures have changed from 1968 to 2008. The daily high temperature reading on January 1 was collected for 1968 and 2008 for 51 randomly selected locations in the continental U.S. Then the difference between the two readings (temperature in 2008 - temperature in 1968) was calculated for each of the 51 different locations. The average of these 51 values was 1.1 degrees with a standard deviation of 4.9 degrees. We are interested in finding whether these data provide strong evidence of temperature warming in the continental U.S.

(a) Are the two data sets of 51 observations dependent or independent of each other? Based on this, what type of analysis should be conducted to test if these data provide strong evidence of temperature warming in the continental U.S?

(b) Write the hypotheses in symbols and in words.

(c) Calculate the test statistic and find the p-value. (Reminder: check assumptions and conditions.)

(d) What do you conclude? Interpret your conclusion in context.

(e) What type of error might we have made? Explain in context what the error means.

(f) Based on the result of this hypothesis test, would you expect a confidence interval for the average difference between the temperature measurements from 1968 and 2008 to include 0? Explain your reasoning.

5.2 During Exercise 5.1, we considered the differences between the temperature readings in January 1 of 1968 and 2008 at 51 locations in the continental U.S.

(a) Calculate a 90% confidence interval for the average difference between the temperature measurements between 1968 and 2008.

(b) Interpret this interval in context.

(c) Does this interval agree with the conclusion of the hypothesis test from Exercise 5.1? Explain.

5.8.2 Difference of two means

5.3 In 1964 the Office of the Surgeon General released their first report linking smoking to various health issues, including cancer. Research done by an ad agency surveyed the number of cigarettes smoked by 80 smokers the day before the Surgeon Generals report came out. The sample average was 13.5 and the standard deviation was 3.2. A year after the report was released, in a random sample of 85 smokers, the average number of cigarettes smoked per day was 12.6 with a standard deviation of 2.9. Is there strong evidence that the average number of cigarettes smoked per day decreased after the Surgeon General's report?

	Before	After
n	80	85
$\bar{x}$	13.5	12.6
s	3.2	2.9

(a) Write the hypotheses in symbols and in words.

(b) Calculate the test statistic and find the p-value. (Reminder: check assumptions and conditions.)

(c) What do you conclude? Interpret your conclusion in context.

(d) Does this imply that the Surgeon General's report was the cause of this decrease? Explain.

(e) What type of error might we have made? Explain.

5.4 Based on the data given in Exercise 5.3, construct a 90% confidence interval for the difference between the average number of cigarettes smoked per day before and after the Surgeon General's report was released. Interpret this interval in context. Also comment on if the confidence interval agrees with the conclusion of the hypothesis test from Exercise 5.3.

5.5 The National Assessment of Educational Progress tested 13 year old students in reading in 2004 and 2008. A random sample of 1,000 students who took the test in 2004 yielded an average score of 257 with a standard deviation of 39. A random sample of 1,000 students who took the test in 2008 yielded an average score of 260 with a standard deviation of 38. Construct a 90% confidence interval for the difference between the average scores in 2004 and 2008. Interpret this interval in context. (Reminder: check assumptions and conditions.) [23]

5.6 Exercise 5.5 provides data on the average math scores from tests conducted by the National Assessment of Educational Progress in 2004 and 2008.

(a) Do these data provide strong information that the average reading score for 13 year old students has changed between 2004 and 2008? Use a 10% significance level.

(b) What type of error might we have made? Explain.

(c) Does the conclusion of your hypothesis test agree with the confidence interval from Exercise 5.5?

5.7 Women are said to be obese if they have greater than 35% body fat, while the cutoff for men being termed obese is 25%. The cutoff for women is higher since women store extra fat in their bodies to be used during childbearing. The third National Health and Nutrition Examination Survey collected body fat percentage (BF) and lean mass data from 13,601 subjects ages 20 to 79.9. A summary table for these data is given below. Note that BF and lean mass are given as *mean* $\pm$ *standard error*. Test the hypothesis that women on average have a higher body fat percentage than men using these data. You may assume that all assumptions and conditions for inference are satisfied. [24]

Gender	n	BF (%)	Lean mass (kg)
Men	6580	23.9 ± 0.07	61.8 ± 0.12
Women	7021	35.0 ± 0.09	44.0 ± 0.08

Test the hypothesis that women have higher average body fat percentages than men using a 1% significance level.

5.8 Exercise 5.7 also provides information on the amount of lean mass of men and women who were surveyed. Calculate a 95% confidence interval for the difference between the lean mass amounts of men and women, and interpret the interval in context. Also comment on whether or not this interval suggests a significant difference between the average lean mass amounts of men and women.

5.8.3 Single population proportion

5.9 Suppose that 8% of college students are vegetarians. Determine if the following statements are true or false, and explain your reasoning.

(a) The distribution of the sample proportions of vegetarians in random samples of size 60 is nearly normal since $n \geq 50$.

(b) The distribution of the sample proportions of vegetarian college students in random samples of size 50 is right skewed.

(c) A random sample of 125 college students where 12% are vegetarians would be considered unusual.

(d) A random sample of 250 college students where 12% are vegetarians would be considered unusual.

(e) The standard error would be reduced by one-half if we increased the sample size from 125 to 250.

5.10 Suppose that the proportion of the adult population who jog is 0.15. Determine if the following statements are true or false, and explain your reasoning.

(a) The distribution of the proportions of joggers in random samples of size 40 is right skewed.

(b) The distribution of the proportions of joggers in random samples of size 80 is nearly normal since $n \geq 50$.

(c) A random sample of 150 where 20% are joggers would be considered unusual.

(d) A random sample of 300 where 20% are joggers would be considered unusual.

(e) The standard error would be reduced by one-half if we increased the sample size from 150 to 600.

5.11 Ninety percent of orange tabby cats are male. Determine if the following statements are true or false, and explain your reasoning.

(a) The distribution of sample proportions of samples of size 30 is left skewed.

(b) Doubling the sample size will reduce the standard error of the sample proportion by one-half.

(c) The distribution of sample proportions of samples of size 140 is approximately normal.

(d) Doubling the sample size will reduce the standard error of the sample proportion by a factor of $\sqrt{2}$.

(e) The distribution of sample proportions of samples of size 70 is approximately normal.

5.12 In a poll conducted by Survey USA on July 12, 2010, 70% of the 119 respondents between the ages of 18 and 34 said they would vote in the 2010 general election for Prop 19, which would change California law to legalize marijuana and allow it to be regulated and taxed. At a 95% confidence level, this sample has an 8% margin of error. Based on this information, determine if the following statements are true or false, and explain your reasoning.

(a) We are 95% confident that between 62% and 78% of the California voters in this sample support support Prop 19.

(b) We are 95% confident that between 62% and 78% of all California voters between the ages of 18 and 34 support Prop 19.

(c) If we considered many random samples of 119 California voters between the ages of 18 and 34, and we calculated the sample proportions of those who support Prop 19, 95% of them will be between 62% and 78%.

(d) In order to decrease the margin of error to 4%, we would need to quadruple (multiply by 4) the sample size.

(e) Based on this confidence interval, there is sufficient evidence to conclude that a majority of California voters between the ages of 18 and 34 support Prop 19.

5.13 We are interested in estimating the proportion of graduates at a mid-sized university who found a job within one year of completing their undergraduate degree. We conduct a survey and find out that 348 of the 400 randomly sampled graduates found jobs. The graduating class under consideration included approximately 4500 students.

(a) Describe the population parameter of interest. What is the value of the point estimate of this parameter?

(b) Construct a 95% confidence interval for the proportion of graduates who found a job within one year of completing their undergraduate degree at this university. (Reminder: check assumptions and conditions.)

(c) Explain what this interval means in the context of this question.

(d) What does "95% confidence" mean?

(e) Without doing any calculations, describe what would happen to the confidence interval if we decided to use a higher confidence level.

(f) Without doing any calculations, describe what would happen to the confidence interval if we used a larger sample.

5.14 After implementing a study abroad program, a university conducted a study to find out what percent of students had already traveled to another country. The survey showed that 42 out of 100 sampled students had previously visited abroad.

(a) Describe the population parameter of interest. What is the value of the point estimate of this parameter?

(b) Construct a 90% confidence interval for the proportion of students at this university who have traveled abroad. (Reminder: check assumptions and conditions.)

(c) Interpret this interval in context.

(d) What does "90% confidence" mean?

5.15 It is believed that large doses of acetaminophen (the active ingredient in over the counter pain relievers like Tylenol) may cause damage to the liver. A researcher wants to conduct a study to estimate the proportion of acetaminophen users who have liver damage. For participating in this study, she will pay each subject $20 and provide a free medical consultation if the patient has liver damage.

(a) If she wants to limit the margin of error of her 98% confidence interval to 2%, what is the minimum amount of money she needs to set aside to pay her subjects?

(b) The amount you calculated in part (a) is substantially over her budget so she decides to use fewer subjects. How will this affect the width of her confidence interval.

5.16 We are interested in estimating the proportion of students at a university who smoke. Out of a random sample of 200 students from this university, 40 students smoke.

(a) Construct a 95% confidence interval for the proportion of students at this university who smoke, and interpret this interval in context.

(b) Construct a 99% confidence interval for the proportion of students at this university who do not smoke, and interpret this interval in context.

(c) Compare the widths of the 95% and 99% confidence intervals. Which one is wider? Can you explain why this is the case?

(d) If we wanted the margin of error to be no larger than 2% for at a 95% confidence level for the proportion of students who smoke, how big of a sample would we need?

5.17 In January 2011, The Marist Poll published a report stating that 66% of adults nationally think licensed drivers should be required to re-take their road test once they reach 65 years of age. It was also reported that interviews were conducted on 1,018 American adults, and that the margin of error was 3% using a 95% confidence level. [25]

(a) Verify the margin of error reported by The Marist Poll. The data collected was based on simple random sampling.

(b) Does the poll contain strong evidence that more than two thirds of the population think that licensed drivers should be required to re-take their road test once they turn 65?

5.18 Exercise 5.14 provides the result of a campus wide survey where 42 of 100 randomly sampled students have traveled abroad. A comprehensive follow-up survey was conducted one year after the study abroad program was implemented. It sampled all students at the university and found that 51% of the university's students had traveled to another country. Do you think there has been an increase in the number of students who traveled abroad since the first survey? Conduct a hypothesis test to check. Would you come to the same conclusion using your confidence interval from Exercise 5.14?

5.19 A national survey conducted in 2011 among a simple random sample of 1,507 adults shows that 56% of Americans think the Civil War is still relevant to American politics and political life. [26]

(a) Conduct a hypothesis test to determine if these data provide strong evidence that majority of the Americans think the Civil War is still relevant.

(b) Interpret the p-value in context.

(c) Calculate a 90% confidence interval for the proportion of Americans who think the Civil War is still relevant. Interpret the interval in context and comment on whether or not the confidence interval agrees with the conclusion of the hypothesis test.

5.20 A college review magazine states that in many business schools there is a certain stigma that marketing is a less stressful major and so most students (>50%) majoring in marketing also major in finance, economics or accounting to be able to show employers that their quantitative skills are also strong. In order to test this claim, an education researcher collects a simple random sample of 80 undergraduate students majoring in marketing at various business schools and finds that 50 of them have a double major.

(a) Conduct a hypothesis test to determine if these data provide strong evidence supporting this magazine's claim that majority of marketing students have a double major.

(b) Interpret the p-value in context.

(c) Calculate a 90% confidence interval for the proportion of marketing students who have a double major. Interpret the interval in context and comment on whether or not the confidence interval agrees with the conclusion of the hypothesis test.

5.21 Among a simple random sample of 331 American adults who do not have a four-year college degree and are not currently enrolled in school, 48% said they decided to not go to college because they could not afford school. [27]

(a) A newspaper article states that only a minority of the Americans who decide not to go to college do so because they cannot afford it, and uses the point estimate from this survey as evidence. Conduct a hypothesis test to determine if these data provide strong evidence supporting this statement. Use a 10% significance level.

(b) Would you expect a confidence interval for the proportion of American adults who decide to not go to college because they cannot afford it to include 0.5? Explain.

5.22 A *Washington Post* article reports that "support for a government-run health-care plan to compete with private insurers has rebounded from its summertime lows and wins clear majority support from the public." More specifically, the article says "seven in 10 Democrats back the plan, while almost nine in 10 Republicans oppose it. Independents divide 52 percent against, 42 percent in favor of the legislation." There were were 819 Democrats, 566 Republicans and 783 Independents surveyed. [28]

(a) A political pundit on TV claims that a majority of Independents oppose the public option health care plan. Do these data provide strong evidence to support this statement?

(b) Would you expect a confidence interval for the proportion of Independents who oppose the public option plan to include 0.5? Explain.

5.23 Exercise 5.21 presents the results of a poll where 48% of 331 Americans who decide to not go to college do so because they cannot afford it.

(a) Construct an 80% confidence interval for the proportion of Americans who decide to not go to college do so because they cannot afford it. Interpret it in context and comment on whether or not the confidence interval agrees with the conclusion of the hypothesis test from Exercise 5.21.

(b) Suppose we wanted the margin of error for the 80% confidence level to be about 1.5%. How large of a survey would you recommend?

5.24 Exercise 5.22 presents the results of a poll evaluating support for the public option health care plan in 2009.

(a) Construct an 90% confidence interval for the proportion of Independents who support the public option health care plan. Interpret it in context and comment on whether or not the confidence interval agrees with the conclusion of the hypothesis test from Exercise 5.22?

(b) If we wanted to limit the margin of error of a 90% confidence interval to 1%, about how many Independents would we need to survey?

5.25 A report by the Centers for Disease Control and Prevention states that 30% of Americans are habitually getting less than six hours of sleep a night – less than the recommended seven to nine hours. New York is known as "the city that never sleeps". In a simple random sample of 300 New Yorkers, it was found that 105 of them get less than six hours of sleep a night.

(a) Do these data provide strong evidence that the rate of sleep deprivation for New Yorkers is higher than the rate of sleep deprivation in the population at large?

(b) Interpret the p-value in context.

5.26 Some people claim that they can tell the difference between a diet soda and a regular soda in the first sip. A researcher wanting to test this claim randomly sampled 80 such people. He then filled 80 plain white cups with soda, half diet and half regular through random assignment, and asked each person to take one sip from their cup and identify the soda as "diet" or "regular". 53 participants correctly identified the soda.

(a) Do these data provide strong evidence that these people are able to detect the difference between diet and regular soda, in other words, are the results significantly better than just random guessing?

(b) Interpret the p-value in context.

5.27 A large survey conducted five years ago at a university showed that 18% of the university students smoked. A more recent survey (simple random sample) found that 40 of 200 students at the university smoked.

(a) Do the data provide strong evidence that the percentage of students who smoke has changed over the last five years?

(b) What type of error might we have made?

5.28 The corporate management at a paper company has reason to believe one of its (supposed) star managers, Michael, is making misleading claims about his regional market share. Michael recently claimed that his office is the sole paper provider to 45% of its regions businesses. When the senior management conducted its own survey, they found that 36% of 180 randomly sampled businesses purchased their paper from Michaels office.

(a) Does this provide strong evidence that Michael is misleading the upper management about his offices performance? Conduct a full hypothesis test.

(b) What type of error might we have made?

5.29 Statistics show that traditionally about 65% of students in a particular rural school district go out of state for college. The school board would like to see if this number will increase next year. To check, they randomly sample 250 college-bound high school students and discover 172 of these students say they will be going to school out of state.

(a) Do these data provide strong evidence that the percentage of students in this rural school district who go out of state for college has increased?

(b) Interpret the p-value in context.

5.30 A law firm associate interviewed 300 randomly selected residents from counties where it is believed mining pollution has caused elevated mercury levels in the water supply. She finds that 38 individuals have higher blood levels of mercury than what the EPA accepts as reasonable; only about 7% of the general population has such high levels of mercury.

(a) Does this provide strong evidence that individuals in these counties are at greater risk of having elevated mercury levels?

(b) Can the result of your hypothesis test be used to prove that mining pollution has caused elevated mercury levels in the water supply?

5.8.4 Difference of two proportions

5.31 In Exercise 5.29, we were introduced to data from a rural school district where 172 out of 250 college-bound high school students planned to go out of state for college. In a similar survey conducted in an urban school district, 450 out of 930 randomly sampled college-bound high school students planned to go out.

(a) Calculate a 95% confidence interval for the difference between the proportions of high school students from the rural and the urban district who go out of state for college. (Reminder: check conditions and assumptions.)

(b) Interpret the confidence interval and describe its practical implications.

5.32 Exercise 5.31 presents the results of two surveys conducted in a rural and an urban school district.

(a) Conduct a hypothesis test to determine if there is strong evidence to suggest that a higher proportion of students from the rural district go out of state for college.

(b) Note any changes you would make to your setup, calculations, and conclusion in part (a) if you were looking for any difference rather than using a one-sided test.

(c) Which of the tests do you think is most appropriate? Explain.

5.33 Exercise 5.22 presents the results of a poll evaluating support for the public option health care plan in 2009. 70% of 819 Democrats and 42% of 783 Independents support the public option.

(a) Do these data provide strong evidence that a higher proportion of Democrats than Independents support the public option plan?

(b) What type of error might we have made?

(c) Would you expect a confidence interval for the difference between the two proportions to include 0? Explain your reasoning. If you answered no, would you expect the confidence interval for $(p_D - p_I)$ to be positive or negative?

(d) Calculate a 95% confidence interval for the difference between $(p_D - p_I)$ and interpret it in context.

(e) True or false: If we had picked a random Democrat and a random Independent at the time of this poll, it is more likely that the Democrat would support the public option than the Independent.

5.34 According to a report on sleep deprivation by the Centers for Disease Control and Prevention, the proportion of California residents who reported insufficient rest or sleep during each of the preceding 30 days is 8.0%, while this proportion is 8.8% for Oregon residents. These data are based on simple random samples of 11,545 California and 4,691 Oregon residents. [29]

(a) What kind of study is this?

(b) Conduct a hypothesis test to determine if these data provide strong evidence the rate of sleep deprivation is different for the two states.

(c) Explain what type of error we might have made in this hypothesis test.

(d) Would you expect a confidence interval for the difference between the two proportions to include 0? Explain your reasoning.

5.35 Using the data provided in Exercise 5.34, construct and interpret a 95% confidence interval for the difference between the population proportions. If we had instead conducted a hypothesis test to check whether the proportions were equal, what conclusion would your confidence interval support?

5.36 A study published in 2001 asked 1924 male and 3666 female undergraduate college students their favorite color. A 95% confidence interval for the difference between the proportions of males and females whose favorite color is black ($p_{male} - p_{female}$) was calculated to be (0.02, 0.06). Based on this information, determine if the following statements are true or false, and explain your reasoning. [30]

(a) We are 95% confident that the true proportion of males whose favorite color is black is 2% lower to 6% higher than the true proportion of females whose favorite color is black.

(b) We are 95% confident that the true proportion of males whose favorite color is black is 2% to 6% higher than the true proportion of females whose favorite color is black.

(c) 95% of random samples will produce 95% confidence intervals that include the true difference between the population proportions of males and females whose favorite color is black.

Continue to parts (d) and (e) on the next page.

(d) We can conclude that there is a significant difference between the proportions of males and females whose favorite color is black and that the difference between the two sample proportions was too large to plausibly be due to chance.

(e) The 95% confidence interval for ($p_{female} - p_{male}$) cannot be calculated with only the information given in this exercise.

5.37 Researchers studying the effectiveness of an anti-anxiety medication randomly sampled 100 patients diagnosed with general anxiety disorder (GAD). Through random assignment, half of the patients received a placebo and the other half received the anti-anxiety medication. The treatment was considered a *success* if the patient experienced a reduction in their anxiety level. A 95% confidence interval for the difference between the proportion of success in the two groups ($p_{medication} - p_{placebo}$) was (-0.02, 0.24). Based on this information, determine if the following statements are true or false, and explain your reasoning.

(a) We are 95% confident that the true proportion of success among the population of people who take the medication is 2% lower to 24% higher than the true proportion of success among the population of people who take a placebo.

(b) 95% of random samples will produce a difference between -0.02 and 0.24.

(c) The 95% confidence interval for ($p_{placebo} - p_{medication}$) cannot be calculated with only the information given in this exercise.

(d) The 95% confidence interval for ($p_{placebo} - p_{medication}$) would be (-0.24, 0.02).

(e) We can conclude that the medication was effective and the difference between the two sample proportions was too large to plausibly be due to chance.

5.38 A 2010 survey asked 827 randomly sampled registered voters in California "Do you support? Or do you oppose? Drilling for oil and natural gas off the Coast of California? Or do you not know enough to say?" Below is the distribution of responses, separated based on whether or not the respondent graduated from college. [31]

(a) What percent of college graduates and what percent of the non-college graduates in this sample do not know enough to have an opinion on drilling for oil and natural gas off the Coast of California?

(b) Conduct a hypothesis test to determine if the data provide strong evidence that the proportion of college graduates who do not have an opinion on this issue is different than that of non-college graduates.

	College Grad	
	Yes	No
Support	154	132
Oppose	180	126
Do not know	104	131
Total	438	389

5.39 Exercise 5.38 presents the results of a poll evaluating support for drilling for oil and natural gas off the coast of California.

(a) What percent of college graduates and what percent of the non-college graduates in this sample support drilling for oil and natural gas off the Coast of California?

(b) Conduct a hypothesis test to determine if the data provide strong evidence that the proportion of college graduates who support off-shore drilling in California is different than that of non-college graduates.

5.40 A news article reports that "Americans have differing views on two potentially inconvenient and invasive practices that airports could implement to uncover potential terrorist attacks." This news piece was based on a survey conducted among a random sample of 1,137 adults nationwide, interviewed by telephone November 7-10, 2010, where one of the questions on the survey was "Some airports are now using 'full-body' digital x-ray machines to electronically screen passengers in airport security lines. Do you think these new x-ray machines should or should not be used at airports?" Below is a summary of responses based on party affiliation. [32]

		Party Affiliation		
		Republican	Democrat	Independent
	Should	264	299	351
Answer	Should not	38	55	77
	Don't know/No answer	16	15	22
	Total	318	369	450

Conduct an appropriate hypothesis test evaluating whether there is a difference in the proportion of Republicans and Democrats who think the full-body scans should be applied in airports. After making your conclusion, describe what type of error we might have committed with this test.

5.41 Exercise 5.40 presents the results of a poll on public opinion on the use of full body scans at airports.

(a) Calculate and interpret a 90% confidence interval for the difference between the proportion of Republicans and Democrats who support full-body scans.

(b) Does this prove that a there is no difference in opinion on the use of full-body scans between Republicans and Democrats? Explain.

5.8.5 When to retreat

5.42 The Stanford University Heart Transplant Study was conducted to determine whether an experimental heart transplant program increased lifespan. Each patient entering the program was designated officially a heart transplant candidate, meaning that he was gravely ill and would most likely benefit from a new heart. Some patients got a transplant (treatment group) and some did not (control group). The table below displays how many patients survived and died in each group. [10]

	control	treatment
alive	4	24
dead	30	45

A hypothesis test would reject the conclusion that the survival rate is the same in each group, and so we might like to construct a confidence interval. Explain why we cannot construct such an interval using our large sample techniques. What might go wrong if we constructed the confidence interval despite this problem?

5.8.6 Testing for goodness of fit using chi-square

5.43 A professor using an open-source introductory statistics book predicts that 60% of the students will purchase a hard copy of the book, 25% will print it out from the web, and 15% will read it online. At the end of the semester she asks her students to complete a survey where they indicate what format of the book they used. Of the 126 students, 71 said they bought a hard copy of the book, 30 said they printed it out from the web, and 25 said they read it online.

(a) State the hypotheses for testing if the professor's predictions were inaccurate.

(b) How many students did the professor expect to buy the book, print the book, and read the book exclusively online?

(c) This is an appropriate setting for a chi-square test. List the assumptions and conditions required for a test and verify they are satisfied (as is always necessary).

(d) Calculate the chi-squared statistic, the degrees of freedom associated with it, and the p-value.

(e) Based on the p-value calculated in part (d), what is the conclusion of the hypothesis test? Interpret your conclusion in context.

5.44 A Gallup Poll released in December 2010 asked 1019 adults living in the Continental U.S. about their belief in the origin of humans. These results, along with results from a more comprehensive poll from 2001 (that we will assume to be exactly accurate), are summarized in the table below: [33]

	Year	
Response	2010	2001
Humans evolved, with God guiding (1)	38%	37%
Humans evolved, but God had no part in process (2)	16%	12%
God created humans in present form (3)	40%	45%
Other / No opinion (4)	6%	6%

(a) Calculate the actual number of respondents in 2010 that fall in each response category.

(b) State hypotheses for the following research question: have beliefs on the origin of human life changed since 2001?

(c) Calculate the expected number of respondents in each category if the null hypothesis is true.

(d) Conduct a chi-square test and state your conclusion (reminder: verify assumptions).

5.8.7 Testing for independence in two-way tables

5.45 Exercise 5.40 introduces data on views on full-body scans at airports and party affiliation. The differences in each political group may be due to chance. Answer each of the following questions under the hypothesis that party affiliation and support of full-body scans are independent. Complete the following computations under the null hypothesis of independence between an individual's party affiliation and her support of full-body scans.

(a) How many Republicans would you expect to not support the use of full-body scans?

(b) How many Democrats would you expect to support the use of full-body scans?

(c) How many Independents would you expect to not know or not answer?

5.46 Exercise 5.31 presents the results of two surveys conducted in a rural and an urban district. In the rural district, 172 out of 250 college-bound high school students went out of state for college, and in the urban district, 450 out of 930 students did.

(a) Create a two-way table presenting the results of these two surveys.

(b) If in fact there was no difference between the true proportions of students from the rural and urban district who went out of state for college, how many students from each district would be expected to go out of state for college?

5.47 Researchers studying the link between prenatal vitamin use and autism surveyed the mothers of a random sample of children aged 24 - 60 months with autism or with typical development. The table below shows the number of mothers in each group who did and did not use prenatal vitamins during the three months before pregnancy (periconceptional period). [34]

		Autism		
		Autism	Typical development	Total
Periconceptional	No	111	70	181
prenatal vitamin	Yes	143	159	302
	Total	254	229	483

(a) State appropriate hypotheses to test for independence of use of prenatal vitamins during the three months before pregnancy and autism.

(b) Complete the hypothesis test and state an appropriate conclusion. (Reminder: verify any necessary assumptions for the test.)

(c) A New York Times article reporting on this study was titled "Prenatal Vitamins May Ward Off Autism". Do you find the title of this article to be appropriate? If not, explain your reasoning, and suggest a more appropriate title. [35]

5.48 A 2011 survey asked 806 randomly sampled adult Facebook users about their Facebook privacy settings. One of the questions on the survey was, "Do you know how to adjust your Facebook privacy settings to control what people can and cannot see?" The responses are cross-tabulated based on gender. [36]

		Gender		
		Male	Female	Total
Response	Yes	288	378	666
	No	61	62	123
	Not sure	10	7	17
	Total	359	447	806

(a) State appropriate hypotheses to test for independence of gender and whether or not Facebook users know how to adjust their privacy settings.

(b) Complete the hypothesis test and state an appropriate conclusion. (Reminder: verify any necessary assumptions for the test.)

5.49 Exercise 5.38 provides a table summarizing the responses of a random sample of 438 college graduates and 389 non-graduates on the topic of oil drilling. Complete a chi-square test for these data to check whether there is a statistically significant differences in responses from college graduate and non-graduates.

5.50 A December 2010 survey asked 500 randomly sampled Los Angeles residents which shipping carrier they prefer to use for shipping holiday gifts. The table below shows the distribution of responses by age group as well as the expected counts for each cell (shown in parentheses).

		Age			
		18-34	35-54	55+	Total
Shipping Method	USPS	72 (81)	97 (102)	76 (62)	245
	UPS	52 (53)	76 (68)	34 (41)	162
	FedEx	31 (21)	24 (27)	9 (16)	64
	Something else	7 (5)	6 (7)	3 (4)	16
	Not sure	3 (5)	6 (5)	4 (3)	13
	Total	165	209	126	500

(a) State the null and alternative hypotheses for testing for independence of age and preferred shipping method for holiday gifts among Los Angeles residents.

(b) Are the assumptions and conditions for inference satisfied?

(c) Is a chi-squared test appropriate for testing for independence of age and preferred shipping method for holiday gifts?

Chapter 6

Small sample inference

Large samples are sometimes unavailable, so it is useful to study methods that apply to small samples. Moving from large samples to small samples creates a number of problems that prevent us from applying the normal model directly, though the general ideas will be similar to those from Chapters 4 and 5. The approach is as follows:

- Determine what test statistic is useful.
- Identify the distribution of the test statistic under the condition the null hypothesis was true.
- Apply the ideas of Chapter 4 under the new distribution.

This is the same approach we used in Chapter 5.

6.1 Small sample inference for the mean

We applied a normal model to the sample mean in Chapter 4 when (1) the observations were independent, (2) the sample size was at least 50, and (3) the data were not strongly skewed. The findings in Section 4.4 also suggested we could relax condition (3) when we considered ever larger samples.

In this section, we examine the distribution of the sample mean for any sample size. To this end, we must strengthen the condition about the distribution of the data. Specifically, our data must meet two criteria:

(1) The observations are independent.

(2) The population distribution is nearly normal.

If we are not confident that the data come from a nearly normal distribution, then we cannot apply the methods of this section. Just like before, we can relax this condition when the sample size becomes larger.

Let's review our motives for requiring a large sample (for this paragraph, we will assume the independence and skew conditions are met). First, a large sample would ensure that the sampling distribution of the mean was nearly normal. Second, it also gave us support that the estimate of the standard error was reliable. Both of these issues seemed to be satisfactorily addressed when the sample size was larger than 50. Now, we'll think about how these issues (the shape of the sampling distribution, and the accuracy of the standard error estimate) change under small samples.

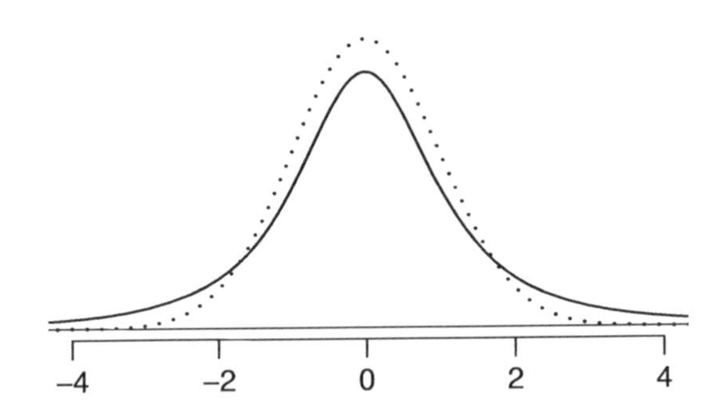

Figure 6.1: Comparison of a t distribution (solid line) and a normal distribution (dotted line).

6.1.1 The normality condition

If the individual observations are independent and come from a nearly normal population distribution, a special case of the Central Limit Theorem ensures the distribution of the sample means will be nearly normal.

Central Limit Theorem for normal data
The sampling distribution of the mean is nearly normal when the sample observations are independent and come from a nearly normal distribution. This is true for any sample size.

While this seems like a very helpful special case, there is one small problem. It is inherently difficult to verify normality in small data sets.

Caution: Checking the normality condition
We should exercise caution when verifying the normality condition for small samples. It is important to not only examine the data but also think about where the data come from. For example, ask: would I expect this distribution to be symmetric, and am I confident that outliers are rare?

6.1.2 Introducing the t distribution

We will address the uncertainty of the standard error estimate by using a new distribution: the t distribution. A t distribution, shown as a solid line in Figure 6.1, has a bell shape. However, its tails are thicker than the normal model's. This means observations are more likely to fall beyond two standard deviations from the mean than under the normal distribution[1]. These extra thick tails are exactly the correction we need to resolve our problem with estimating the standard error.

The t distribution, always centered at zero, has a single parameter: degrees of freedom. The **degrees of freedom (df)** describe the exact bell shape of the t distribution. Several

[1] The standard deviation of the t distribution is actually a little more than 1. However, it is useful to always think of the t distribution as having a standard deviation of 1 in all of our applications.

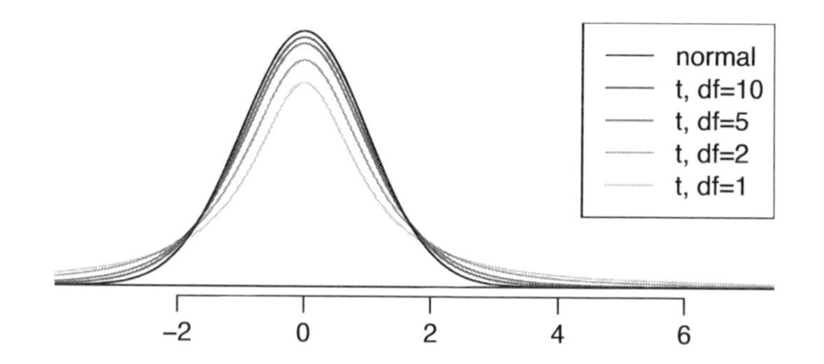

Figure 6.2: The larger the degrees of freedom, the more closely the t distribution resembles the standard normal model.

t distributions are shown in Figure 6.2. When there are more degrees of freedom, the t distribution looks very much like the standard normal distribution.

Degrees of freedom
The degrees of freedom describe the shape of the t distribution. The larger the degrees of freedom, the more closely the distribution approximates the normal model.

When the degrees of freedom is about 50 or more, the t distribution is nearly indistinguishable from the normal distribution. In Section 6.1.4, we relate degrees of freedom to sample size.

6.1.3 Working with the t distribution

We will find it very useful to become familiar with the t distribution, because it plays a very similar role to the normal distribution during inference. We use a **t table** in place of the normal probability table, which is partially shown in Table 6.3. A larger table is presented in Appendix C.2 on page 364.

Each row in the t table represents a t distribution with different degrees of freedom. The columns correspond to tail probabilities. For instance, if we know we are working with the t distribution with $df = 18$, we can examine row 18, which is highlighted in Table 6.3. If we want the value in this row that identifies the cutoff for an upper tail of 10%, we can look in the column where *one tail* is 0.100. This cutoff is 1.33. If we had wanted the cutoff for the lower 10%, we would use -1.33. Just like the normal distribution, all t distributions are symmetric.

Example 6.1 What proportion of the t distribution with 18 degrees of freedom falls below -2.10?

Just like a normal probability problem, we first draw the picture in Figure 6.4 and shade the area below -2.10. To find this area, we identify the appropriate row: $df = 18$. Then we identify the column containing the absolute value of -2.10; it is the third column. Because we are looking for just one tail, we examine the top line of the table, which shows that a one tail area for a value in the third row corresponds to 0.025. About 2.5% of the distribution falls below -2.10. In the next example we encounter a case where the exact t value is not listed in the table.

one tail		0.100	0.050	0.025	0.010	0.005
two tails		0.200	0.100	0.050	0.020	0.010
df	1	3.08	6.31	12.71	31.82	63.66
	2	1.89	2.92	4.30	6.96	9.92
	3	1.64	2.35	3.18	4.54	5.84
	⋮	⋮	⋮	⋮	⋮	
	17	1.33	1.74	2.11	2.57	2.90
	18	*1.33*	*1.73*	*2.10*	*2.55*	*2.88*
	19	1.33	1.73	2.09	2.54	2.86
	20	1.33	1.72	2.09	2.53	2.85
	⋮	⋮	⋮	⋮	⋮	
	400	1.28	1.65	1.97	2.34	2.59
	500	1.28	1.65	1.96	2.33	2.59
	∞	1.28	1.64	1.96	2.33	2.58

Table 6.3: An abbreviated look at the t table. Each row represents a different t distribution. The columns describe the tail areas at each standard deviation. The row with $df = 18$ has been *highlighted*.

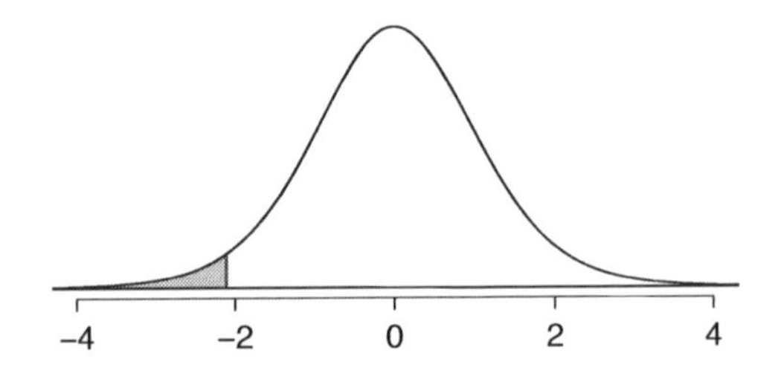

Figure 6.4: The t distribution with 18 degrees of freedom. The area below -2.10 has been shaded.

● **Example 6.2** A t distribution with 20 degrees of freedom is shown in the left panel of Figure 6.5. Estimate the proportion of the distribution falling above 1.65.

We identify the row in the t table using the degrees of freedom: $df = 20$. Then we look for 1.65; it is not listed. It falls between the first and second columns. Since these values bound 1.65, their tail areas will bound the tail area corresponding to 1.65. We identify the one tail area of the first and second columns, 0.050 and 0.10, and we conclude that between 5% and 10% of the distribution is more than 1.65 standard deviations above the mean. If we like, we can identify the precise area using statistical software: 0.0573.

● **Example 6.3** A t distribution with 2 degrees of freedom is shown in the right panel of Figure 6.5. Estimate the proportion of the distribution falling more than 3 units from the mean (above or below).

As before, first identify the appropriate row: $df = 2$. Next, find the columns that capture 3; because $2.92 < 3 < 4.30$, we use the second and third columns. Finally, we find bounds for the tail areas by looking at the two tail values: 0.05 and 0.10. We use the two tail values because we are looking for two (symmetric) tails.

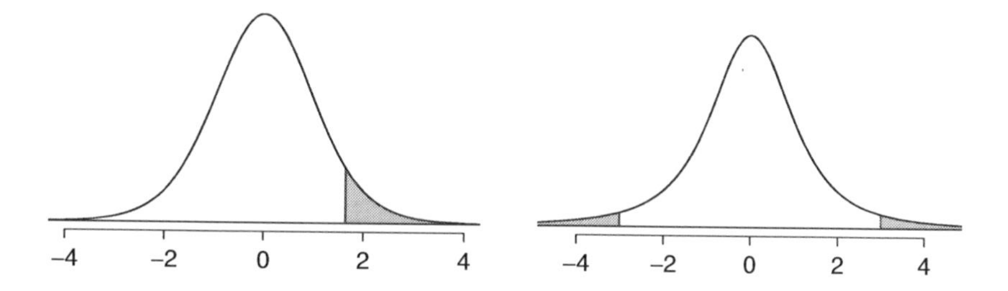

Figure 6.5: Left: The t distribution with 20 degrees of freedom, with the area above 1.65 shaded. Right: The t distribution with 2 degrees of freedom, and the area further than 3 units from 0 has been shaded.

⊙ **Exercise 6.4** What proportion of the t distribution with 19 degrees of freedom falls above -1.79 units? Answer in the footnote[2].

6.1.4 The t distribution as a solution to the standard error problem

When estimating the mean and standard error from a small sample, the t distribution is a more accurate tool than the normal model.

> **TIP: When to use the t distribution**
> When observations are independent and nearly normal, we can use the t distribution for inference of the sample mean.

We use the t distribution instead of the normal model because we have extra uncertainty in the estimate of the standard error. To proceed with the t distribution for inference about a single mean, we must check two conditions.

- Independence of observations: We verify this condition exactly as before. We either collect a simple random sample from less than 10% of the population or, if it was an experiment or random process, carefully ensure to the best of our abilities that the observations were independent.
- Observations come from a nearly normal distribution: This second condition is more difficult to verify since we are usually working with small data sets. Instead we often (i) take a look at a plot of the data for obvious departures from the normal model and (ii) consider whether any previous experiences alert us that the data may not be nearly normal.

When examining a sample mean and estimated standard error from a sample of n independent and nearly normal observations, we will use a t distribution with $n-1$ degrees of freedom (df). For example, if the sample size was 19, then we would use the t distribution with $df = 19 - 1 = 18$ degrees of freedom and proceed exactly as we did in Chapter 4, except that *now we use the t table.*

We can relax the normality condition for the observations when the sample size becomes large. For instance, a slightly skewed data set might be acceptable if there were at least 15 observations. For a strongly skewed data set, we might require 30 or 40 observations. For an extremely skewed data set, perhaps 100 or more.

[2]We finding the shaded area *above* -1.79 (we leave the picture to you). The small left tail is between 0.025 and 0.05, so the larger upper region must have an area between 0.95 and 0.975.

6.1.5 One sample confidence intervals with small n

Dolphins are at the top of the oceanic food chain, which causes dangerous substances such as mercury to concentrate in their organs and muscles. This is an important problem for both dolphins and other animals, like humans, who occasionally eat them. For instance, this is particularly relevant in Japan where school meals have included dolphin at times.

Figure 6.6: A Risso's dolphin.

Photo by Mike Baird (http://www.bairdphotos.com/).

Here we identify a confidence interval for the average mercury content in dolphin muscle using a sample of 19 Risso's dolphins from the Taiji area in Japan[3]. The data are summarized in Table 6.7. The minimum and maximum observed values can be used to evaluate whether or not there are any extreme outliers or obvious skew.

n	$\bar{x}$	s	minimum	maximum
19	4.4	2.3	1.7	9.2

Table 6.7: Summary of mercury content in the muscle of 19 Risso's dolphins from the Taiji area. Measurements are in μg/wet g (micrograms of mercury per wet gram of muscle).

⊙ **Exercise 6.5** Are the independence and normality conditions satisfied for this data set? Answer in the footnote[4].

In the normal model, we used $z^\star$ and the standard error to determine the width of a confidence interval. When we have a small sample, we try the t distribution instead of the normal model:

$$\bar{x} \pm t^\star_{df} SE$$

$t^\star_{df}$
Multiplication factor for t conf. interval

The sample mean and estimated standard error are computed just as before ($\bar{x} = 4.4$ and

[3]Taiji was featured in the movie *The Cove*, and it is a significant source of dolphin and whale meat in Japan. Thousands of dolphins pass through the Taiji area annually, and we will assume these 19 dolphins represent a random sample from those dolphins. Data reference: Endo T and Haraguchi K. 2009. High mercury levels in hair samples from residents of Taiji, a Japanese whaling town. Marine Pollution Bulletin 60(5):743-747.

[4]The observations are a random sample and consist of less than 10% of the population, therefore independence is reasonable. The summary statistics in Table 6.7 do not suggest any strong skew or outliers, which is encouraging. Based on this evidence – and that we don't have any clear reasons to believe the data are not roughly normal – the normality assumption is reasonable.

$SE = s/\sqrt{n} = 0.528$), while the value $t^\star_{df}$ is a change from our previous formula. Here $t^\star_{df}$ corresponds to the appropriate cutoff from the t distribution with df degrees of freedom, which is identified below.

> **Degrees of freedom for a single sample**
> If our sample has n observations and we are examining a single mean, then we use the t distribution with $df = n - 1$ degrees of freedom.

Applying the rule in our current example, we should use the t distribution with $df = 19 - 1 = 18$ degrees of freedom. To build a 95% confidence interval, we will use the abbreviated t table on page 244 where each tail has 2.5% (both tails total to 5%), which is the third column. Then we identify the row with 18 degrees of freedom to obtain $t^\star_{18} = 2.10$. Generally the value of $t^\star_{df}$ is slightly larger than what we would expect under the normal model with $z^\star$.

Finally, we can substitute all our values into the confidence interval equation to create the 95% confidence interval for the average mercury content in muscles from Risso's dolphins that pass through the Taiji area:

$$\bar{x} \pm t^\star_{18} SE \quad \rightarrow \quad 4.4 \pm 2.10 * 0.528 \quad \rightarrow \quad (3.87, 4.93)$$

We are 95% confident the average mercury content of muscles in Risso's dolphins is between 3.87 and 4.93 μg/wet gram. This falls below the US safety limit, which is 0.5 μg per wet gram[5].

> **Finding a t confidence interval for the mean**
> Based on a sample of n independent and nearly normal observations, a confidence interval for the population mean is
>
> $$\bar{x} \pm t^\star_{df} SE$$
>
> where $\bar{x}$ is the sample mean, $t^\star_{df}$ corresponds to the confidence level and degrees of freedom, and SE is the standard error as estimated by the sample.

⊙ **Exercise 6.6** The FDA's webpage provides some data on mercury content of fish[6]. Based on a sample of 15 croaker white fish (Pacific), a sample mean and standard deviation were computed as 0.287 and 0.069 ppm (parts per million), respectively. The 15 observations ranged from 0.18 to 0.41 ppm. We will assume these observations are independent. Based on the summary statistics of the data, do you have any objections to the normality condition of the individual observations? Answer in the footnote[7].

[5] http://www.ban.org/ban-hg-wg/Mercury.ToxicTimeBomb.Final.PDF

[6] http://www.fda.gov/Food/FoodSafety/Product-SpecificInformation/Seafood/FoodbornePathogensContaminants/Methylmercury/ucm115644.htm

[7] There are no extreme outliers; all observations are within 2 standard deviations of the mean. If there is skew, it is not evident. There are no red flags for the normal model based on this (limited) information, and we do not have reason to believe the mercury content is not nearly normal in this type of fish.

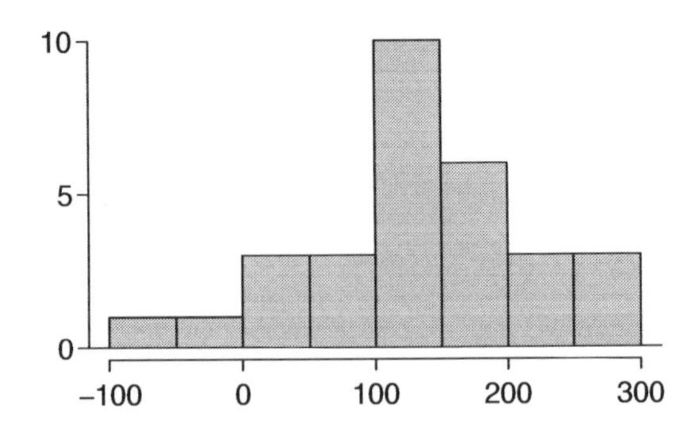

Figure 6.8: Sample distribution of improvements in SAT scores after taking the SAT course.

● **Example 6.7** Estimate the standard error of $\bar{x} = 0.287$ ppm from the statistics in Exercise 6.6. If we are to use the t distribution to create a 90% confidence interval for the actual mean of the mercury content, identify the degrees of freedom we should use and also find $t^\star_{df}$.

$SE = \frac{0.069}{\sqrt{15}} = 0.0178$ and $df = n - 1 = 14$. Looking in the column where two tails is 0.100 (since we want a 90% confidence interval) and row $df = 14$, we identify $t^\star_{14} = 1.76$.

⊙ **Exercise 6.8** Based on the results of Exercise 6.6 and Example 6.7, compute a 90% confidence interval for the average mercury content of croaker white fish (Pacific). Answer in the footnote[8].

6.1.6 One sample t tests with small n

An SAT preparation company claims that its students' scores improve by over 100 points on average after their course. A consumer group would like to evaluate this claim, and they collect data on a random sample of 30 students who took the class. Each of these students took the SAT before and after taking the company's course, and we would like to examine the differences in these scores to evaluate the company's claim. (This was originally paired data, so we use the differences; see Section 5.1 for more details on paired data.) The distribution of the difference in scores, shown in Figure 6.8, has mean 135.9 and standard deviation 82.2. Do these data provide convincing evidence to back up the company's claim?

⊙ **Exercise 6.9** Set up hypotheses to evaluate the company's claim. Use μ to represent the true average difference in student scores. Answer in the footnote[9].

⊙ **Exercise 6.10** Are the conditions to use the t distribution method satisfied? Answer in the footnote[10].

[8] Use $\bar{x} \pm t^\star_{14} SE$: $0.287 \pm 1.76 * 0.0178$. This corresponds to $(0.256, 0.318)$. We are 90% confident that the average mercury content of croaker white fish (Pacific) is between 0.256 and 0.318 ppm.

[9] This is a one-sided test. H_0: student scores do not improve by more than 100 after taking the company's course. $\mu \leq 100$ (or simply $\mu = 100$). H_A: students scores improve by more than 100 points on average after taking the company's course. $\mu > 100$.

[10] This is a random sample from less than 10% of the company's students (assuming they have more than 300 former students), so the independence condition is reasonable. The normality condition also seems reasonable based on Figure 6.8. We can use the t distribution method.

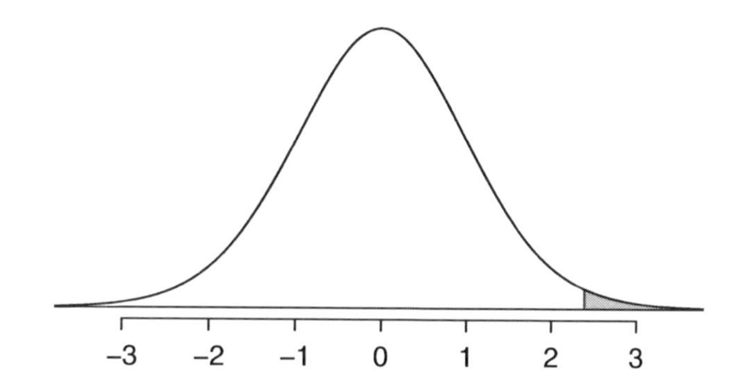

Figure 6.9: The t distribution with 29 degrees of freedom.

Just as we did for the normal case, we standardize the sample mean using the Z score to identify the test statistic. However, we will write T instead of Z, because we have a small sample and are basing our inference on the t distribution:

T
T score
(like Z score)

$$T = \frac{\bar{x} - \text{null value}}{SE} = \frac{135.9 - 100}{82.2/\sqrt{30}} = 2.39$$

If the null hypothesis was true, the test statistic T would follow a t distribution with $df = n - 1 = 29$ degrees of freedom. We can draw a picture of this distribution and mark the observed T, as in Figure 6.9. The shaded right tail represents the p-value: the probability of observing such strong evidence in favor of the SAT company's claim, if the average student improvement is really only 100.

⊙ **Exercise 6.11** Use the t table in Appendix C.2 on page 364 to identify the p-value. What do you conclude? Answer in the footnote[11].

⊙ **Exercise 6.12** Because we rejected the null hypothesis, does this mean that taking the company's class improves student scores by more than 100 points on average? Answer in the footnote[12].

6.2 The t distribution for the difference of two means

It is useful to be able to compare two means. For instance, a teacher might like to test the notion that two versions of an exam were equally difficult. She could do so by randomly assigning each version to students. If she found that the average scores on the exams were so different that we cannot write it off as chance, then she may want to award extra points to students who took the more difficult exam.

In a medical context, we might investigate whether embryonic stem cells (ESCs) can improve heart pumping capacity in individuals who have suffered a heart attack. We could look for evidence of greater heart health in the ESC group against a control group.

[11]We use the row with 29 degrees of freedom. The value $T = 2.39$ falls between the third and fourth columns. Because we are looking for a single tail, this corresponds to a p-value between 0.01 and 0.025. The p-value is guaranteed to be less than 0.05 (the default significance level), so we reject the null hypothesis. The data provide convincing evidence to support the company's claim that student scores improve by more than 100 points following the class.

[12]This is an observational study, so we cannot make this causal conclusion. For instance, maybe SAT test takers tend to improve their score over time even if they don't take a special SAT class, or perhaps only the most motivated students take such SAT courses.

The ability to make conclusions about a difference in two means, $\mu_1 - \mu_2$, is often useful. If the sample sizes are small and the data are nearly normal, the t distribution can be applied to the sample difference in means, $\bar{x}_1 - \bar{x}_2$, to make inference about the difference in population means.

6.2.1 Sampling distributions for the difference in two means

In the example of two exam versions, the teacher would like to evaluate whether there is convincing evidence that the difference in average scores is not due to chance.

It will be useful to extend the t distribution method from Section 6.1 to apply to a new point estimate:

$$\bar{x}_1 - \bar{x}_2$$

Just as we did in Section 5.2, we verify conditions for each sample separately and then verify that the samples are also independent. For instance, if the teacher believes students in her class are independent, the exam scores are nearly normal, and the students taking each version of the exam were independent, then we can use the t distribution for the sampling distribution of the point estimate, $\bar{x}_1 - \bar{x}_2$.

The formula for the standard error of $\bar{x}_1 - \bar{x}_2$, introduced in Section 5.2, remains useful for small samples:

$$SE_{\bar{x}_1 - \bar{x}_2} = \sqrt{SE_{\bar{x}_1}^2 + SE_{\bar{x}_2}^2} = \sqrt{\frac{s_1^2}{n_1} + \frac{s_2^2}{n_2}} \qquad (6.13)$$

Because we will use the t distribution, we will need to identify the appropriate degrees of freedom. This can be done using computer software. An alternative technique is to use the smaller of $n_1 - 1$ and $n_2 - 1$, which is the method we will apply in the examples and exercises[13].

Using the t distribution for a difference in means
The t distribution can be used for the (standardized) difference of two means if (1) each sample meets the conditions for the t distribution and (2) the samples are independent. We estimate the standard error of the difference of two means using Equation (6.13).

6.2.2 Two sample t test

Summary statistics for each exam version are shown in Table 6.10. The teacher would like to evaluate whether this difference is so large that it provides convincing evidence that Version B was more difficult (on average) than Version A.

⊙ **Exercise 6.14** Construct a two-sided hypothesis test to evaluate whether the observed difference in sample means, $\bar{x}_A - \bar{x}_B = 5.3$, might be due to chance. Answer in the footnote[14].

[13] This technique for degrees of freedom is conservative with respect to a Type 1 Error; it is more difficult to reject the null hypothesis using this *df* method.

[14] Because the professor did not expect one exam to be more difficult prior to examining the test results, she should use a two-sided hypothesis test. H_0: the exams are equally difficult, on average. $\mu_A - \mu_B = 0$. H_A: one exam was more difficult than the other, on average. $\mu_A - \mu_B \neq 0$.

Version	n	$\bar{x}$	s	min	max
A	30	79.4	14	45	100
B	27	74.1	20	32	100

Table 6.10: Summary statistics of scores for each exam version.

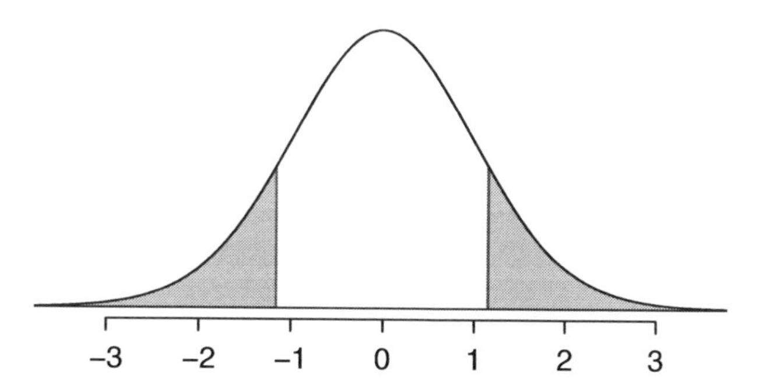

Figure 6.11: The t distribution with 26 degrees of freedom. The shaded right tail represents values with $T \geq 1.15$. Because it is a two-sided test, we also shade the corresponding lower tail.

⊙ **Exercise 6.15** To evaluate the hypotheses in Exercise 6.14 using the t distribution, we must first verify assumptions. (a) Does it seem reasonable that the scores are independent? (b) What about the normality condition for each group? (c) Do you think each group would be independent? Answer in the footnote[15].

After verifying the conditions for each sample and confirming the samples are independent of each other, we are ready to conduct the test using the t distribution. In this case, we are estimating the true difference in average test scores using the sample data, so the point estimate is $\bar{x}_A - \bar{x}_B = 5.3$. The standard error of the estimate can be calculated using Equation (6.13):

$$SE = \sqrt{\frac{s_A^2}{n_A} + \frac{s_B^2}{n_B}} = \sqrt{\frac{14^2}{30} + \frac{20^2}{27}} = 4.62$$

Finally, we construct the test statistic:

$$T = \frac{\text{point estimate} - \text{null value}}{SE} = \frac{(79.4 - 74.1) - 0}{4.62} = 1.15$$

If we have a computer handy, we can identify the degrees of freedom as 45.97. Otherwise we use the smaller of $n_1 - 1$ and $n_2 - 1$: $df = 26$.

⊙ **Exercise 6.16** Identify the p-value, shown in Figure 6.11. Use $df = 26$. Answer in the footnote[16].

[15](a) It is probably reasonable to conclude the scores are independent. (b) The summary statistics suggest the data are roughly symmetric about the mean, and it doesn't seem unreasonable to suggest the data might be normal. (c) It seems reasonable to suppose that the samples are independent since the exams were handed out randomly.

[16]We examine row $df = 26$ in the t table. Because this value is smaller than the value in the left column, the p-value is at least 0.200 (two tails!). Because the p-value is so large, we do not reject the null hypothesis. That is, the data do not convincingly show that one exam version is more difficult than the other, and the teacher is not convinced that she should add points to the Version B exam scores.

	n	$\bar{x}$	s
ESCs	9	3.50	5.17
control	9	-4.33	2.76

Table 6.12: Summary statistics of scores, split by exam version.

In Exercise 6.16, we could have used $df = 45.97$. However, this value is not listed in the table. In such cases, we use the next lower degrees of freedom (unless the computer also provides the p-value). For example, we could have used $df = 45$ but not $df = 46$.

Do embryonic stem cells (ESCs) help improve heart function following a heart attack? Table 6.12 contains summary statistics for an experiment to test ESCs in sheep that had a heart attack. Each of these sheep was randomly assigned to the ESC or control group, and the change in their hearts' pumping capacity was measured. A positive value generally corresponds to increased pumping capacity, which suggests a stronger recovery. We will consider this study in the exercises and examples below.

⊙ **Exercise 6.17** Set up hypotheses that will be used to test whether there is convincing evidence that ESCs actually increase the amount of blood the heart pumps. Answer in the footnote[17].

● **Example 6.18** The raw data from the ESC experiment described in Exercise 6.17 may be viewed in Figure 6.13. Using 8 degrees of freedom for the t distribution, evaluate the hypotheses.

We first compute the point estimate of the difference along with the standard error:

$$\bar{x}_{esc} - \bar{x}_{control} = 7.88$$

$$SE = \sqrt{\frac{5.17^2}{9} + \frac{2.76^2}{9}} = 1.95$$

The p-value is depicted as the shaded right tail in Figure 6.14, and the test statistic is computed as follows:

$$T = \frac{7.88 - 0}{1.95} = 4.03$$

We use the smaller of $n_1 - 1$ and $n_2 - 1$ (each are the same) for the degrees of freedom: $df = 8$. Finally, we look for $T = 4.03$ in the t table; it falls to the right of the last column, so the p-value is smaller than 0.005 (one tail!). Because the p-value is less than 0.005 and therefore also smaller than 0.05, we reject the null hypothesis. The data provide convincing evidence that embryonic stem cells improve the heart's pumping function in sheep that have suffered a heart attack.

[17] We first setup the hypotheses:

H_0: The stem cells do not improve heart pumping function. $\mu_{esc} - \mu_{control} = 0$.

H_A: The stem cells do improve heart pumping function. $\mu_{esc} - \mu_{control} > 0$.

Before we move on, we must first verify that the t distribution method can be applied. Because the sheep were randomly assigned their treatment and, presumably, were kept separate from one another, the independence assumption is verified for each sample as well as for between samples. The data are very limited, so we can only check for obvious outliers in the raw data in Figure 6.13. Since the distributions are (very) roughly symmetric, we will assume the normality condition is acceptable. Because the conditions are satisfied, we can apply the t distribution.

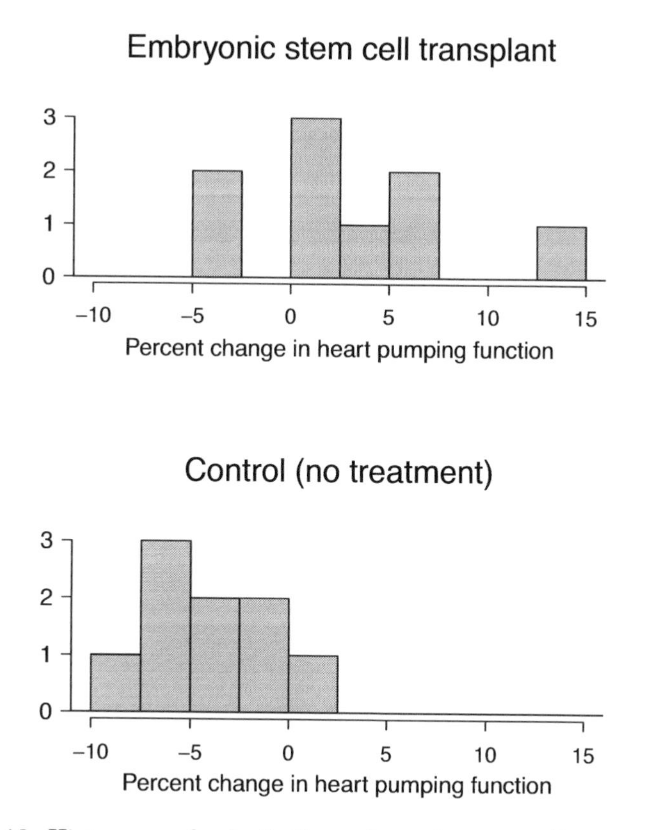

Figure 6.13: Histograms for both the embryonic stem cell group and the control group. Higher values are associated with greater improvement.

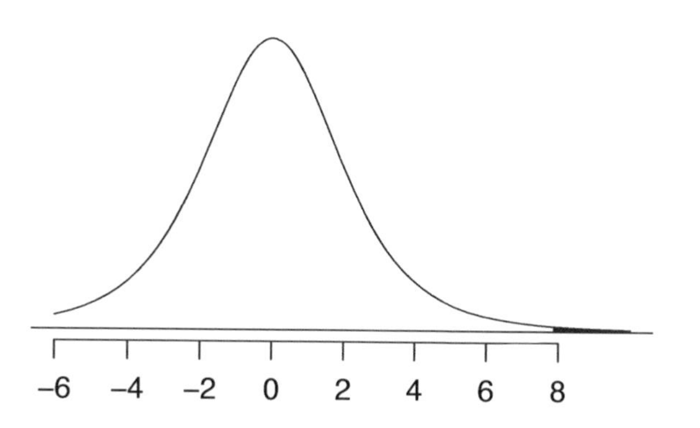

Figure 6.14: Distribution of the sample difference of the mean improvements if the null hypothesis was true. The shaded area represents the p-value.

6.2.3 Two sample t confidence interval

Based on the results of Exercise 6.17, you found significant evidence that ESCs actually help improve the pumping function of the heart. But how large is this improvement? To answer this question, we can use a confidence interval.

⊙ **Exercise 6.19** In Exercise 6.17, you found that the point estimate, $\bar{x}_{esc} - \bar{x}_{control} = 7.88$, has a standard error of 1.95. Using $df = 8$, create a 99% confidence interval for the improvement due to ESCs. Answer in the footnote[18].

6.2.4 Pooled standard deviation estimate (special topic)

Occasionally, two populations will have standard deviations that are so similar, that they can be treated as identical. For example, historical data or a well-understood biological mechanism may justify this strong assumption. In such cases, we can make our t distribution approach slightly more precise by using a pooled standard deviation.

The **pooled standard deviation** of two groups is a way to use data from both samples to better estimate the standard deviation and standard error. If s_1 and s_2 are the standard deviations of groups 1 and 2 and there are good reasons to believe that the population standard deviations are equal, then we can obtain an improved estimate of the group variances by pooling their data:

$$s_{pooled}^2 = \frac{s_1^2 * (n_1 - 1) + s_2^2 * (n_2 - 1)}{n_1 + n_2 - 2}$$

where n_1 and n_2 are the sample sizes, as before. To utilize this new statistic, we substitute s_{pooled}^2 in place of s_1^2 and s_2^2 in the standard error formula, and we use an updated formula for the degrees of freedom:

$$df = n_1 + n_2 - 2$$

The benefits of pooling the standard deviation are realized through obtaining a better estimate of the population standard deviations and using a larger degrees of freedom parameter for the t distribution. Both of these changes may permit a better model of the sampling distribution of $\bar{x}_1^2 - \bar{x}_2^2$.

> **Caution: Pooling standard deviations should be done only after careful research**
> A pooled standard deviation is only appropriate when background research indicates the population standard deviations are nearly equal. When the sample size is large and the condition may be adequately checked with data, the benefits of pooling the standard deviations greatly diminishes.

[18]We know the point estimate, 7.88, and the standard error, 1.95. We also verified the conditions for using the t distribution in Exercise 6.17. Thus, we only need identify $t_8^\star$ to create a 99% confidence interval: $t_8^\star = 3.36$. Thus, the 99% confidence interval for the improvement from ESCs is given by

$$7.88 \pm 3.36 * 1.95 \quad \rightarrow \quad (1.33, 14.43)$$

That is, we are 99% confident that the true improvement in heart pumping function is somewhere between 1.33% and 14.43%.

6.3 Small sample hypothesis testing for a proportion (special topic)

In this section we develop inferential methods for a single proportion that are appropriate when the sample size is too small to apply the normal model to $\hat{p}$. Just like the other small sample techniques, the methods introduced here can be applied to large samples.

6.3.1 When the success-failure condition is not met

People providing an organ for donation sometimes seek the help of a special "medical consultant". These consultants assist the patient in all aspect of the surgery, with the goal of reducing the possibility of complications during the medical procedure and recovery. Patients might choose a consultant based in part on the historical complication rate of the consultant's clients. One consultant tried to attract patients by noting the average complication rate for liver donor surgeries in the US is about 10%, but her clients have only had 3 complications in the 62 liver donor surgeries she has facilitated. She claims this is strong evidence that her work meaningfully contributes to reducing complications (and therefore she should be hired!).

⊙ **Exercise 6.20** We will let p represent the true complication rate for liver donors working with this consultant. Estimate p using the data, and label this value $\hat{p}$.

● **Example 6.21** Is it possible to assess the consultant's claim using the data provided?

No. The claim is that there is a causal connection, but the data are observational. Patients who hire this medical consultant may have lower complication rates for other reasons.

While it is not possible to assess this causal claim, it is still possible to test for an association using these data. For this question we ask, Could the low complication rate of $\hat{p} = 3/62 = 0.048$ be due to chance?

⊙ **Exercise 6.22** Write out hypotheses in both plain and statistical language to test for the association between the consultant's work and the true complication rate, p, for this consultant's clients. Answer in the footnote[19].

● **Example 6.23** In the examples based on large sample theory, we modeled $\hat{p}$ using the normal distribution. Why is this not appropriate here?

The independence assumption may be reasonable if each of the surgeries is from a different surgical team. However, the success-failure condition is not satisfied. Under the null hypothesis, we would anticipate seeing $62 * 0.10 = 6.2$ complications, not the 10 required for the normal approximation.

The uncertainty associated with the sample proportion should not be modeled using the normal distribution. However, we would still like to assess the hypotheses from Exercise 6.22 in absence of the normal framework. To do so, we need to evaluate the possibility

[19] H_0: There is no association between the consultant's contributions and the clients' complication rate. In statistical language, $p = 0.10$. H_A: Patients who work with the consultant tend to have a complication rate lower than 10%, i.e. $p < 0.10$.

of a sample value ($\hat{p}$) this far below the null value, $p_0 = 0.10$. This possibility is usually measured with a p-value.

The p-value is computed based on the null distribution, which is the distribution of the test statistic if the null hypothesis is true. Supposing the the null hypothesis is true, we can compute the p-value by identifying the chance of observing a test statistic that favors the alternative hypothesis at least as strongly as the observed test statistic. In other words, we compute the tail area (or areas) to identify the p-value.

6.3.2 Generating the null distribution and p-value by simulation

We want to identify the sampling distribution of the test statistic ($\hat{p}$) if the null hypothesis was true. In other words, we want to see how the sample proportion changes due to chance alone. Then we plan to use this information to decide whether there is enough evidence to reject the null hypothesis.

Under the null hypothesis, 10% of liver donors have complications during or after surgery. Suppose this rate was really no different for the consultant's clients. If this was the case, we could *simulate* 62 clients to get a sample proportion for the complication rate from the null distribution.

Each client can be simulated using a deck of cards. Take one red card, nine black cards, and mix them up. Then drawing a card is one way of simulating the chance a patient has a complication *if the true complication rate is 10%* for the data. If we do this 62 times and compute the proportion of patients with complications in the simulation, $\hat{p}_{sim}$, then this sample proportion is exactly a sample from the null distribution.

An undergraduate student was paid $2 to complete this simulation. There were 5 simulated cases with a complication and 57 simulated cases without a complication, i.e. $\hat{p}_{sim} = 5/62 = 0.081$.

● **Example 6.24** Is this one simulation enough to determine whether or not we should reject the null hypothesis from Exercise 6.22? Explain.

No. To assess the hypotheses, we need to see a distribution of many $\hat{p}_{sim}$, not just a *single* draw from this sampling distribution.

One simulation isn't enough to get a sense of the null distribution; many simulation studies are needed. Roughly 10,000 seems sufficient. However, paying someone to simulate 10,000 studies by hand is a waste of time and money. Instead, simulations are typically programmed into a computer, which is fast and cheap.

Figure 6.15 shows the results of 10,000 simulated studies. The proportions that are equal to or less than $\hat{p} = 0.048$ are shaded. The shaded areas represent sample proportions under the null distribution that provide at least as much evidence as $\hat{p}$ favoring the alternative hypothesis. There were 1222 simulated sample proportions with $\hat{p}_{sim} \leq 0.048$. We use these to construct the null distribution's left-tail area and find the p-value:

$$\text{left tail} = \frac{\text{Number of observed simulations with } \hat{p}_{sim} \leq \ 0.048}{10000} \tag{6.25}$$

Of the 10,000 simulated $\hat{p}_{sim}$, 1222 were equal to or smaller than $\hat{p}$. Since the hypothesis test is one-sided, the estimated p-value is equal to this tail area: 0.1222.

⊙ **Exercise 6.26** Based on the estimated p-value of 0.1222, should the null hypothesis be rejected? Give a brief explanation.

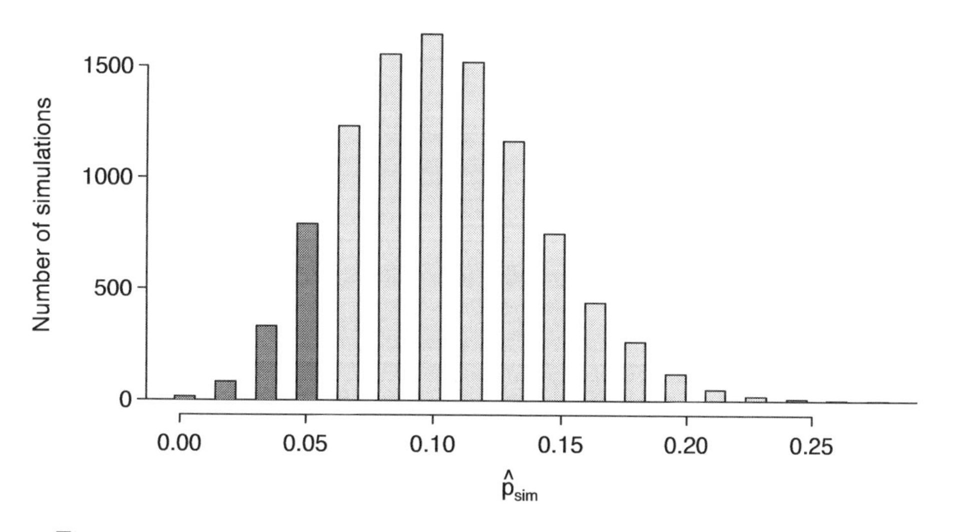

Figure 6.15: The null distribution for $\hat{p}$, created from 10,000 simulated studies. The left tail, representing the p-value for the hypothesis test, contains 12.22% of the simulations.

⊙ **Exercise 6.27** Because the estimated p-value is 0.1222, which is larger than the significance level 0.05, we do not reject the null hypothesis. Explain what this means in plain language in the context of the problem. Answer in the footnote[20].

⊙ **Exercise 6.28** Does the conclusion in Exercise 6.27 imply there is no real association between the surgical consultant's work and fewer complications? Explain. Answer in the footnote[21].

One-sided hypothesis test for p with a small sample
The p-value is always derived by analyzing the null distribution of the test statistic. The normal model poorly approximates the null distribution for $\hat{p}$ when the success-failure condition is not satisfied. As a substitute, we generate the null distribution using simulated sample proportions ($\hat{p}_{sim}$) and use this distribution to compute the tail area, i.e. the p-value.

How do we compute the p-value when the test is two-sided? We continue to use the same rule as before: double the single tail area, which remains a reasonable approach even when the sampling distribution is assymmetric. However, this can result in p-values larger than 1 when the point estimate is very near the mean in the null distribution; in such cases, we write that the p-value is 1. Also, very large p-values computed in this way (e.g. 0.85), may also be slightly inflated.

Exercises 6.26 and 6.27 said the p-value is *estimated*. It is not exact because the simulated null distribution itself is not exact, only a close approximation. However, we can generate an exact null distribution and p-value using the binomial model from Section 3.4.

[20]There isn't sufficiently strong evidence to support an association between the consultant's work and fewer surgery complications.

[21]No. It might be that the consultant's work is associated with a reduction but that there isn't enough data to convincingly show this connection.

6.3.3 Generating the exact null distribution and p-value

The number of successes in n independent cases can be described using the binomial model, which was introduced in Section 3.4. Recall that the probability of observing exactly k successes is given by

$$P(k \text{ successes}) = \binom{n}{k} p^k (1-p)^{n-k} = \frac{n!}{k!(n-k)!} p^k (1-p)^{n-k} \tag{6.29}$$

where p is the true probability of success. The expression $\binom{n}{k}$ is read as n *choose* k, and the exclamation points represent factorials. For instance, 3! is equal to $3 * 2 * 1 = 6$, 4! is equal to $4 * 3 * 2 * 1 = 24$, and so on (see Section 3.4).

The tail area of the null distribution is computed by adding up the probability in Equation (6.29) for each k that provides at least as strong of evidence favoring the alternative hypothesis as the data. If the hypothesis test is one-sided, then the p-value is represented by a single tail area. If the test is two-sided, compute the single tail area and double it to get the p-value, just as we have done in the past.

● **Example 6.30** Compute the exact p-value to check the consultant's claim that her clients' complication rate is below 10%.

Exactly $k = 3$ complications were observed in the $n = 62$ cases cited by the consultant. Since we are testing against the 10% national average, our null hypothesis is $p = 0.10$. We can compute the p-value by adding up the cases where there are 3 or fewer complications:

$$\begin{aligned}
\text{p-value} &= \sum_{j=0}^{3} \binom{n}{j} p^j (1-p)^{n-j} \\
&= \sum_{j=0}^{3} \binom{62}{j} 0.1^j (1-0.1)^{62-j} \\
&= \binom{62}{0} 0.1^0 (1-0.1)^{62-0} + \binom{62}{1} 0.1^1 (1-0.1)^{62-1} \\
&\quad + \binom{62}{2} 0.1^2 (1-0.1)^{62-2} + \binom{62}{3} 0.1^3 (1-0.1)^{62-3} \\
&= 0.0015 + 0.0100 + 0.0340 + 0.0755 \\
&= 0.1210
\end{aligned}$$

This exact p-value is very close to the p-value based on the simulations (0.1222), and we come to the same conclusion. We do not reject the null hypothesis, and there is not statistically significant evidence to support the association.

If it were plotted, the exact null distribution would look almost identical to the simulated null distribution shown in Figure 6.15 on page 257.

6.4 Hypothesis testing for two proportions (special topic)

Cardiopulmonary resuscitation (CPR) is a procedure commonly used on individuals suffering a heart attack when other emergency resources are not available. This procedure is helpful in maintaining some blood circulation, but the chest compressions involved can also cause internal injuries. Internal bleeding and other injuries complicate additional treatment efforts following arrival at a hospital. For instance, blood thinners may be used to help release a clot that is causing the heart attack. However, the blood thinner would have negative repercussions on any internal injuries. Here we consider an experiment[22] for patients who underwent CPR for a heart attack and were subsequently admitted to a hospital. These patients were randomly divided into a treatment group where they received a blood thinner or the control group where they did not receive the blood thinner. The outcome variable of interest was whether the patients survived for at least 24 hours.

● **Example 6.31** Form hypotheses for this study in plain and statistical language. Let p_c represent the true survival proportion in the control group and p_t represent the survival proportion for the treatment group.

We are interested in whether the blood thinners are helpful or hurtful, so this should be a two-sided test.

H_0: Blood thinners do not have an overall effect on survival, i.e. the survival proportions are the same in each group. $p_t - p_c = 0$.

H_A: Blood thinners do have an impact on survival. $p_t - p_c \neq 0$.

6.4.1 Large sample framework for a difference in two proportions

There were 50 patients in the experiment who did not receive the blood thinner and 40 patients who did. The study results are shown in Table 6.16.

	Survived	Died	Total
Control	11	39	50
Treatment	14	26	40
Total	25	65	90

Table 6.16: Results for the CPR study. Patients in the treatment group were given a blood thinner, and patients in the control group were not.

⊙ **Exercise 6.32** What is the sample survival proportion of the control group? Of the treatment group? Provide a point estimate of the difference in survival proportions of the two groups: $\hat{p}_t - \hat{p}_c$. Answer in the footnote[23].

According to the point estimate, there is a 13% increase in the survival proportion when patients who have undergone CPR outside of the hospital are treated with blood thinners. However, we wonder if this difference could be due to chance. We'd like to investigate this using a large sample framework, but we first need to check the conditions for such an approach.

[22] *Efficacy and safety of thrombolytic therapy after initially unsuccessful cardiopulmonary resuscitation: a prospective clinical trial*, by Böttiger et al., The Lancet, 2001.

[23] $\hat{p}_t - \hat{p}_c = 0.35 - 0.22 = 0.13$

● **Example 6.33** Can the point estimate of the difference in survival proportions be adequately modeled using a normal distribution?

We will assume the patients are independent, which is probably reasonable. The success-failure condition is also satisfied. There were at least 10 successes and 10 failures in each group.

While we can apply a normal framework as an approximation to find a p-value, we might keep in mind that there were just 11 successes in one group and 14 in the other. Below we conduct an analysis relying on the large sample normal theory. We will follow up with a small sample analysis and compare the results.

● **Example 6.34** Assess the hypotheses presented in Example 6.31 using a large sample framework. Use a significance level of $\alpha = 0.05$.

We suppose the null distribution of the sample difference follows a normal distribution with mean 0 (the null value) and a standard deviation equal to the standard error of the estimate. Because the null hypothesis in this case would be that the two proportions are the same, we compute the standard error using the pooled standard error formula from Equation (5.30) on page 211:

$$SE = \sqrt{\frac{p(1-p)}{n_t} + \frac{p(1-p)}{n_c}} \approx \sqrt{\frac{0.278(1-0.278)}{40} + \frac{0.278(1-0.278)}{50}} = 0.095$$

where we have used the pooled estimate $\left(\hat{p} = \frac{11+14}{50+40} = 0.278\right)$ in place of the true proportion, p.

The null distribution with mean zero and standard deviation 0.095 is shown in Figure 6.17. We compute the tail areas to identify the p-value. To do so, we use the Z score of the point estimate:

$$Z = \frac{(\hat{p}_t - \hat{p}_c) - \text{null value}}{SE} = \frac{0.13 - 0}{0.095} = 1.37$$

If we look this Z score up in Appendix C.1, we see that the right tail has area 0.0853. The p-value is twice the single tail area: 0.176. This p-value does not provide convincing evidence that the blood thinner helps. Thus, there is insufficient evidence to conclude whether or not the blood thinner helps or hurts. (Remember, we never "accept" the null hypothesis – we can only reject or fail to reject.)

The p-value given here, 0.176, relies on the normal approximation. We know that as the samples sizes are large, this approximation is quite good. However, when the sample sizes are relatively small – the success failure condition is either not satisfied or is just barely satisfied – the approximation may only be adequate. Next we develop a small sample technique, apply it to these data, and compare our results. In general, the small sample method we develop may be used for any size sample, small or large, and should be considered as more accurate than the corresponding large sample technique.

6.4.2 Simulating a difference under the null distribution

The ideas in this section were first introduced in the optional Section 1.8 on page 36. For the interested reader, this earlier section provides a more in-depth discussion.

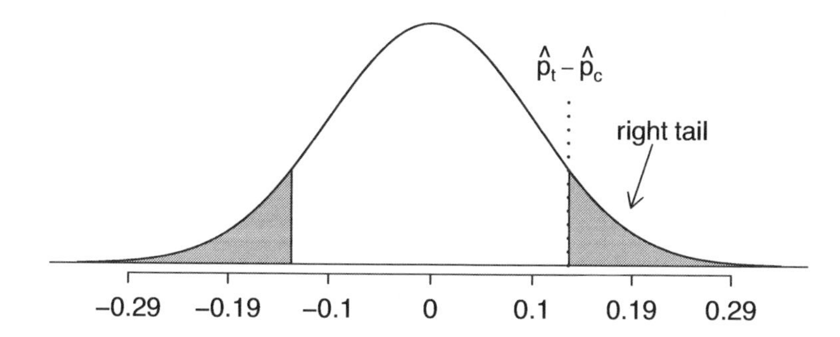

Figure 6.17: The null distribution of the point estimate, $\hat{p}_t - \hat{p}_c$, under the large sample framework is a normal distribution with mean 0 and standard deviation equal to the standard error, in this case $SE = 0.095$. The p-value is represented as sample differences in the null distribution that provide greater evidence for the alternative hypothesis than what the observed data present. The p-value is represented by the shaded areas.

Suppose the null hypothesis is true. Then the blood thinner has no impact on survival and the 13% difference was due to chance. In this case, we can simulate *null* differences that are due to chance using a *randomization technique*[24]. By randomly assigning "fake treatment" and "fake control" stickers to the patients' files, we could get a new grouping – one that is completely due to chance. The expected difference between the two proportions under this simulation is zero.

We run this simulation by taking 40 `treatmentFake` and 50 `controlFake` labels and randomly assigning them to the patients. The label counts of 40 and 50 correspond to the number of treatment and control assignments in the actual study. We use a computer program to randomly assign these labels to the patients, and we organize the simulation results into Table 6.18.

	Survived	Died	Total
`controlFake`	15	35	50
`treatmentFake`	10	30	40
Total	25	65	90

Table 6.18: Simulated results for the CPR study under the null hypothesis. The labels were randomly assigned and are independent of the outcome of the patient.

⊙ **Exercise 6.35** What is the difference in death rates between the two fake groups in Table 6.18? How does this compare to the observed 13% in the real groups?

The difference computed in Exercise 6.35 represents a draw from the null distribution of the sample differences. Next we generate many more simulations to build up the null distribution, much like we did in Section 6.3.2 to build a null distribution for one sample proportion.

[24]The test procedure we employ in this section is formally called a **permutation test**.

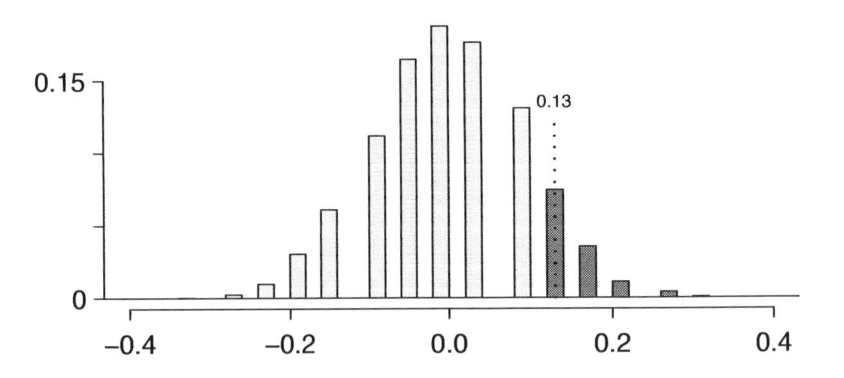

Figure 6.19: An approximation of the null distribution of the point estimate, $\hat{p}_t - \hat{p}_c$. The p-value is twice the right tail area.

Caution: Simulation in the two proportion case requires that the null difference is zero
The technique described here to simulate a difference from the null distribution relies on an important condition in the null hypothesis: there is no connection between the two variables considered. In some special cases, the null difference might not be zero, and more advanced methods (or a large sample approximation, if appropriate) would be necessary.

6.4.3 Null distribution for the difference in two proportions

We build up an approximation to the null distribution by repeatedly creating tables like the one shown in Table 6.18 and computing the sample differences. The null distribution from 10,000 simulations is shown in Figure 6.19.

● **Example 6.36** Compare Figures 6.17 and 6.19. How are they similar? How are they different?

The shapes are similar, but the simulated results show that the continuous approximation of the normal distribution is not very good. We might wonder, how close are the p-values?

⊙ **Exercise 6.37** The right tail area is about 0.13. (It is only a coincidence that we also have $\hat{p}_t - \hat{p}_c = 0.13$.) Since this is a two-sided test, what is the p-value?

⊙ **Exercise 6.38** The p-value is computed by doubling the right tail area: 0.26. How does this value compare with the large sample approximation for the p-value? Answer in the footnote[25].

In general, small sample methods produce more accurate results since they rely on fewer assumptions. However, they often require some extra work or simulations. For this reason, many statisticians use small sample methods only when conditions for large sample methods are not satisfied.

[25]The approximation in this case is fairly poor (p-values: 0.174 vs. 0.26), though we come to the same conclusion. The data do not provide convincing evidence showing the blood thinner hurts or helps patients.

6.5 Exercises

6.5.1 Small sample inference for the mean

6.1 An independent random sample is selected from an approximately normal population with unknown standard deviation. The sample is small ($n < 50$). Find the degrees of freedom and the critical t value ($t^\star$) for the given confidence level.

(a) $n = 42$, CL $= 90\%$

(b) $n = 21$, CL $= 98\%$

(c) $n = 29$, CL $= 95\%$

(d) $n = 12$, CL $= 99\%$

6.2 A 90% confidence interval for a population mean is (65,77). The population distribution is approximately normal and the population standard deviation is unknown. Given that this confidence interval is calculated based on an independent random sample of size 25, calculate the sample mean, the margin of error, and the sample standard deviation.

6.3 For a given confidence level, $t^\star_{df}$ is larger than $z^\star$. Explain how $t^\star_{df}$ being slightly larger than $z^\star$ affects the width of the confidence interval.

6.4 An independent random sample is selected from an approximately normal population with unknown standard deviation. The sample is small ($n < 50$). Find the p-value for the given set of hypotheses and t values. Also determine if the null hypothesis would be rejected at $\alpha = 0.05$.

(a) $H_A : \mu > \mu_0$, $n = 11$, $T = 1.91$

(b) $H_A : \mu < \mu_0$, $n = 17$, $T = -3.45$

(c) $H_A : \mu \neq \mu_0$, $n = 38$, $T = 0.83$

(d) $H_A : \mu > \mu_0$, $n = 47$, $T = 2.13$

6.5 New York is known as "the city that never sleeps". A random sample of 25 New Yorkers were asked how much sleep they get per night. Statistical summaries of these data are shown below. Do these data provide strong evidence that New Yorkers sleep less than 8 hours a night on average?

n	$\bar{x}$	s	min	max
25	7.73	0.77	6.17	9.78

(a) Write the hypotheses in symbols and in words.

(b) Calculate the test statistic, T. (Reminder: check conditions and assumptions.)

(c) Find and interpret the p-value in context. Drawing a picture may be helpful.

(d) What is the conclusion of the hypothesis test?

(e) If you were to construct a confidence interval that corresponded to this hypothesis test, would you expect 8 hours to be in the interval?

6.6 A college newspaper article claims that students at this college spend more than an hour per day on average on social networking sites. The article is based on a survey conducted at this college on a random sample of 45 college students who use social networking sites, which yielded a sample mean of 68.2 minutes with a standard deviation of 21 minutes. A histogram of the data is shown below. Do these data provide strong evidence that students at this college who use social networking sites spend on average more than an hour (60 minutes) per day on such sites?

(a) Write the hypotheses in symbols and in words.

(b) Calculate the test statistic, T. (Reminder: check conditions and assumptions.)

(c) Find and interpret the p-value in context. Drawing a picture may be helpful.

(d) What is the conclusion of the hypothesis test?

(e) If you were to construct a confidence interval that corresponded to this hypothesis test, would you expect 60 minutes to be in the interval?

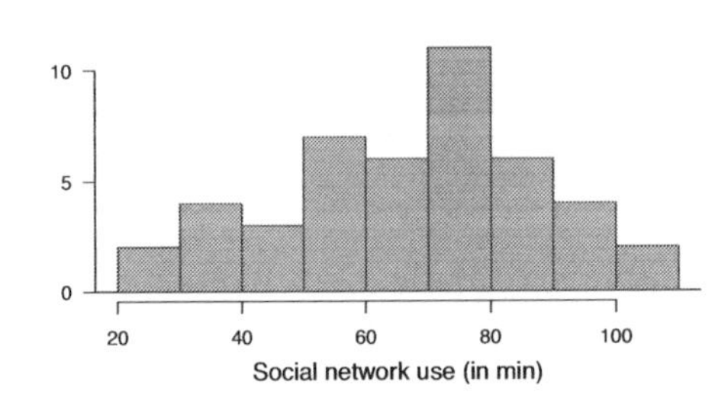

6.7 Exercise 6.5 provides summary statistics on the number of hours of sleep 25 randomly sampled New Yorkers get per night.

(a) Calculate a 90% confidence interval for the number of hours of New Yorkers sleep on average and interpret this interval in context.

(b) Using your confidence interval, would you reject the notion that New Yorkers sleep an average of 8 hours per night?

6.8 Exercise 6.6 provides the mean (68.2 minutes) and standard deviation (21 minutes) for the time spent on social networking sites of 45 randomly sampled students (under the condition that they use social networking sites) at an unnamed college

(a) Calculate a 90% confidence interval for the true average amount of time students who use social networking sites at this college spend on social networking sites per day.

(b) Using your confidence interval, would you reject the notion that students at this college who use social networking sites spend an average of 60 minutes on these sites?

6.9 Chain restaurants in California are required to display calorie counts of each menu item. Prior to October 2008, when a law went into effect that required calorie information on menus, the average calorie intake of a diner at a particular restaurant was 1900 calories. Suppose a nutritionist randomly samples 30 diners at this restaurant and finds an average calorie intake of 1806 calories with a standard deviation of 310 calories.

(a) Do these data provide strong evidence that the average calorie intake has changed after calorie counts started to be displayed on the menus at this restaurant? You may assume that the distribution of the data is nearly normal.

(b) Calculate a 95% confidence interval for the average calorie intake of diners at this restaurant.

(c) Does the conclusion of your hypothesis test agree with the confidence interval you calculated?

6.10 Fueleconomy.gov, the official U.S. government source for fuel economy information, allows users to share gas mileage information on their vehicles. The histogram below shows the distribution of gas mileage (in miles per gallon, MPG) data from 25 users who drive a 2009 Toyota Prius. The sample mean is 50.3 MPG and the standard deviation is 6.8 MPG. Note that these data are user estimates and since the source data cannot be verified, the accuracy of these estimates are not guaranteed. [37]

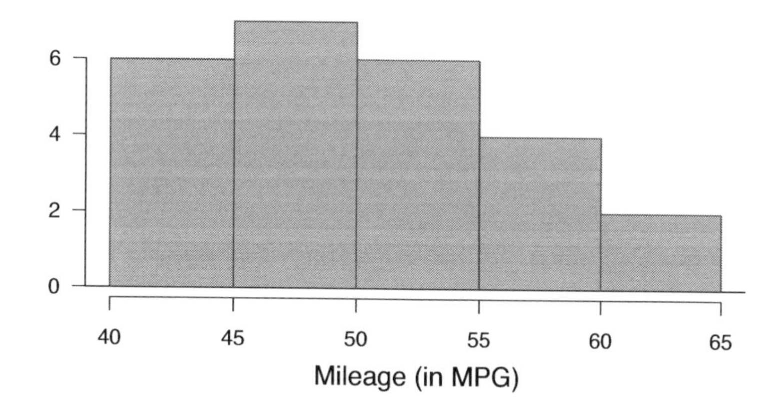

(a) We would like to use this data to evaluate the average mileage of all 2009 Prius drivers. Do you think this is reasonable? Why or why not?

(b) The EPA claims that a 2009 Prius gets 46 MPG. Do these data provide strong evidence against this estimate for drivers who participate on fueleconomy.gov? Use a 10% significance level.

(c) Calculate a 90% confidence interval for the average gas mileage of a 2009 Prius by drivers who participate on fueleconomy.gov.

(d) Does the conclusion of your hypothesis test agree with the confidence interval you calculated?

6.11 You are given the following hypotheses:

$$H_0 : \mu = 60$$
$$H_A : \mu < 60$$

We know that the sample standard deviation is 8 and the sample size is 20. For what sample mean would the p-value be equal to 0.05? Assume that all assumptions and conditions necessary for inference are satisfied.

6.12 A 95% confidence interval for a population mean, μ, is given as (18.985, 21.015). This confidence interval is based on a simple random sample of 36 observations. Calculate the sample mean, $\bar{x}$, and the sample standard deviation, s. Assume that all assumptions and conditions necessary for inference are satisfied.

6.5.2 The t distribution for the difference of two means

6.13 Average income varies from one region of the country to another, and it often reflects both lifestyles and regional living expenses. Suppose a new graduate is considering a job in two locations, Cleveland, OH and Sacramento, CA, and she wants to see whether the average income in one of these cities is higher than the other. She would like to conduct a t test based on two small samples from the 2000 Census, but she first must consider whether the conditions are met to implement the test. Below are a histograms for each city. Should she move forward with the t test? Explain.

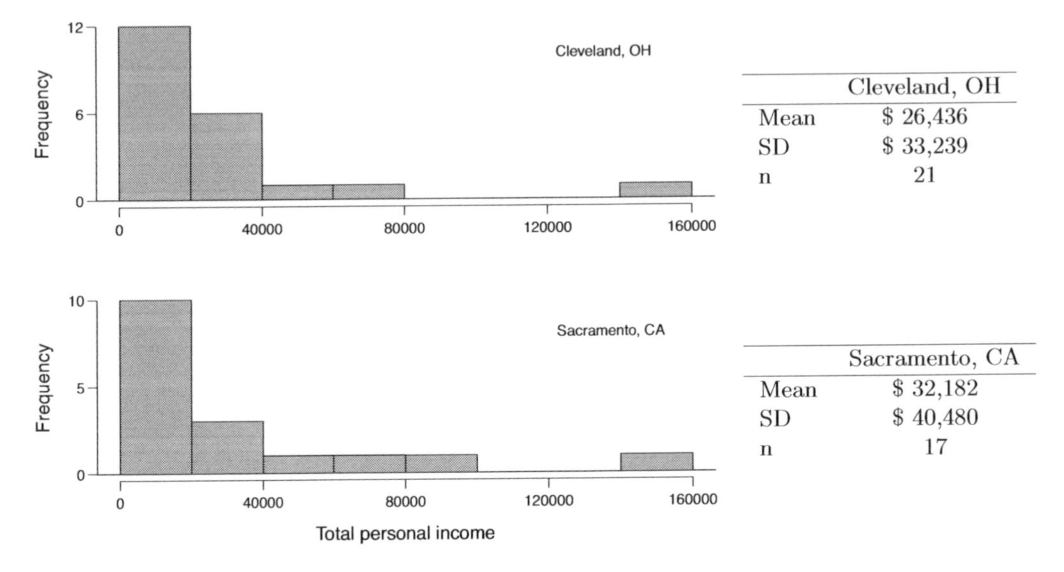

	Cleveland, OH
Mean	\$ 26,436
SD	\$ 33,239
n	21

	Sacramento, CA
Mean	\$ 32,182
SD	\$ 40,480
n	17

6.14 The first Oscar award for best actor and best actresses were given out in 1929. The histograms below show the age distribution for all the best actor and best actress winners from 1929 to 2011. Summary statistics for these distributions are also provided. Is a t test appropriate for testing whether the difference in the sample means for age might be due to chance? Explain.

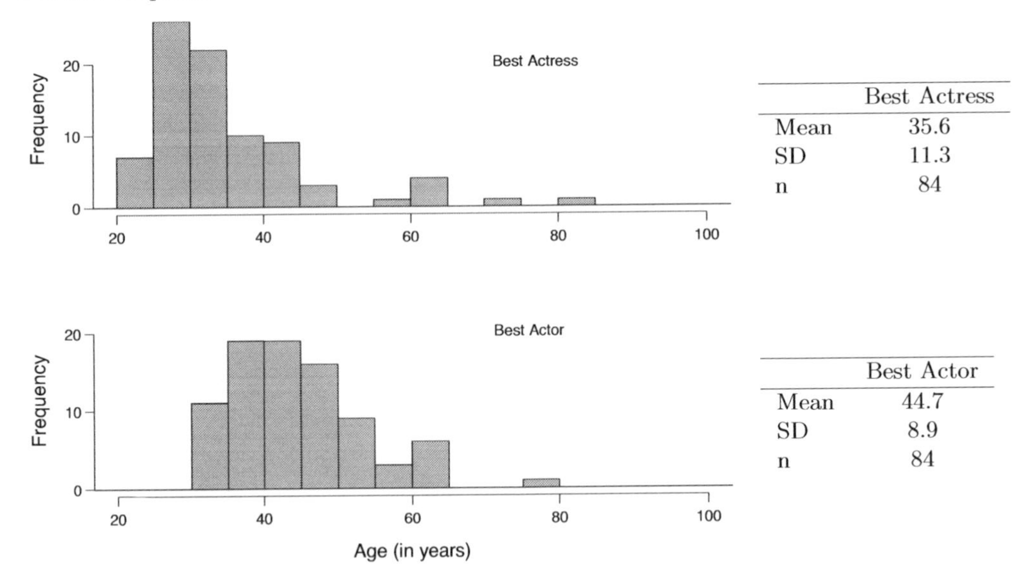

	Best Actress
Mean	35.6
SD	11.3
n	84

	Best Actor
Mean	44.7
SD	8.9
n	84

6.15 Two independent random samples are selected from normal populations with unknown standard deviations. Both samples are small ($n < 50$). Find the p-value for the given set of hypotheses and t values. Also determine if the null hypothesis would be rejected at $\alpha = 0.05$. Remember that a reasonable choice of degrees of freedom for the two-sample case is the minimum of $n_1 - 1$ and $n_2 - 1$. The "exact" df is something we cannot compute from the given information.

(a) $H_A : \mu_1 > \mu_2$, $n_1 = 23$, $n_2 = 25$, $T = 3.16$

(b) $H_A : \mu_1 \neq \mu_2$, $n_1 = 38$, $n_2 = 37$, $T = 2.72$

(c) $H_A : \mu_1 < \mu_2$, $n_1 = 45$, $n_2 = 41$, $T = -1.83$

(d) $H_A : \mu_1 \neq \mu_2$, $n_1 = 11$, $n_2 = 15$, $T = 0.28$

6.16 Two independent random samples are selected from normal populations with unknown standard deviations. Both samples are small ($n < 50$). Find the degrees of freedom and the critical t value ($t^{\star}_{df}$) for the given confidence level.

(a) $n_1 = 16$, $n_2 = 16$, CL = 90%.

(b) $n_1 = 36$, $n_2 = 41$, CL = 95%

(c) $n_1 = 8$, $n_2 = 10$, CL = 99%

(d) $n_1 = 23$, $n_2 = 27$, CL = 98%

6.17 A weight loss pill claims to accelerate weight loss when accompanied with exercise and diet. Diet researchers from a consumer advocacy group decided to test this claim using an experiment. 42 subjects were randomly assigned to two groups: 21 took the pill and 21 only received a placebo. Both groups underwent the same diet and exercise regiment. In the group that received the pill the average weight loss was 20 lbs with a standard deviation of 4 lbs. In the placebo group the average weight loss was 18 lbs with a standard deviation of 5 lbs.

(a) Calculate a 95% confidence interval for the difference between the two means and interpret it in context.

(b) Based on your confidence interval, is there significant evidence that the weight loss pill is effective?

(c) Does this prove that the weight loss pill is effective?

6.18 A company has two factories in which they manufacture engines. Once a month they randomly select 10 engines from each factory and test if there is a difference in performance in engines made in the two factories. This month the average output of the motors from Factory 1 is 120 horsepower with a standard deviation of 5 horsepower, and the average output of the motors from Factory 2 is 132 horsepower with a standard deviation of 4 horsepower.

(a) Calculate a 95% confidence interval for the difference in the average horsepower for engines coming from the two factories and interpret it in context.

(b) Based on your confidence interval, is there a significant evidence that there is a difference in performance in engines made in the two factories? If so, can you tell which factory produces motors with lower performance? Explain.

(c) Recently upgrades were made in Factory 2. Do these data prove that these upgrades enhanced the performance in engines made in this factory? Explain.

6.19 Chicken farming is a multi-billion dollar industry, and any methods that increase the growth rate of young chicks can reduce consumer costs while increasing company profits, possibly by millions of dollars. An experiment was conducted to measure and compare the effectiveness of various feed supplements on the growth rate of chickens. Newly hatched chicks were randomly allocated into six groups, and each group was given a different feed supplement. Their weights in grams after six weeks are given along with feed types in the data set called `chickwts`. Below are some summary statistics from this data set along with box plots showing the distribution of weights by feed type. [38]

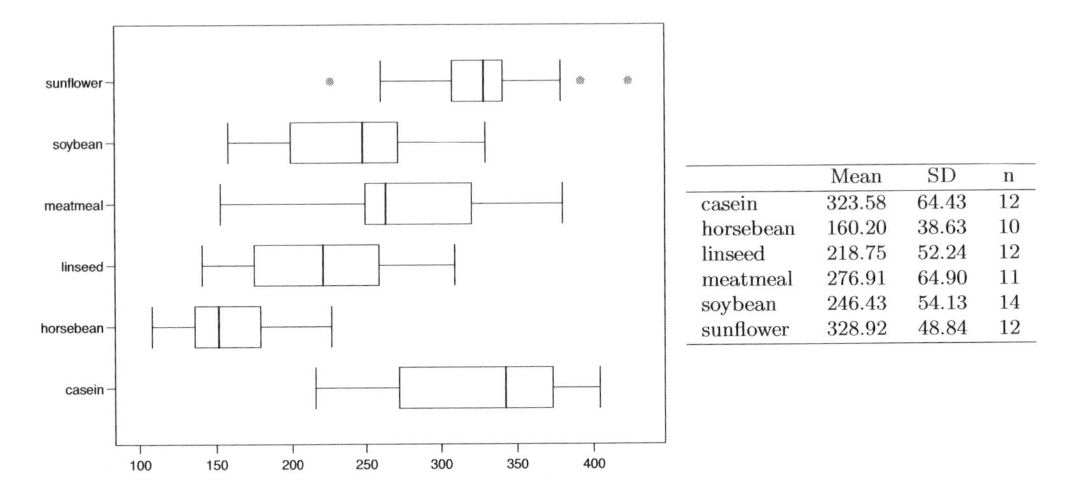

	Mean	SD	n
casein	323.58	64.43	12
horsebean	160.20	38.63	10
linseed	218.75	52.24	12
meatmeal	276.91	64.90	11
soybean	246.43	54.13	14
sunflower	328.92	48.84	12

(a) Describe the distributions of weights of chickens that were fed linseed and horsebean.

(b) Do these data provide strong evidence that the average weights of chickens that were fed linseed and horsebean are different? Use a 5% significance level.

(c) What type of error might we have committed? Explain.

(d) Would your conclusion change if we used $\alpha = 0.01$?

6.20 Casein is a common weight gain supplement for humans. Does it have the same effect on chickens? Using data provided in Exercise 6.19, test the hypothesis that the average weight of chickens that were fed casein is different than the average weight of chickens that were fed soybean. Assume that conditions for inference are satisfied.

6.21 Each year the US Environmental Protection Agency (EPA) releases fuel economy data on cars manufactured in that year. Below are summary statistics on fuel efficiency (in miles/gallon) from random samples of cars with manual and automatic transmissions manufactured in 2010. Do these data provide strong evidence of a difference between the average fuel efficiency of cars with manual and automatic transmissions in terms of their average city mileage? Assume that conditions for inference are satisfied. [39]

	City MPG		Hwy MPG		
	Mean	SD	Mean	SD	n
Manual	21.08	4.29	29.31	4.63	26
Automatic	15.62	2.76	21.38	3.73	26

6.22 An organization is studying whether women have caught up to men in starting pay after attending college. They randomly sampled 28 women, who earned an average of $38,293.78 out of college with a standard deviation of $5,170.22. Twenty-four men were also randomly sampled, and their earnings averaged $41,981.82 with a standard deviation of $3,195.42. Using a significance level of 0.02, do these data provide strong evidence that women have not yet caught up to men in terms of pay? If so, can we make a causal conclusion? If so, explain why. If not, provide an example of why the causal interpretation would not be valid.

6.23 Exercise 6.21 provides data on fuel efficiency of cars manufactured in 2010. Use these statistics to calculate a 95% confidence interval for the difference between average highway mileage of manual and automatic cars and interpret this interval in context.

6.24 Exercise 6.22 provides summary statistics on the starting salaries of men and women who recently graduated from a college. Based on this information, calculate a 98% confidence interval for the difference between the average starting salaries of such men and women and interpret this interval in context.

6.25 A group of researchers hypothesize that the presence of distracting stimuli during eating increases the amount of food consumed by a person and could thereby contribute to overeating and obesity. To test their hypothesis, the researchers monitored food intake for a group of 44 patients who were randomized into two equal groups. The treatment group ate lunch while playing solitaire, and the control group ate lunch without any added distractions. Patients in the control group ate 57.1 grams of biscuits, with a standard deviation of 45.1 grams, and patients in the treatment group ate 27.1 grams of biscuits, with a standard deviation of 26.4 grams. Do these data provide strong evidence that the amount of biscuits consumed by the patients in the treatment and control groups are different? Assume that assumptions and conditions for inference are satisfied. [40]

6.26 The researchers from Exercise 6.25 also investigated the effects of being distracted by a game on how much people eat. The 22 patients in the treatment group who ate their lunch while playing solitaire were asked to do a serial-order recall of the food lunch items they ate. The average number of items recalled by the patients in this group was 4.9, with a standard deviation of 1.8. The average number of items recalled by the patients in the control group (no distraction) was 6.1, with a standard deviation of 1.8. Do these data provide strong evidence that the average number of food items recalled by the patients in the treatment and control groups are different?

6.5.3 Small sample hypothesis testing for a proportion

6.27 A popular uprising that started on January 25, 2011 in Egypt led to the 2011 Egyptian Revolution. Polls show that about 69% of American adults followed the news about the political crisis and demonstrations in Egypt closely during the first couple weeks following the start of the uprising. Among a random sample of 30 high school students, it was found that only 17 of them followed the news about Egypt closely during this time. [41]

(a) Write the hypotheses for testing if the proportion of high school students who followed the news about Egypt is different than the proportion of American adults who did.

(b) Calculate the proportion of high schoolers in this sample who followed the news about Egypt closely during this time.

(c) For large sample theory, we modeled $\hat{p}$ using the normal distribution. Why should we be cautious about this approach for these data?

(d) Since the normal approximation may not be as reliable here as a small sample approach, we evaluate the hypotheses using a simulation. Describe how to perform such a simulation and, once you had results, how to estimate the p-value.

(e) Below is a histogram showing the distribution of $\hat{p}_{sim}$ in 10,000 simulations under the null hypothesis. Estimate the p-value using the plot and determine the conclusion of the hypothesis test.

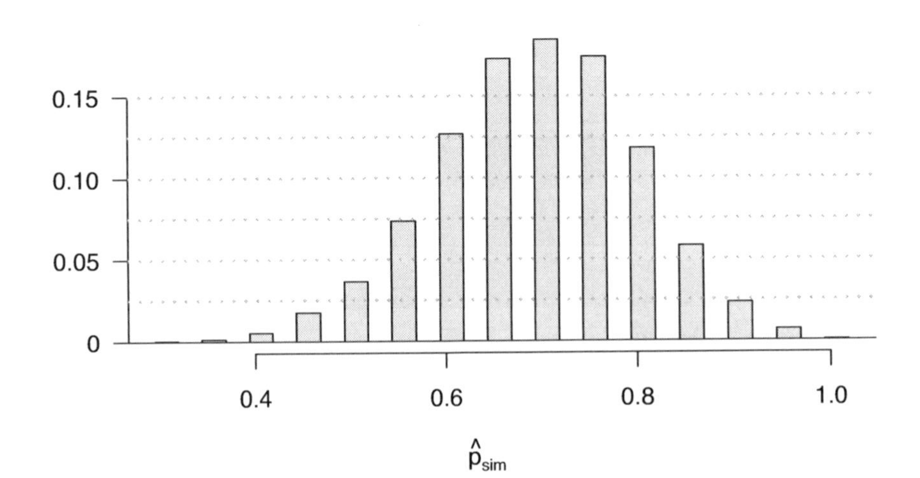

6.28 Assisted Reproductive Technology (ART) is a collection of techniques that help facilitate pregnancy (e.g. in vitro fertilization). A 2008 report by the Centers for Disease Control and Prevention estimated that ART has been successful in leading to a live birth in 31% of cases [42]. A new infertility clinic claims that their success rate is higher than average. A random sample of 30 of their patients yielded a success rate of 40%.

(a) Write the hypotheses to test if the success rate for ART at this clinic is significantly higher than the average success rate.

(b) For large sample theory, we modeled $\hat{p}$ using the normal distribution. Why is this not appropriate here?

(c) The normal approximation would be less reliable here, so we use a simulation strategy. Describe a setup for a simulation that would be appropriate in this situation and how the p-value can be calculated using the simulation results.

(d) Below is a histogram showing the distribution of $\hat{p}_{sim}$ in 10,000 simulations under the null hypothesis. Estimate the p-value using the plot and use it to evaluate the hypotheses.

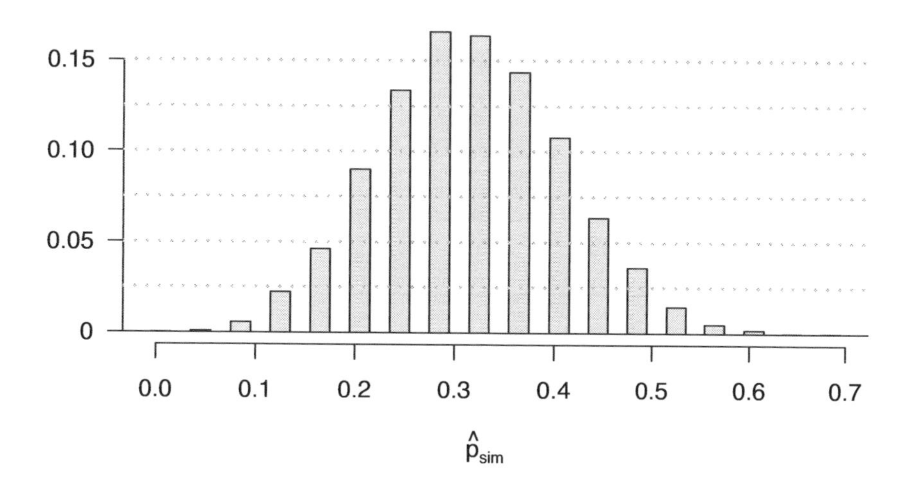

6.5.4 Hypothesis testing for two proportions

6.29 A "social experiment" conducted by a TV program questioned what people do when they see a very obviously bruised woman getting picked on by her boyfriend. On two different occasions at the same restaurant the same couple was depicted, however in one scenario the woman was dressed "provocatively" and in the other scenario the woman was dressed "conservatively". The table below shows how many restaurant diners were present under each scenario, and whether or not they intervened.

		Scenario		
		Provocative	Conservative	Total
Intervene	Yes	5	15	20
	No	15	10	25
	Total	20	25	45

A simulation was conducted to test if people react differently under the two scenarios. In order to conduct the simulation, a researcher wrote yes on 20 index cards and no on 25 index cards to indicate whether or not a diner (represented by each card) intervened. Then he shuffled the cards and dealt them into two groups of size 20 and 25, the provocative and conservative scenarios, respectively. He counted how many diners in each scenario intervened, calculated the difference between the simulated proportions of intervention as $\hat{p}_{pr,sim} - \hat{p}_{con,sim}$. This simulation was repeated 10,000 times using software to obtain 10,000 differences that are due to chance alone. The histogram below shows the distribution of the simulated differences.

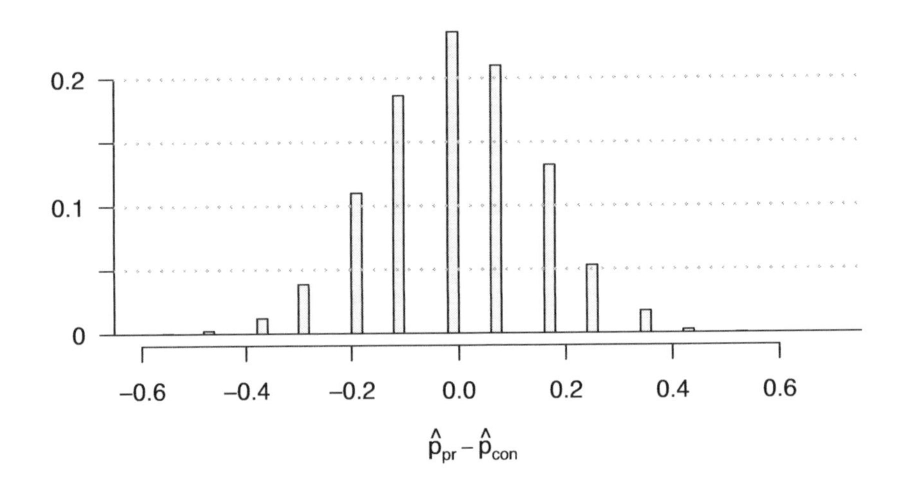

(a) What are the hypotheses?

(b) Calculate the observed difference between the rates of intervention under the two scenarios.

(c) Estimate the p-value using the figure above and determine the conclusion of the hypothesis test.

6.30 An experiment conducted by the *MythBusters*, a science entertainment TV program on the Discovery Channel, tested if a person can be subconsciously influenced into yawning if another person near them yawns. 50 people were randomly assigned to two groups: 34 to a group where a person near them yawned (treatment) and 16 to a group where there wasn't a yawn seed (control). The following table shows the results of this experiment. [43]

		Group		
		Treatment	Control	Total
Result	Yawn	10	4	14
	Not Yawn	24	12	36
	Total	34	16	50

A simulation was conducted to understand the distribution of the test statistic under the assumption of independence: having someone yawn near another person has no influence on if the other person will yawn. In order to conduct the simulation, a researcher wrote yawn on 14 index cards and not yawn on 36 index cards to indicate whether or not a person yawned. Then he shuffled the cards and dealt them into two groups of size 34 and 16 for treatment and control, respectively. He counted how many participants in each simulated group yawned in an apparent response to a nearby yawning person, calculated the difference between the simulated proportions of yawning as $\hat{p}_{trtmt,sim} - \hat{p}_{ctrl,sim}$. This simulation was repeated 10,000 times using software to obtain 10,000 differences that are due to chance alone. The histogram below shows the distribution of the simulated differences.

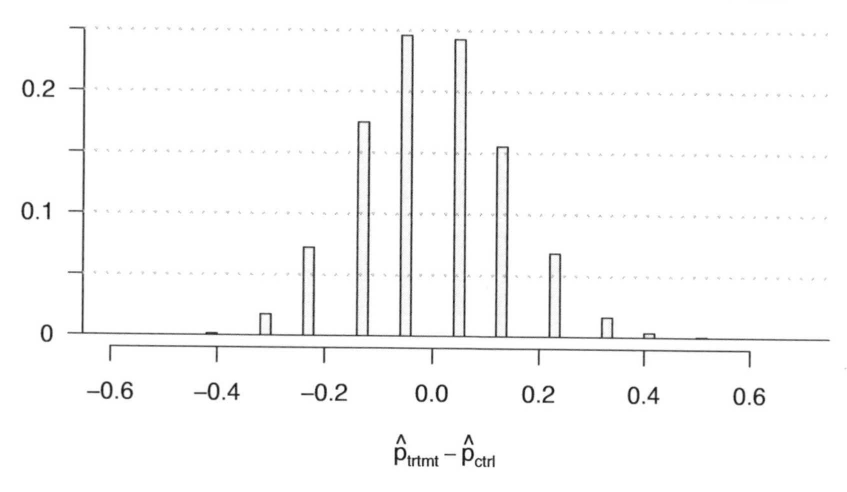

(a) What are the hypotheses?

(b) Calculate the observed difference between the yawning rates under the two scenarios.

(c) Estimate the p-value using the figure above and determine the conclusion of the hypothesis test.

Chapter 7

Introduction to linear regression

Linear regression is a very powerful statistical technique. Many people have some familiarity with regression just from reading the news, where graphs with straight lines are overlaid on scatterplots. Linear models can be used for prediction or to evaluate whether there is a linear relationship between two numerical variables.

Figure 7.1 shows two variables whose relationship can be modeled perfectly with a straight line. The equation for the line is

$$y = 7 + 49.24x$$

Imagine what a perfect linear relationship would mean: you would know the exact value of y, just by knowing the value of x. This is unrealistic in almost any natural process. Consider height and weight of school children for example. Their height, x, gives you some information about their weight, y, but there is still a lot of variability, even for children of the same height.

We often write a linear regression line as

$$y = \beta_0 + \beta_1 x$$

β_0, β_1
Linear model parameters

where β_0 and β_1 represent two parameters that we wish to identify. Usually x represents an explanatory or **predictor** variable and y represents a response. We use the variable x to predict a response y. Usually, we use b_0 and b_1 to denote the point estimates of β_0 and β_1.

Examples of several scatterplots are shown in Figure 7.2. While none reflect a perfect linear relationship, it will be useful to fit approximate linear relationships to each. The lines represent models relating x to y. The first plot shows a relatively strong downward linear trend. The second shows an upward trend that, while evident, is not as strong as the first. The last plot shows a very weak downward trend in the data, so slight we can hardly notice it.

We will soon find that there are cases where fitting a straight line to the data, even if there is a clear relationship between the variables, is not helpful. One such case is shown in Figure 7.3 where there is a very strong relationship between the variables even if the trend is not linear. Such nonlinear trends are beyond the scope of this textbook.

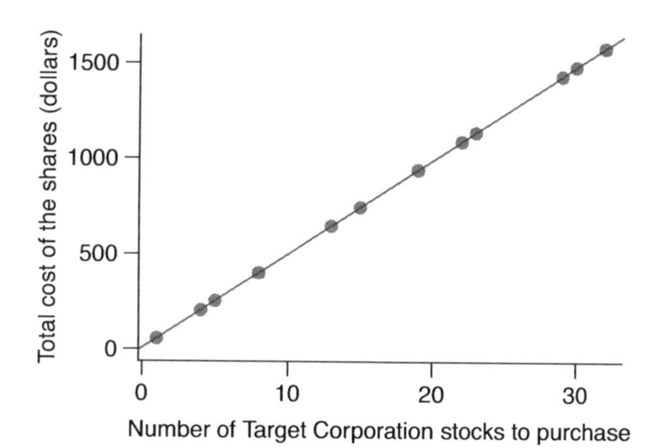

Figure 7.1: Twelve requests were put into a trading company to buy Target Corporation stock (ticker TGT, May 24th, 2011), and the total cost of the shares were reported. Because the cost is computed using a linear formula, the linear fit is perfect.

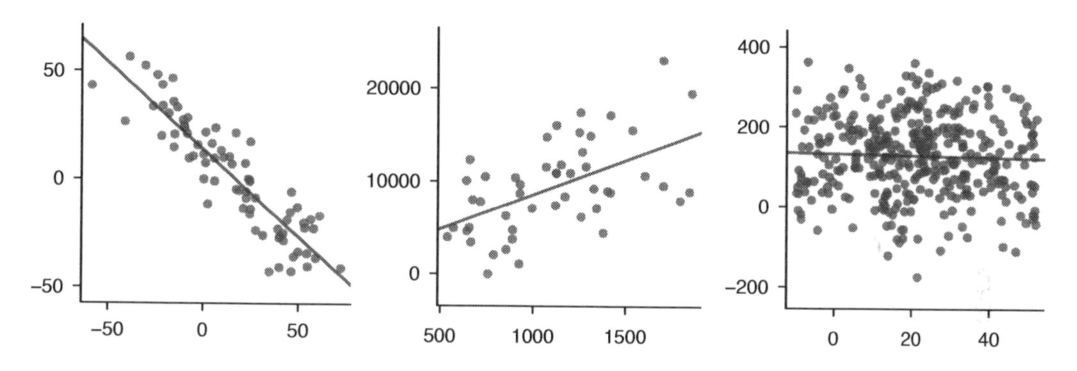

Figure 7.2: Three data sets where a linear model may be useful but is not perfect.

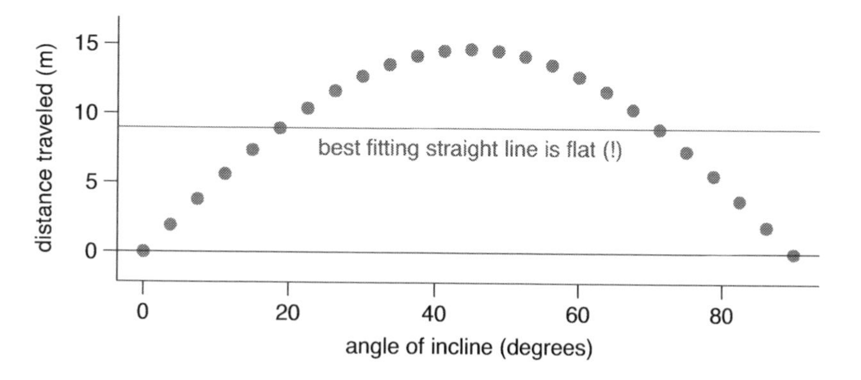

Figure 7.3: A linear model is not useful in this nonlinear case. (These data are from an introductory physics experiment.)

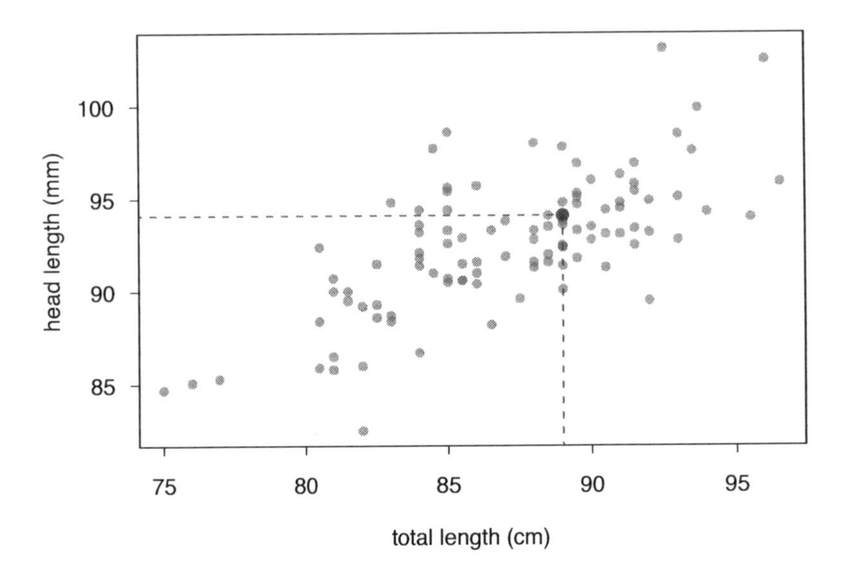

Figure 7.4: A scatterplot showing `headL` against `totalL`. The first possum with a head length of 94.1mm and a length of 89cm is highlighted.

7.1 Line fitting, residuals, and correlation

It is helpful to think deeply about the line fitting process. In this section, we examine criteria for identifying a linear model and introduce a new statistic, *correlation*.

7.1.1 Beginning with straight lines

Scatterplots were introduced in Chapter 1 as a graphical technique to present two numerical variables simultaneously. Such plots permit the relationship between the variables to be examined with ease. Figure 7.4 shows a scatterplot for the `headL` and `totalL` variables from the `possum` data set introduced in Chapter 1. Each point represents a single possum from the data.

The `headL` and `totalL` variables are associated. Possums with an above average total length also tend to have above average head lengths. While the relationship is not perfectly linear, it could be helpful to partially explain the connection between these variables with a straight line.

Straight lines should only be used when the data appear to have a linear relationship, such as the case shown in the left panel of Figure 7.5. The right panel of Figure 7.5 shows a case where a curved band would be more useful in capturing a different set of data.

Caution: Watch out for curved trends
We only consider models based on straight lines in this chapter. If data show a nonlinear trend, like that in the right panel of Figure 7.5, more advanced techniques should be used.

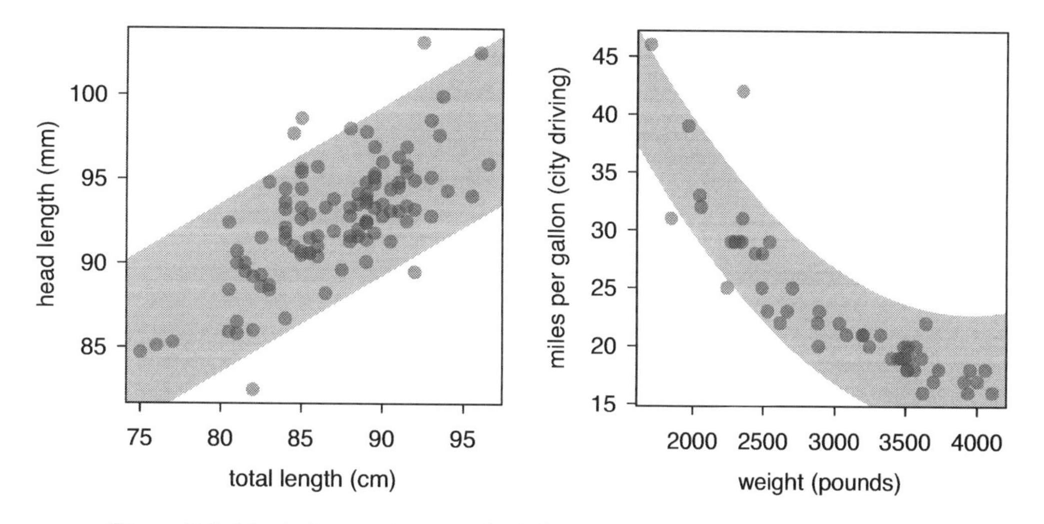

Figure 7.5: Most observations on the left can be captured in a straight band. On the right, we have the `weight` and `mpgCity` variables represented from the `cars` data set, and a curved band does a better job of capturing these cases than a straight band.

7.1.2 Fitting a line by eye

We want to describe the relationship between the `headL` and `totalL` variables using a line. In this example, we will use the total length as the predictor variable, x, to predict a possum's head length, y. We could fit the linear relationship by eye, as in Figure 7.6. The equation for this line is

$$\widehat{y} = 41 + 0.59 * x \tag{7.1}$$

We can use this model to discuss properties of possums. For instance, our model predicts a possum with a total length of 80 cm will have a head length of

$$\widehat{y} = 41 + 0.59 * 80 = 88.2mm$$

A "hat" on y is used to signify that this is an estimate. The model predicts that possums with a total length of 80 cm will have an average head length of 88.2 mm. Without further information about an 80 cm possum, this prediction for head length that uses the average is a reasonable estimate. Generally, linear models predict the average value of y for a particular value of x based on the model.

7.1.3 Residuals

Residuals can be thought of as the leftovers from the model fit:

$$\text{Data} = \text{Fit} + \text{Residual}$$

Each observation will have a residual. If an observation is above the regression line, then its residual, the vertical distance from the observation to the line, is positive. Observations below the line have negative residuals. One goal in picking the right linear model is for these residuals to be as small as possible.

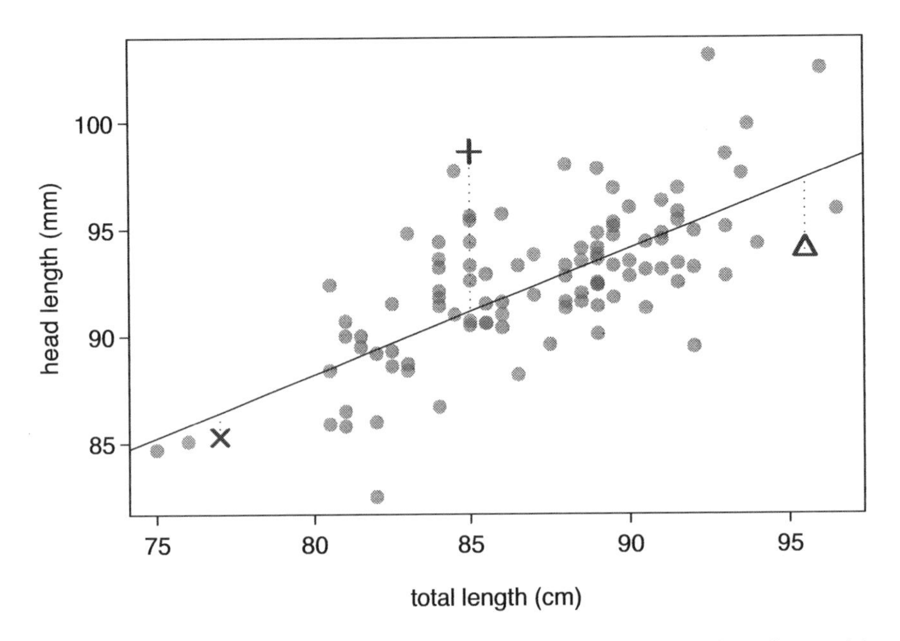

Figure 7.6: A reasonable linear model was fit to represent the relationship between `headL` and `totalL`.

Three observations are noted specially in Figure 7.6. The "×" has a small, negative residual of about -1; the observation marked by "+" has a large residual of about +7; and the observation marked by "△" has a moderate residual of about -4. The size of residuals is usually discussed in terms of its absolute value. For example, the residual for "△" is larger than that of "×" because $|-4|$ is larger than $|-1|$.

Residual: difference between observed and expected

The *residual* of an observation (x_i, y_i) is the difference of the observed response (y_i) and the response we would predict based on the model fit $(\hat{y}_i)$:

$$e_i = y_i - \hat{y}_i$$

We typically identify $\hat{y}_i$ by plugging x_i into the model. The residual of the ith observation is denoted by e_i.

● **Example 7.2** The linear fit shown in Figure 7.6 is given in Equation (7.1). Based on this line, formally compute the residual of the observation $(77.0, 85.3)$. This observation is denoted by "×" on the plot. Check it against the earlier visual estimate, -1.

We first compute the predicted value of point "×" based on the model:

$$\hat{y}_\times = 41 + 0.59 * x_\times = 41 + 0.59 * 77.0 = 86.4$$

Next we compute the difference of the actual head length and the predicted head length:

$$e_\times = y_\times - \hat{y}_\times = 85.3 - 86.43 = -0.93$$

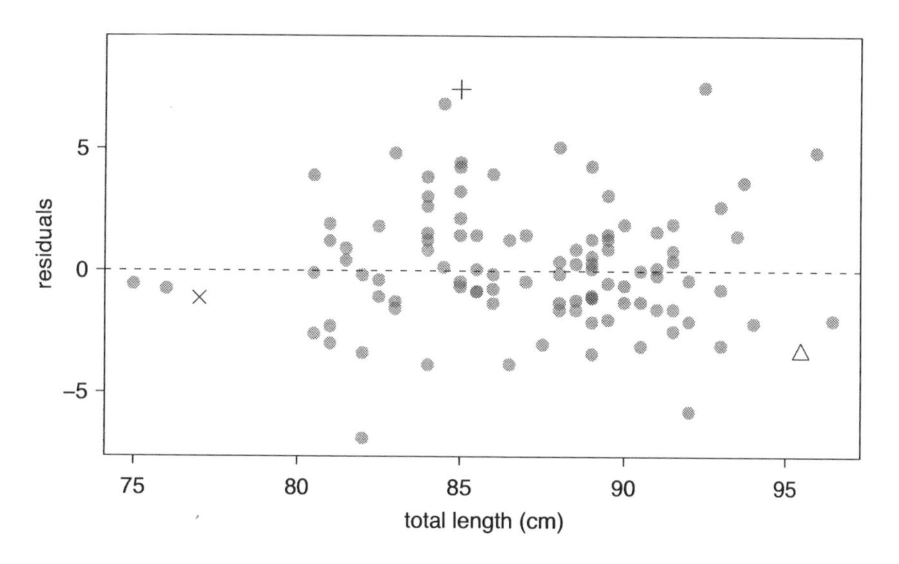

Figure 7.7: Residual plot for the model in Figure 7.6.

This is very close to the visual estimate of -1.

⊙ **Exercise 7.3** If a model underestimates an observation, will the residual be positive or negative? What about if it overestimates the observation? Answer in the footnote[1].

⊙ **Exercise 7.4** Compute the residuals for the observations $(85.0, 98.6)$ ("+" in the figure) and $(95.5, 94.0)$ ("$\triangle$") using the linear model given in Equation (7.1). Answer for "+" is in the footnote[2].

Residuals are helpful in evaluating how well a linear model fits a data set. We often display them in a **residual plot** such as the one shown in Figure 7.7 for the regression line in Figure 7.6. The residuals are plotted at their original horizontal locations but with the vertical coordinate as the residual. For instance, the point $(85.0, 98.6)_+$ had a residual of 7.45, so in the residual plot it is placed at $(85.0, 7.45)$. Creating a residual plot is sort of like tipping the scatterplot over so the regression line is horizontal.

● **Example 7.5** One purpose of residual plots is to identify characteristics or patterns still apparent in data after fitting a model. Figure 7.8 shows three scatterplots with linear models in the first row and residual plots in the second row. Can you identify any patterns remaining in the residuals?

In the first data set (first column), the residuals show no obvious patterns. The residuals appear to be scattered randomly about 0, represented by the dashed line.

---

[1] If a model underestimates an observation, then the model estimate is below the actual. The residual – the actual minus the model estimate – must then be positive. The opposite is true when the model overestimates the observation: the residual is negative.

[2] First compute the predicted value based on the model:

$$\hat{y}_+ = 41 + 0.59 * x_+ = 41 + 0.59 * 85.0 = 91.15$$

Then the residual is given by

$$e_+ = y_+ - \hat{y}_+ = 98.6 - 91.15 = 7.45$$

This was close to the earlier estimate of 7.

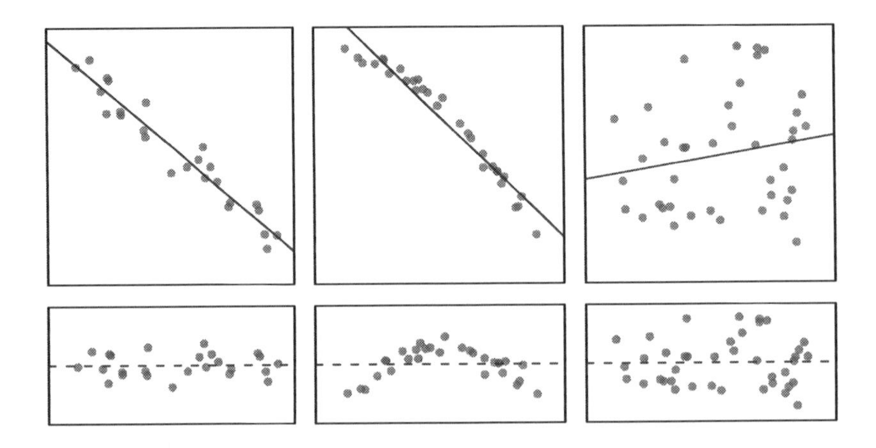

Figure 7.8: Sample data with their best fitting lines (top row) and their corresponding residual plots (bottom row).

The second data set shows a pattern in the residuals. There is some curvature in the scatterplot, which is more obvious in the residual plot. We should not use a straight line to model these data and should use a more advanced technique instead.

The last plot shows very little upwards trend, and the residuals also show no obvious patterns. It is reasonable to try to fit this linear model to the data. However, it is unclear whether there is statistically significant evidence that the slope parameter is different from zero. The point estimate, b_1, is not zero, but we wonder if this could just be due to chance. We will address this sort of question in Section 7.4.

7.1.4 Describing linear relationships with correlation

R
correlation

Correlation: strength of a linear relationship
The **correlation** describes the strength of the linear relationship between two variables and takes values between -1 and 1. We denote the correlation by R.

We compute the correlation using a formula, just as we did with the sample mean and standard deviation. However, this formula is rather complex[3], so we generally perform the calculations on a computer or calculator. Figure 7.9 shows eight plots and their corresponding correlations. Only when the relationship is perfectly linear is the correlation either -1 or 1. If the relationship is strong and positive, the correlation will be near +1. If it is strong and negative, it will be near -1. If there is no apparent linear relationship between the variables, then the correlation will be near zero.

The correlation is intended to quantify the strength of a linear trend. Nonlinear trends, even when strong, sometimes produce correlations that do not reflect the strength of the relationship; see three such examples in Figure 7.10.

[3]Formally, we can compute the correlation for observations (x_1, y_1), (x_2, y_2), ..., (x_n, y_n) using the formula

$$R = \frac{1}{n-1}\sum_{i=1}^{n}\frac{x_i - \bar{x}}{s_x}\frac{y_i - \bar{y}}{s_y}$$

where $\bar{x}$, $\bar{y}$, s_x, and s_y are the sample means and standard deviations for each variable.

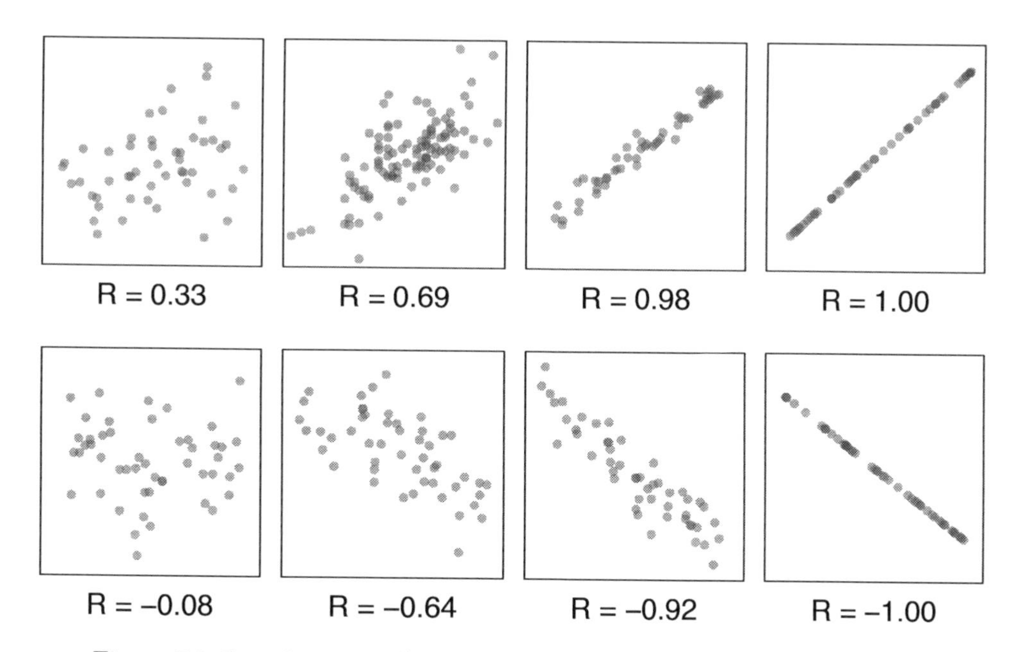

Figure 7.9: Sample scatterplots and their correlations. The first row shows variables with a positive relationship, represented by the trend up and to the right. The second row shows variables with a negative trend, where a large value in one variable is associated with a low value in the other.

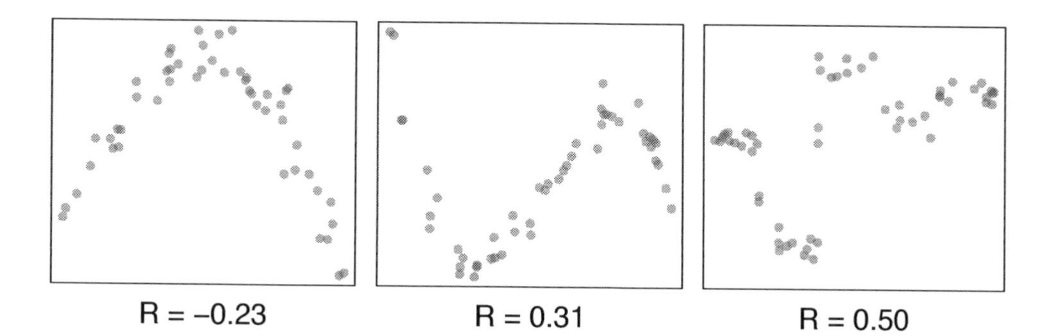

Figure 7.10: Sample scatterplots and their correlations. In each case, there is a strong relationship between the variables. However, the correlation is not very strong, and the relationship is not linear.

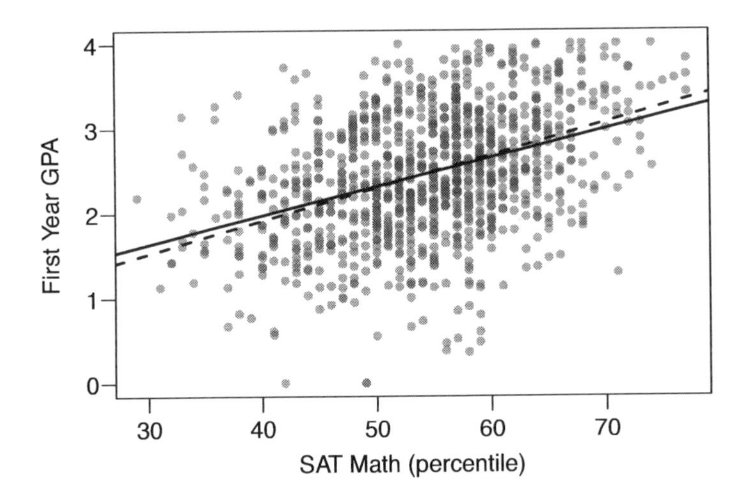

Figure 7.11: SAT math (percentile) and GPA scores for students after one year in college. Two lines are fit to the data, the solid line being the *least squares line*.

⊙ **Exercise 7.6** While no straight line will fit any of the data sets in any of the scatterplots shown in Figure 7.10, try drawing nonlinear curves to each plot. Once you create a curve for each, describe what is important in your fit. Comment in the footnote[4].

7.2 Fitting a line by least squares regression

Fitting linear models by eye is (rightfully) open to criticism since it is based on an individual preference. In this section, we propose *least squares regression* as a more rigorous approach.

This section will use data on SAT math scores and first year GPA from a random sample of students at a college[5]. A scatterplot of the data is shown in Figure 7.11 along with two linear fits. The lines follow a positive trend in the data; students who scored higher math SAT scores also tended to have higher GPAs after their first year in school.

⊙ **Exercise 7.7** Is the correlation positive or negative? Answer in the footnote[6].

7.2.1 An objective measure for finding the best line

We begin by thinking about what we mean by "best". Mathematically, we want a line that has small residuals. Perhaps our criterion could minimize the sum of the residual magnitudes:

$$|e_1| + |e_2| + \cdots + |e_n| \tag{7.8}$$

[4]Possible explanation: The line should be close to most points and reflect overall trends in the data.

[5]These data were collected by Educational Testing Service from an unnamed college. More information: https://www.dartmouth.edu/ chance/course/Syllabi/Princeton96/Class12.html

[6]Positive: because larger SAT scores are associated with higher GPAs, the correlation will be positive. Using a computer, the correlation can be computed: 0.387.

We could use a computer program to find a line that minimizes this criterion (the sum). This does result in a pretty good fit, which is shown as the dashed line in Figure 7.11. However, a more common practice is to choose the line that minimizes the sum of the squared residuals:

$$e_1^2 + e_2^2 + \dots e_n^2 \tag{7.9}$$

The line that minimizes this **least squares criterion** is represented as the solid line in Figure 7.11. This is commonly called the **least squares line**. Three possible reasons to choose Criterion (7.9) over Criterion (7.8) are the following:

1. It is the most commonly used method.
2. Computing the line based on Criterion (7.9) is much easier by hand and in most statistical software.
3. In many applications, a residual twice as large as another is more than twice as bad. For example, being off by 4 is usually more than twice as bad as being off by 2. Squaring the residuals accounts for this discrepancy.

The first two reasons are largely for tradition and convenience, and the last reason explains why Criterion (7.9) is typically most helpful[7].

7.2.2 Conditions for the least squares line

When fitting a least squares line, we generally require

- **Linearity.** The data should show a linear trend. If there is a nonlinear trend (e.g. left panel of Figure 7.12), an advanced regression method from another book or later course should be applied.
- **Nearly normal residuals.** Generally the residuals must be nearly normal. When this condition is found to be unreasonable, it is usually because of outliers or concerns about influential points, which we will discuss in greater depth in Section 7.3. An example of non-normal residuals is shown in the center panel of Figure 7.12.
- **Constant variability.** The variability of points around the least squares line remains roughly constant. An example of non-constant variability is shown in the right panel of Figure 7.12.

⊙ **Exercise 7.10** Should you have any concerns about applying least squares to the `satMath` and `GPA` data in Figure 7.11 on the facing page? Answer in the footnote[8].

7.2.3 Finding the least squares line

For the `satMath` and `GPA` data, we could write the equation as

$$\widehat{GPA} = \beta_0 + \beta_1 * satMath \tag{7.11}$$

[7]There are applications where Criterion (7.8) may be more useful, and there are plenty of other criteria we might consider. However, this book only applies the least squares criterion.

[8]The trend appears to be linear, the data fall around the line with no obvious outliers, and the variance is roughly constant. Least squares regression can be applied to these data.

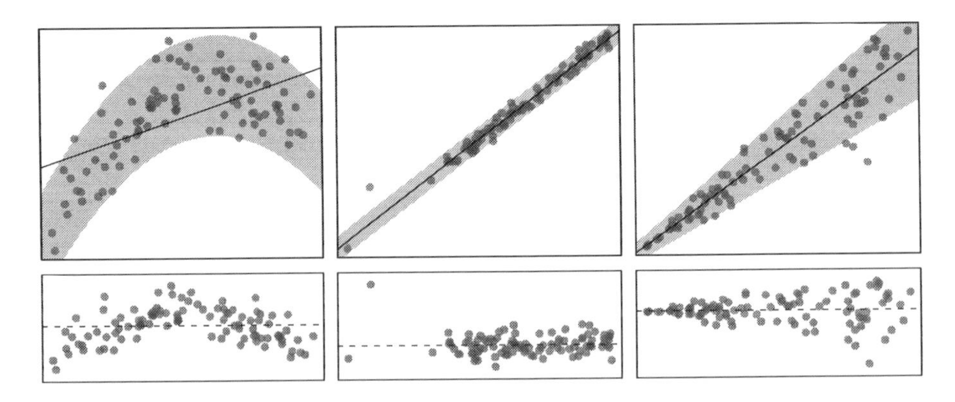

Figure 7.12: Three examples showing when the methods in this chapter are insufficient to apply to the data. In the left panel, a straight line does not fit the data. In the second panel, there are outliers; two points on the left are relatively distant from the rest of the data and one of these points is very far away from the line. In the third panel, the variability of the data around the line increases with larger values of x.

Here the equation is set up to predict `GPA` based on a student's `satMath` score, which would be useful to a college admissions office. These two values, β_0 and β_1, are the *parameters* of the regression line.

Just like in Chapters 4-6, the parameters are estimated using observed data. In practice, this estimation is done using a computer in the same way that other estimates, like a sample mean, can be estimated using a computer or calculator. However, we can also find the parameter estimates by applying two properties of the least squares line:

- If $\bar{x}$ is the mean of the horizontal variable (from the data) and $\bar{y}$ is the mean of the vertical variable, then the point $(\bar{x}, \bar{y})$ is on the least squares line.
- The slope of the least squares line is estimated by

$$b_1 = \frac{s_y}{s_x} R \qquad (7.12)$$

  where R is the correlation between the two variables, and s_x and s_y are the sample standard deviations of the explanatory variable (variable on the horizontal axis) and response (variable on the vertical axis), respectively.

b_0, b_1
Sample estimates of β_0, β_1

We use b_1 to represent the point estimate of the parameter β_1, and we will similarly use b_0 to represent the sample point estimate for β_0.

⊙ **Exercise 7.13** Table 7.13 shows the sample means for the `satMath` and `GPA` variables: 54.395 and 2.468. Plot the point (54.395, 2.468) on Figure 7.11 on page 282 to verify it falls on the least squares line (the solid line).

⊙ **Exercise 7.14** Using the summary statistics in Table 7.13, compute the slope for the regression line of GPA against SAT math percentiles. Answer in the footnote[9].

[9] Apply Equation (7.12) with the summary statistics from Table 7.13 to compute the slope:

$$b_1 = \frac{s_y}{s_x} R = \frac{0.741}{8.450} 0.387 = 0.03394$$

	satMath ("x")	GPA ("y")
mean	$\bar{x} = 54.395$	$\bar{y} = 2.468$
sd	$s_x = 8.450$	$s_y = 0.741$
	correlation:	$R = 0.387$

Table 7.13: Summary statistics for the satMath and GPA data.

You might recall from math class the **point-slope** form of a line (another common form is *slope-intercept*). Given the slope of a line and a point on the line, (x_0, y_0), the equation for the line can be written as

$$y - y_0 = slope * (x - x_0) \tag{7.15}$$

A common exercise to become more familiar with foundations of least squares regression is to use basic summary statistics and point-slope form to produce the least squares line.

TIP: Identifying the least squares line from summary statistics
To identify the least squares line from summary statistics:

- Estimate the slope parameter, b_1, using Equation (7.12)
- Noting that the point $(\bar{x}, \bar{y})$ is on the least squares line, use $x_0 = \bar{x}$ and $y_0 = \bar{y}$ along with the slope b_1 in the point-slope equation:

$$y - \bar{y} = b_1(x - \bar{x})$$

- Simplify the equation.

● **Example 7.16** Using the point $(54.395, 2.468)$ from the sample means and the slope estimate $b_1 = 0.034$ from Exercise 7.14, find the least-squares line for predicting GPA based on satMath.

Apply the point-slope equation using $(54.395, 2.468)$ and the slope, $b_1 = 0.03394$:

$$y - y_0 = b_1(x - x_0)$$
$$y - 2.468 = 0.03394(x - 54.395)$$

Expanding the right side and then adding 2.468 to each side, the equation simplifies:

$$\widehat{GPA} = 0.622 + 0.03394 * satMath$$

Here we have replaced y with $\widehat{GPA}$ and x with $satMath$ to put the equation in context. This form matches the form of Equation (7.11).

We mentioned earlier that a computer is usually used to compute the least squares line. A summary table based on some computer output is shown in Table 7.14 for the satMath and GPA data. The first column of numbers provide estimates for b_0 and b_1, respectively. Compare these to the result from Example 7.16.

	Estimate	Std. Error	t value	Pr(>|t|)
(Intercept)	0.6219	0.1408	4.42	0.0000
satMath	0.0339	0.0026	13.26	0.0000

Table 7.14: Summary of least squares fit for the SAT/GPA data. Compare the parameter estimates in the first column to the results of Example 7.16.

● **Example 7.17** Examine the second, third, and fourth columns in Table 7.14. Can you guess what they represent?

These columns help determine whether the estimates are significantly different from zero and to create a confidence interval for each parameter. The second column lists the standard errors of the estimates, the third column contains t test statistics, and the fourth column lists p-values (2-sided test). We will describe the interpretation of these columns in greater detail in Section 7.4.

7.2.4 Interpreting regression line parameter estimates

Interpreting parameters in a regression model is often one of the most important steps in the analysis.

● **Example 7.18** The slope and intercept estimates for the SAT-GPA data are 0.6219 and 0.0339. What do these numbers really mean?

The intercept $b_0 = 0.6219$ describes the average GPA if a student was at the zeroth percentile score, if the linear relationship held all the way to $satMath = 0$. This interpretation – while perhaps interesting – may not be very meaningful since there are no students enrolled at this college who are very close to the zeroth percentile.

Interpreting the slope parameter is often more realistic and helpful. For each additional SAT percentile score, we would expect a student to have an additional 0.0339 points in their first-year GPA on average. We must be cautious in this interpretation: while there is a real association, we cannot interpret a causal connection between the variables. That is, increasing a student's SAT score may not cause the student's GPA to increase.

Interpreting least squares estimate parameters
The intercept describes the average outcome of y if $x = 0$ *and* the linear model holds. The slope describes the estimated difference in the y variable if the explanatory variable (x) for a case happened to be one unit larger.

7.2.5 Extrapolation is treacherous

When those blizzards hit the East Coast this winter, it proved to my satisfaction that global warming was a fraud. That snow was freezing cold. But in an alarming trend, temperatures this spring have risen. Consider this: On February 6th it was 10 degrees. Today it hit almost 80. At this rate, by August it will be 220 degrees. So clearly folks the climate debate rages on.

Stephen Colbert
April 6th, 2010 [10]

[10] http://www.colbertnation.com/the-colbert-report-videos/269929/

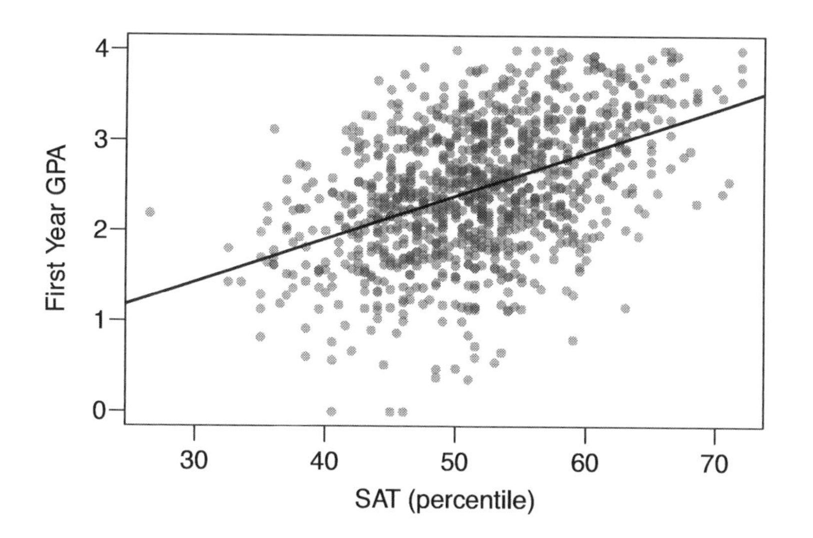

Figure 7.15: A scatterplot of GPA scores versus overall SAT percentile. The least squares regression line is shown.

Linear models are used to approximate the relationship between two variables. However, these models have real limitations. Linear regression is simply a modeling framework. The truth is almost always much more complex than our simple line. For example, we do not know how the data outside of our limited window will behave.

SAT math scores were used to predict freshman GPA in Section 7.2.3. We could also use the overall percentile as a predictor in place of just the math score:

$$\widehat{GPA} = 0.0019 + 0.0477 * satTotal$$

These data are shown in Figure 7.15. The linear model in this case was built for observations between the 26th and the 72nd percentiles. The data meet all conditions necessary for the least squares regression line, so could the model safely be applied to students at the 90th percentile?

● **Example 7.19** Use the model $\widehat{GPA} = 0.0019 + 0.0477 * satTotal$ to estimate the GPA of a student who is at the 90th percentile.

To predict the GPA for a person with an SAT percentile of 90, we plug 90 into the regression equation:

$$0.0019 + 0.0477 * satTotal = 0.0019 + 0.0477 * 90 = 4.29$$

The model predicts a GPA score of 4.29. GPAs only go up to 4.0 (!).

Applying a model estimate to values outside of the realm of the original data is called **extrapolation**. Generally, a linear model is only an approximation of the real relationship between two variables. If we extrapolate, we are making an unreliable bet that the approximate linear relationship will be valid in places where it has not been analyzed.

7.2.6 Using R^2 to describe the strength of a fit

We evaluated the strength of the linear relationship between two variables earlier using correlation, R. However, it is more common to explain the strength of a linear fit using R^2, called **R-squared**. If we are given a linear model, we would like to describe how closely the data cluster around the linear fit.

The R^2 of a linear model describes the amount of variation in the response that is explained by the least squares line. For example, consider the SAT-GPA data, shown in Figure 7.15. The variance of the response variable, GPA, is $s^2_{_{GPA}} = 0.549$. However, if we apply our least squares line, then this model reduces our uncertainty in predicting GPA using a student's SAT score. The variability in the residuals describes how much variation remains after using the model: $s^2_{_{RES}} = 0.433$. In short, there was a reduction of

$$\frac{s^2_{_{GPA}} - s^2_{_{RES}}}{s^2_{_{GPA}}} = \frac{0.549 - 0.433}{0.549} = \frac{0.116}{0.549} = 0.21$$

or about 21% in the data's variation by using information about the SAT scores via a linear model. This corresponds exactly to the R-squared value:

$$R = 0.46 \qquad\qquad R^2 = 0.21$$

⊙ **Exercise 7.20** If a linear model has a very strong negative relationship with a correlation of -0.97, how much of the variation in the response is explained by the explanatory variable? Answer in the footnote[11].

7.3 Types of outliers in linear regression

In this section, we identify (loose) criteria for which outliers are important and influential.

Outliers in regression are observations that fall far from the "cloud" of points. These points are especially important because they can have a strong influence on the least squares line.

⊙ **Exercise 7.21** There are six plots shown in Figure 7.16 along with the least squares line and residual plots. For each scatterplot and residual plot pair, identify any obvious outliers and note how you think they influence the least squares line. Recall that an outlier is any point that doesn't appear to belong with the vast majority of the other points. Answer in the footnote[12].

Examine the residual plots in Figure 7.16. You will probably find that there is some trend in the main clouds of (3) and (4). In these cases, the outlier influenced the slope of

[11] About $R^2 = (-0.97)^2 = 0.94$ or 94% of the variation is explained by the linear model.

[12] Across the top, then across the bottom: (1) There is one outlier far from the other points, though it only appears to slightly influence the line. (2) One outlier on the right, though it is quite close to the least squares line, which suggests it wasn't very influential. (3) One point is far away from the cloud, and this outlier appears to pull the least squares line up on the right; examine how the line around the primary cloud doesn't appear to fit very well. (4) There is a primary cloud and then a small secondary cloud of four outliers. The secondary cloud appears to be influencing the line somewhat strongly, making the least square line fit poorly almost everywhere. There might be an interesting explanation for the dual clouds, which is something that could be investigated. (5) there is no obvious trend in the main cloud of points and the outlier on the right appears to largely control the slope of the least squares line. (6) There is one outlier far from the cloud, however, it falls quite close to the least squares line and does not appear to be very influential.

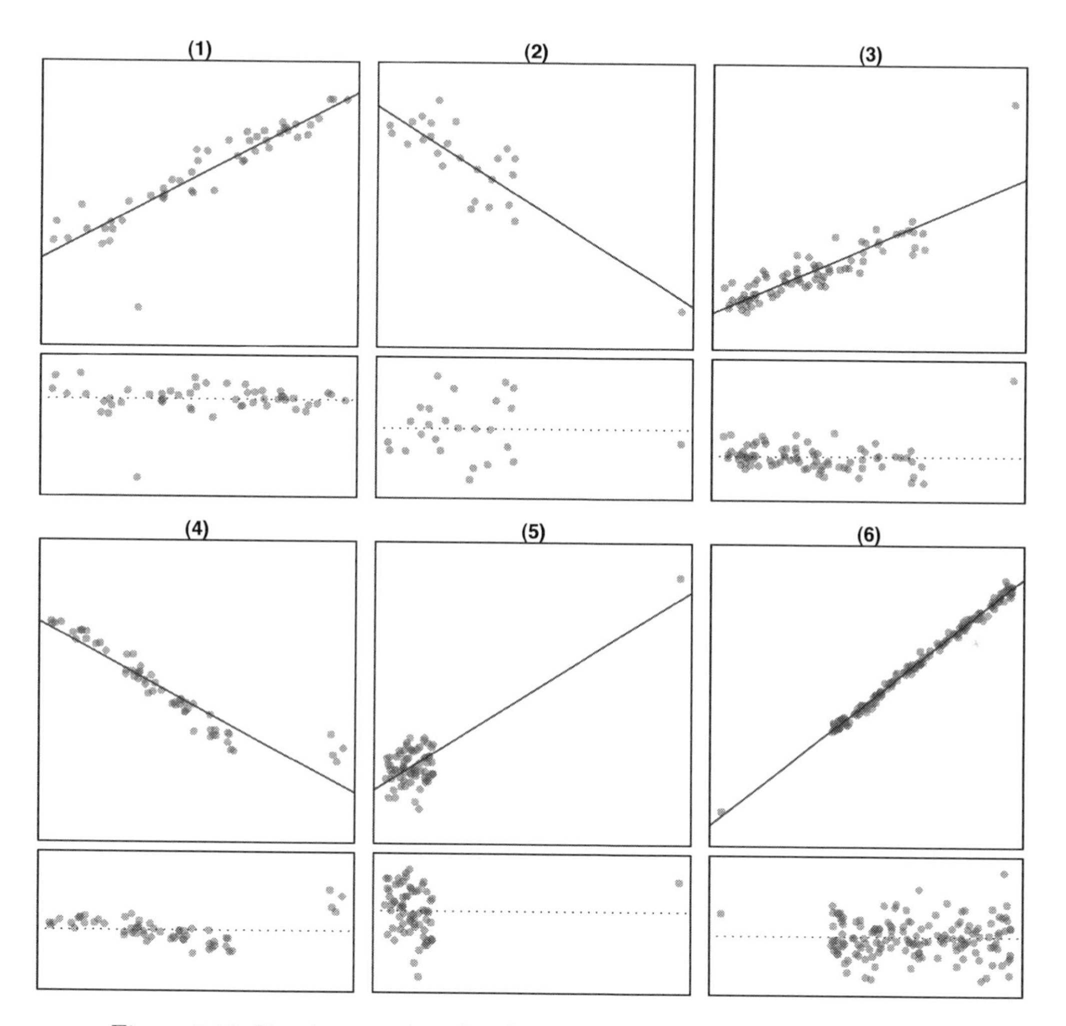

Figure 7.16: Six plots, each with a least squares line and residual plot. All data sets have at least one outlier.

the least squares line. In (5), data with no clear trend were assigned a line with a large trend simply due to one outlier (!).

Leverage
Points that fall, horizontally, away from the center of the cloud tend to pull harder on the line, so we call them points with **high leverage**.

Points that fall horizontally far from the line are points of high leverage; these points can strongly influence the slope of the least squares line. If one of these high leverage points does appear to actually invoke its influence on the slope of the line – as in cases (3), (4), and (5) of Exercise 7.21 – then we call it an **influential point**. Usually we can say a point is influential if, had we fit the line without it, the influential point would have been unusually far from the least squares line.

It is tempting to remove outliers. Don't do this without very good reason. Models that ignore exceptional (and interesting) cases often perform poorly. For instance, if a financial firm ignored the largest market swings – the "outliers" – they would soon go bankrupt by making poorly thought-out investments.

Caution: Don't ignore outliers when fitting a final model
If there are outliers in the data, they should not be removed or ignored without good reason. Whatever final model is fit to the data would not be very helpful if it ignores the most exceptional cases.

7.4 Inference for linear regression

In this section we discuss uncertainty in the estimates of the slope and y-intercept for a regression line. Just as we identified standard errors for point estimates in previous chapters, we first discuss standard errors for these new estimates. However, in the case of regression, we will identify standard errors using statistical software.

7.4.1 Midterm elections and unemployment

Elections for members of the United States House of Representatives occur every two years, coinciding every four years with the U.S. Presidential election. The set of House elections occurring during the middle of a Presidential term are called midterm elections, and are thought to be closely linked with unemployment. In America's two-party system, one political theory suggests the higher the unemployment rate, the worse the President's party will do in the midterm elections.

To assess the validity of this claim, we can compile historical data and look for a connection. We consider every midterm election from 1898 to 2010, with the exception of those elections during the Great Depression. Figure 7.17 shows these data and the least-squares regression line:

$$\begin{aligned}&\text{\% change in House seats for President's party}\\&\qquad = -2.94 - 1.08 * (\text{unemployment rate})\end{aligned}$$

We consider the percent change in the number of seats of the President's party (e.g. percent change in the number of seats for Democrats in 2010) against the unemployment rate.

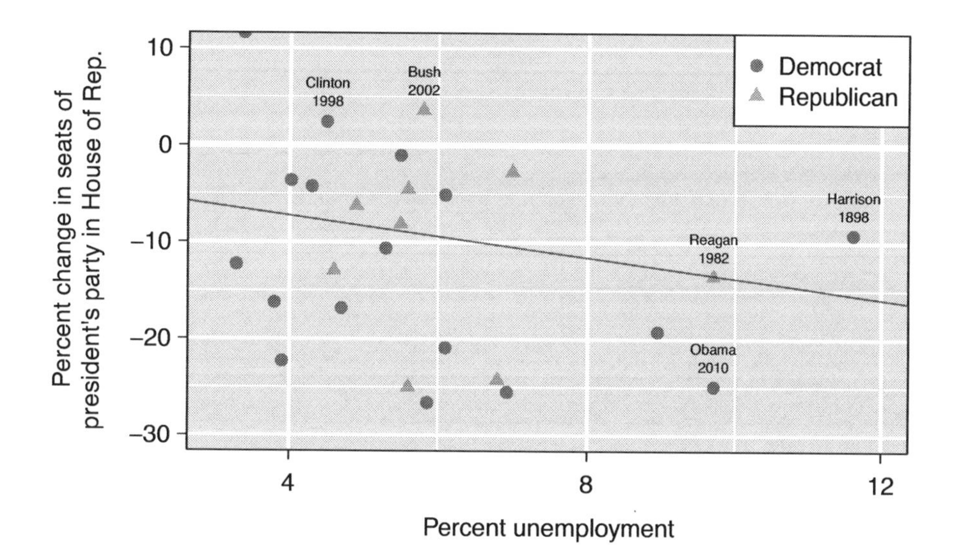

Figure 7.17: The percent change in House seats for the President's party in each election from 1898 to 2010 plotted against the unemployment rate. The two points for the Great Depression have been removed, and a least squares regression line has been fit to the data.

⊙ **Exercise 7.22** The data for the Great Depression (1934 and 1938) were removed because the unemployment rate was 21% and 18%, respectively. Do you agree that they should be removed for this investigation? Why or why not? Two considerations in the footnote[13].

There is a negative slope in the line shown in Figure 7.17. However, this slope (and the y-intercept) are only estimates of the parameter values. We might wonder, is this convincing evidence that the "true" linear model has a negative slope? That is, do the data provide strong evidence that the political theory is accurate? We can frame this investigation into a one-sided statistical hypothesis test:

H_0: $\beta_1 = 0$. The true linear model has slope zero.

H_A: $\beta_1 < 0$. The true linear model has a negative slope. That is, the higher the unemployment, the greater the losses for the President's party in the House of Representatives.

We would reject H_0 in favor of H_A if the data provide strong evidence that the true slope parameter is less than zero. To assess the hypothesis test, we identify a standard error for the estimate, compute an appropriate test statistic, and identify the p-value.

7.4.2 Understanding regression output from software

Just like other point estimates we have seen before, we can compute a standard error for b_1 and a test statistic. We will generally label the test statistic using a T, since it follows the t distribution.

[13]Each of these points would have very high leverage on any least-squares regression line, and years with such high unemployment may not help us understand what would happen in other years where the unemployment is only modestly high. On the other hand, these are exceptional cases, and we would be discarding important information if we exclude them from a final analysis.

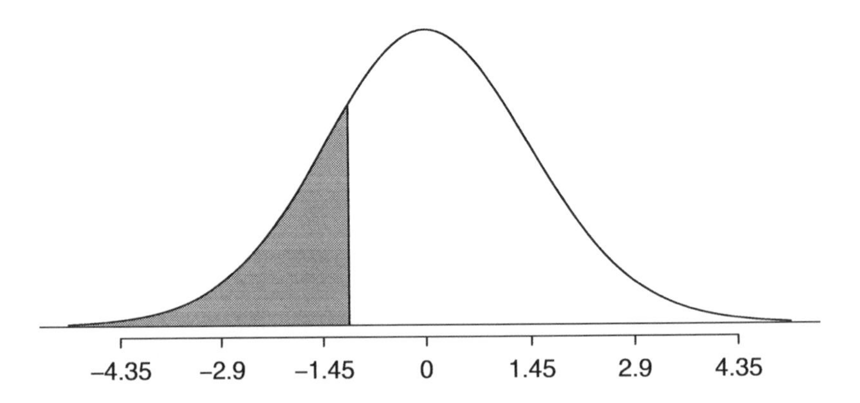

Figure 7.19: The distribution shown here is the sampling distribution for b_1, if the null hypothesis was true. The shaded tail represents the p-value for the hypothesis test evaluating whether there is convincing evidence that higher unemployment corresponds to a greater loss of House seats for the President's party during a midterm election.

We will rely on statistical software to compute the standard error and leave the derivation to a second or third statistics course. Table 7.18 shows software output for the least squares regression line in Figure 7.17. The row labeled *unemp* represents the information for the slope, which is the coefficient of the unemployment variable.

	Estimate	Std. Error	t value	Pr(>|t|)
(Intercept)	-2.9417	9.0851	-0.32	0.7488
unemp	-1.0805	1.4513	-0.74	0.4635
				$df = 25$

Table 7.18: Output from statistical software for the regression line modeling the midterm election losses for the President's party as a response to unemployment.

● **Example 7.23** There are two rows in Table 7.18: one for the y-intercept estimate and one for the slope estimate. What do the first and second columns represent?

The entries in the first column represent the least squares estimate, and the values in the second column correspond to the standard errors of each estimate.

We previously used a t test statistic for hypothesis testing on small (or large!) samples. Regression is very similar. In the hypotheses we consider, the null value for the slope is 0, so we can compute the test statistic using the T (or Z) score formula:

$$T = \frac{\text{estimate} - \text{null value}}{\text{SE}} = \frac{-1.0808 - 0}{1.4513} = -0.74$$

We can look for the one-sided p-value – shown in Figure 7.19 – using the probability table for the t distribution in Appendix C.2 on page 364.

● **Example 7.24** Table 7.18 offers the degrees of freedom for the test statistic T: $df = 25$. Identify the p-value for the hypothesis test.

Looking in the 25 degrees of freedom row in Appendix C.2, we see that the absolute value of the test statistic is smaller than any value listed, which means the tail area and therefore also the p-value is larger than 0.100 (one tail!). Because the p-value is so large, we fail to reject the null hypothesis. That is, the data do not provide convincing evidence that a higher unemployment rate tends to correspond to larger losses for the President's party in the House of Representatives in midterm elections.

We could have identified the t test statistic from the software output in Table 7.18, shown in the second row (unemp) and third column under (t value). The entry in the second row and last column in Table 7.18 represents the p-value for the *two-sided* hypothesis test where the null value is zero. Under close examination, we can see a null hypothesis of 0 was also used to compute the p-value in the *(Intercept)* row. However, we are more often interested in evaluating the significance of the slope since the slope represents a connection between the two variables while the intercept represents an estimate of the average outcome if the x-value was zero.

Inference for regression
We usually rely on statistical software to identify point estimates and standard errors for parameters of a regression line. After verifying conditions hold for fitting a line, we can use the methods learned in Section 6.1 for the t distribution to create confidence intervals for regression parameters or to evaluate hypothesis tests.

Caution: Don't carelessly use the p-value from regression output
The last column in regression output is often used to list p-values for one particular hypothesis: a two-sided test where the null value is zero. If your test is one-sided and the point estimate is in the direction of H_A, then you can halve the software's p-value to get the one-tail area. If neither of these scenarios match your hypothesis test, be cautious about using the software output to obtain the p-value.

● **Example 7.25** Examine Figure 7.15 on page 287, which relates freshman GPA and SAT scores. How sure are you that the slope is statistically significantly different from zero? That is, do you think a formal hypothesis test would reject the claim that the true slope of the line should be zero? Why or why not?

While the relationship between the variables is not perfect, it is difficult to deny that there is some increasing trend in the data. This suggests the hypothesis test will reject the null claim that the slope is zero.

⊙ **Exercise 7.26** Table 7.20 shows statistical software output from fitting the least squares regression line shown in Figure 7.15. Use this output to formally evaluate the following hypotheses. H_0: The true coefficient for `satTotal` is zero. H_A: The true coefficient for `satTotal` is not zero. Answer in the footnote[14].

[14] We look in the second row corresponding to the `satTotal` variable. We see the point estimate of the slope of the line is 0.0477, the standard error of this estimate is 0.0029, and the t test statistic is 16.38. The p-value corresponds exactly to the two-sided test we are interested in: 0.0000. This output doesn't mean the p-value is exactly zero, only that when rounded to four decimal places it is zero. That is, the p-value is so small that we can reject the null hypothesis and conclude that GPA and SAT scores are positively correlated and the true slope parameter is indeed greater than 0, just as we believed in Example 7.25.

	Estimate	Std. Error	t value	Pr(>|t|)
(Intercept)	0.0019	0.1520	0.01	0.9899
satTotal	0.0477	0.0029	16.38	0.0000

Table 7.20: Summary of least squares fit for the SAT/GPA data.

TIP: Always check assumptions
If conditions for fitting the regression line do not hold, then the methods presented here should not be applied. The standard error or distribution assumption of the point estimate – assumed to be normal when applying the t test statistic – may not be valid.

7.4.3 An alternative test statistic

We considered the t test statistic as a way to evaluate the strength of evidence for a hypothesis test in Section 7.4.2. However, we could focus on R^2. Recall that R^2 described the proportion of variability in the response variable (y) explained by the explanatory variable (x). If this proportion is large, then this suggests a linear relationship exists between the variables. If this proportion is small, then the evidence provided by the data may not be convincing.

This concept – considering the amount of variability in the response variable explained by the explanatory variable – is a key component in some statistical techniques. A method called **analysis of variance (ANOVA)** relies on this general principle and is a common topic in statistics. The method states that if enough variability is explained away by the explanatory variable, then we would conclude the variables are connected. On the other hand, we might not be convinced if only a little variability is explained. We will discuss this method further in Section 8.4.

7.5 Exercises

7.5.1 Line fitting, residuals, and correlation

7.1 The scatterplots shown below each have a superimposed regression line. If we were to construct a residual plot (residuals versus x) for each, what would those plots look like?

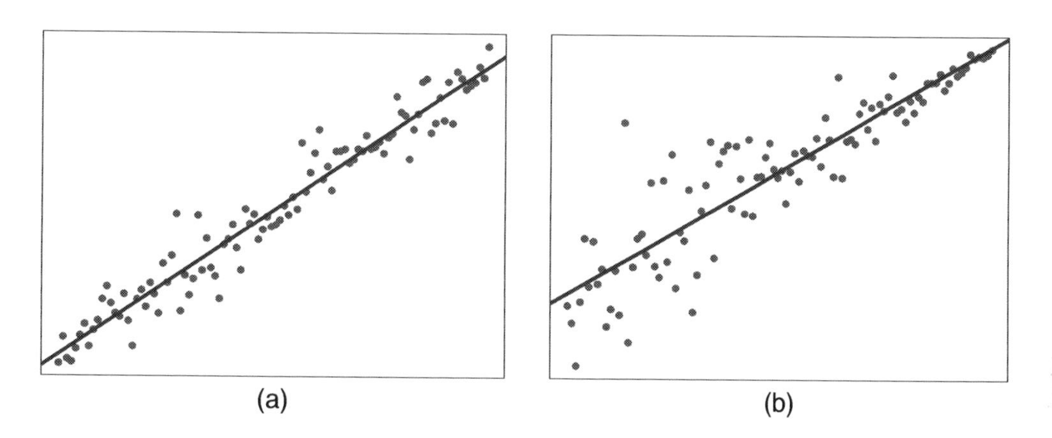

7.2 Shown below are two plots of residuals remaining after fitting a linear model to two different sets of data. Describe any apparent trends in these plots and determine if a linear model would be appropriate for these data. Explain your reasoning.

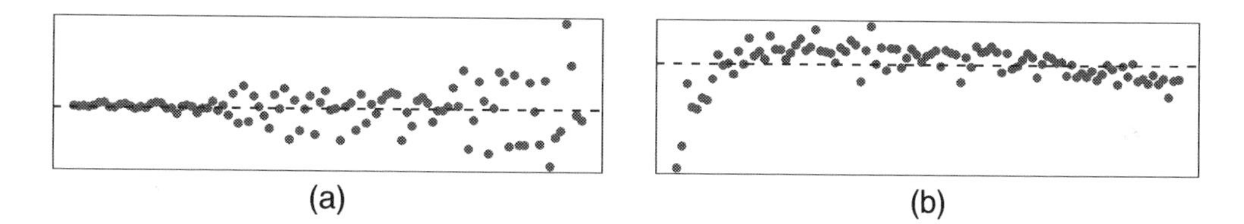

7.3 For each of the six plots, identify the strength of the relationship (e.g. weak, moderate, or strong) in the data and whether fitting a linear model would be reasonable.

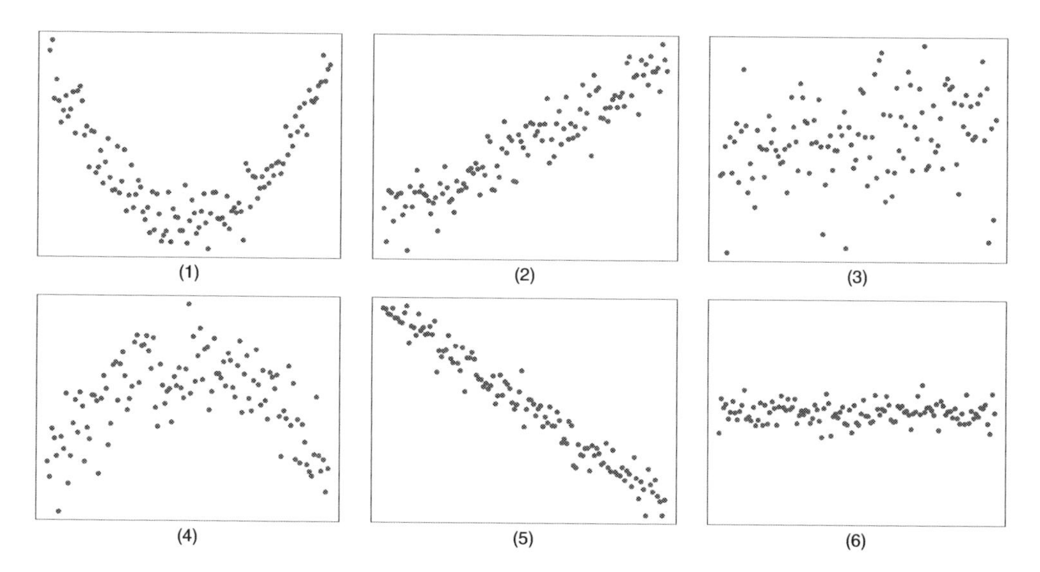

7.4 For each of the six plots, identify the strength of the relationship (e.g. weak, moderate, or strong) in the data and whether fitting a linear model would be reasonable.

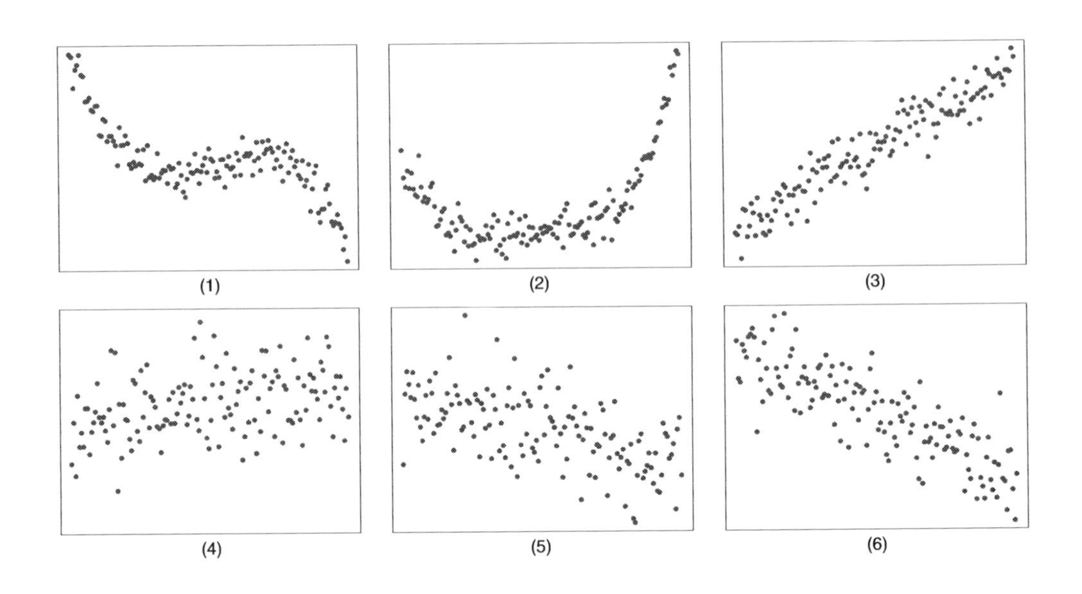

7.5 Below are two scatterplots based on grades recorded during several years for a Statistics course at a university. The first plot shows a scatterplot for final exam grade versus first exam grade for 233 students, and the second scatterplot shows the final exam grade versus second exam grade.

(a) Based on these graphs, which of the two exams has the strongest correlation with the final exam grade? Explain.

(b) Can you think of a reason why the correlation between the exam you chose in part (a) and the final exam is higher?

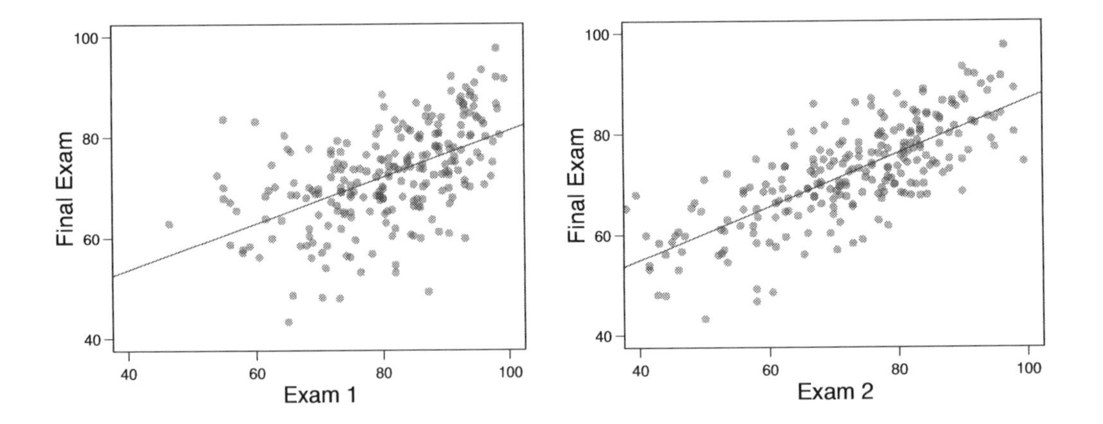

7.6 The Great Britain Office of Population Census and Surveys once collected data on a random sample of 170 married couples in Britain, recording the age (in years) and heights (converted here to inches) of the husbands and wives [44]. The scatterplot on the left shows the wife's age plotted against her husband's age, and the plot on the right shows wife's height plotted against husband's height.

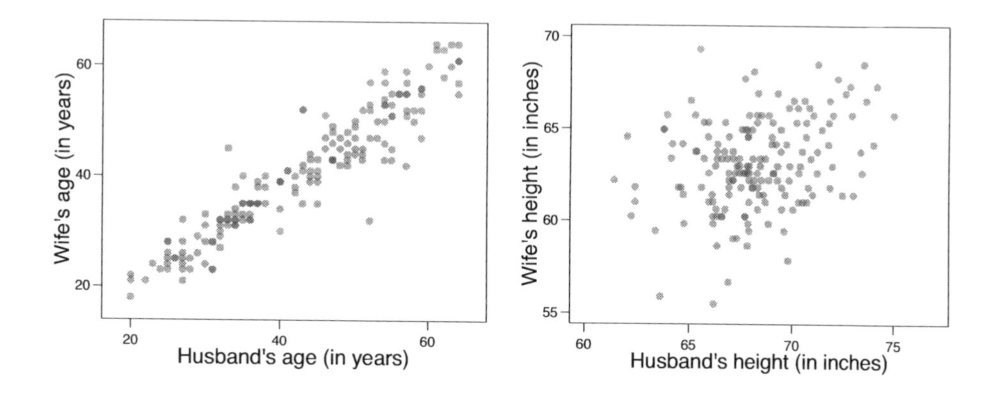

(a) Describe the relationship between husbands' and wives' ages.

(b) Describe the relationship between husbands' and wives' heights.

(c) Which plot shows a stronger correlation? Explain your reasoning.

(d) Data on heights were originally collected in centimeters, and then converted to inches. Does this conversion affect the correlation between husbands' and wives' heights?

7.7 Match the calculated correlations to the corresponding scatterplot.

(a) $R = -0.73$

(b) $R = 0.35$

(c) $R = -0.02$

(d) $R = 0.92$

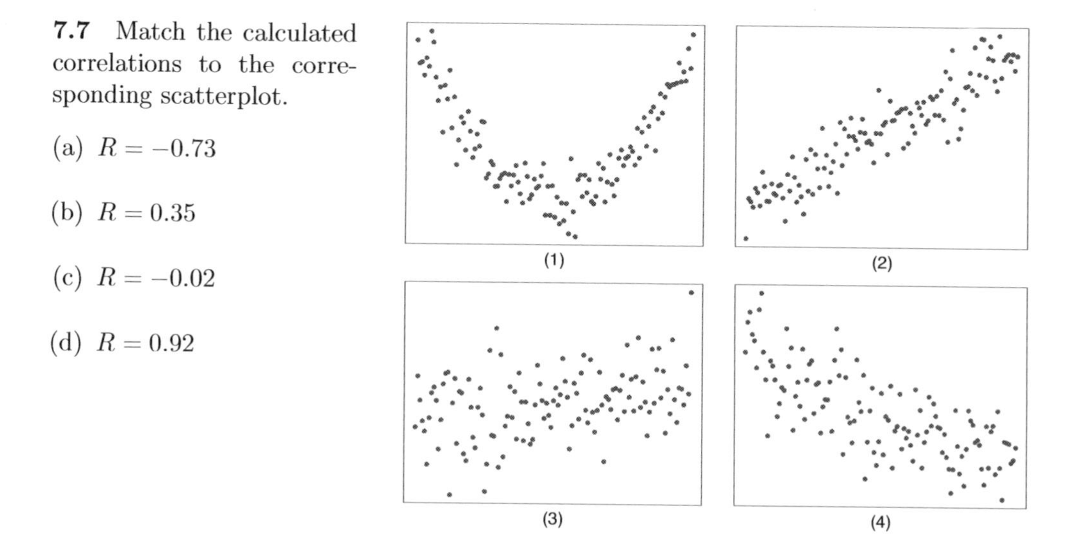

7.8 Match the calculated correlations to the corresponding scatterplot.

(a) $R = 0.35$

(b) $R = -0.52$

(c) $R = -0.06$

(d) $R = -0.80$

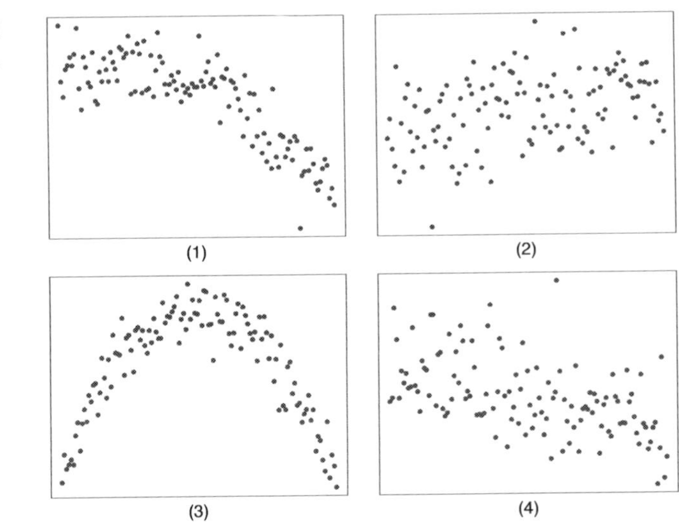

7.9 63 college students were asked to fill out a survey where they were asked about their height, fastest speed they ever drove at, and gender. Below is a scatterplot displaying the relationship between height and fastest speed.

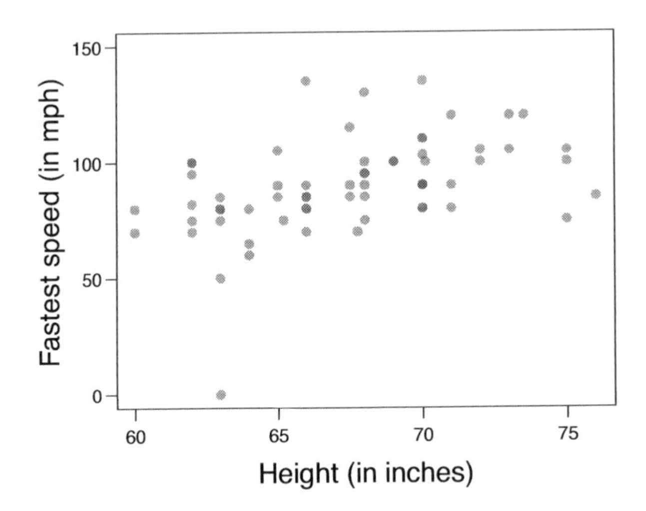

(a) Describe the relationship between height and fastest speed.

(b) Why do you think these variables are positively associated?

(c) Below is another scatterplot displaying the relationship between height and fastest speed. In this plot, female students are represented with triangles and male students are represented with circles. How does gender play a role in the relationship between height and fastest speed.

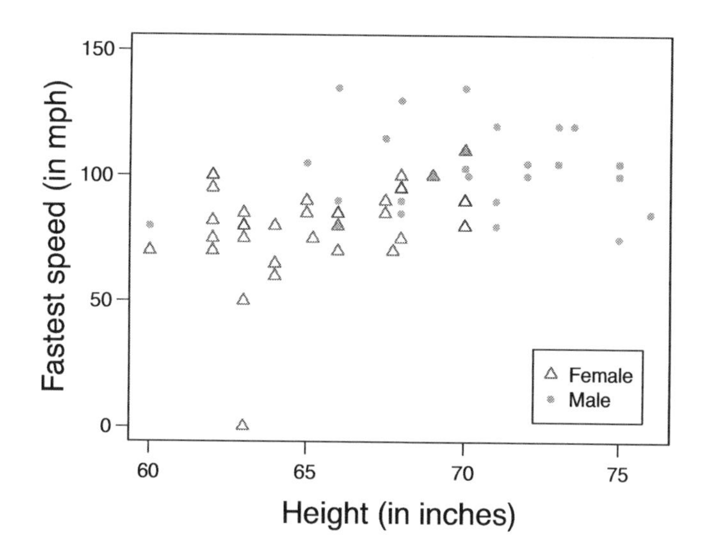

7.10 The scatterplots below show the relationship between height, diameter, and volume of timber in 31 felled black cherry trees. Note that the diameter of the tree is measured at 4 feet 6 inches above the ground. [45]

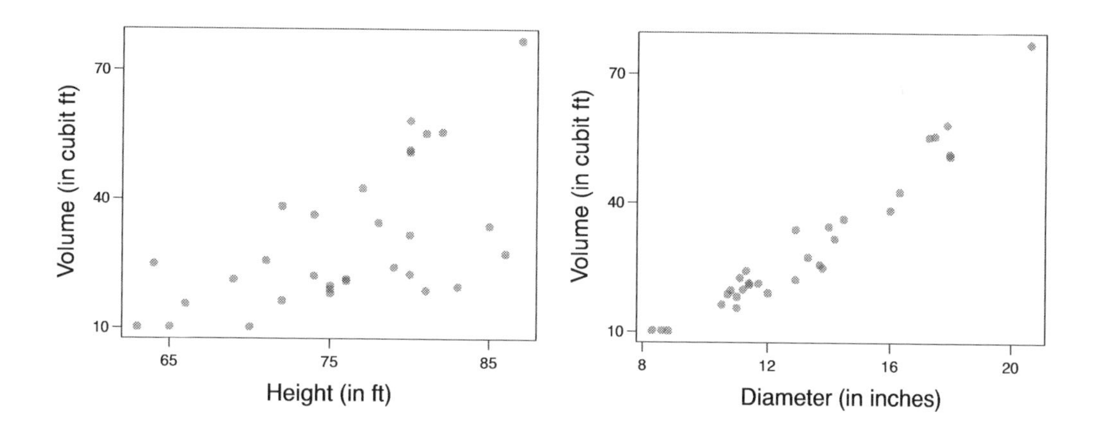

(a) Describe the relationship between volume and height of these trees.

(b) Describe the relationship between volume and diameter of these trees.

(c) Suppose you have height and diameter measurements for another black cherry tree. Which of these variables would you prefer to use to predict the volume of timber in this tree using a simple linear regression model? Explain your reasoning.

7.11 The Coast Starlight Amtrak train runs from Seattle to Los Angeles. The scatterplot below displays the distance between each stop (in miles) and the amount of time it takes to travel from one stop to another (in minutes).

(a) Describe the relationship between distance and travel time.

(b) How would the relationship change if travel time was instead measured in hours, and distance was instead measured in kilometers?

(c) Correlation between travel time (in miles) and distance (in minutes) is $R = 0.636$. What is the correlation between travel time (in kilometers) and distance (in hours).

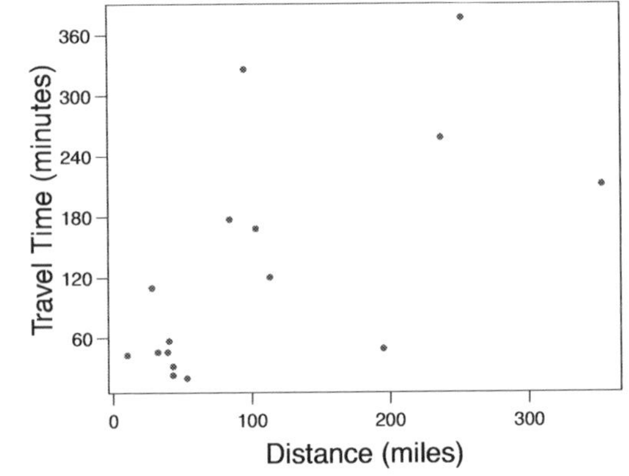

7.12 A study conducted at University of Denver investigated whether babies take longer to learn to crawl in cold months, when they are often bundled in clothes that restrict their movement, than in warmer months [46]. Infants born during the study year were split into twelve groups, one for each birth month. We consider the average crawling age of babies in each group against the average temperature when the babies are six months old (that's when babies often begin trying to crawl). Temperature is measured in degrees Fahrenheit (°F) and age is measured in weeks.

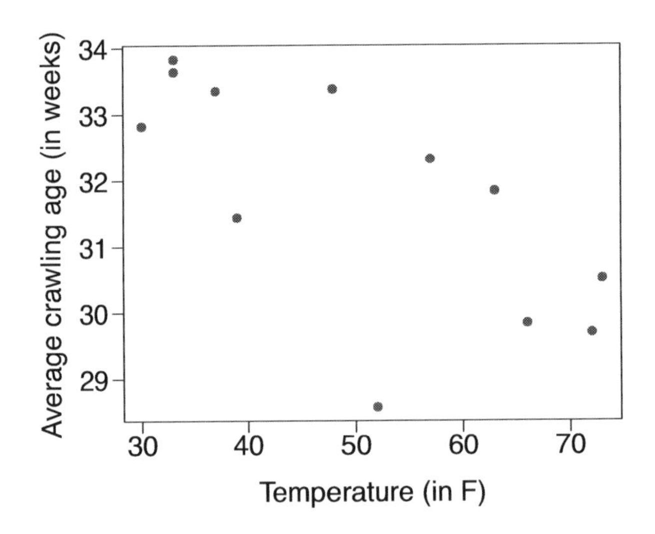

(a) Describe the relationship between temperature and crawling age.

(b) How would the relationship change if temperature was measured in degrees Celsius (°C) and age was measured in months instead of in weeks?

(c) The correlation between temperature in Fahrenheit and age in weeks was $R = -0.70$. If we converted the temperature to Celsius and age to months, what would the correlation be?

7.13 Researchers studying anthropometry collected body girth measurements and skeletal diameter measurements, as well as age, weight, height and gender for 507 physically active individuals [21]. The scatterplot below shows the relationship between height and shoulder girth (over deltoid muscles), both measured in centimeters.

(a) Describe the relationship between shoulder girth and height.

(b) How would the relationship change if shoulder girth was measured in inches while the units of height remained in centimeters?

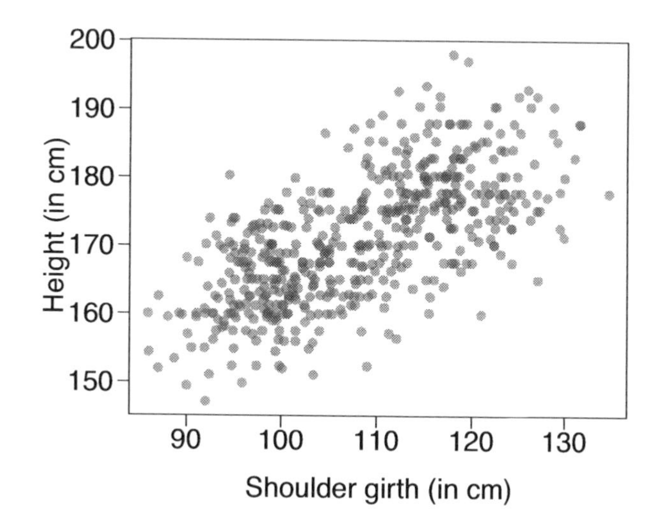

7.14 The scatterplot below shows the relationship between weight measured in kilograms and hip girth measured in centimeters from the data described in Exercise 7.13.

(a) Describe the relationship between hip girth and weight.

(b) How would the relationship change if weight was measured in pounds while the units for hip girth remained in centimeters?

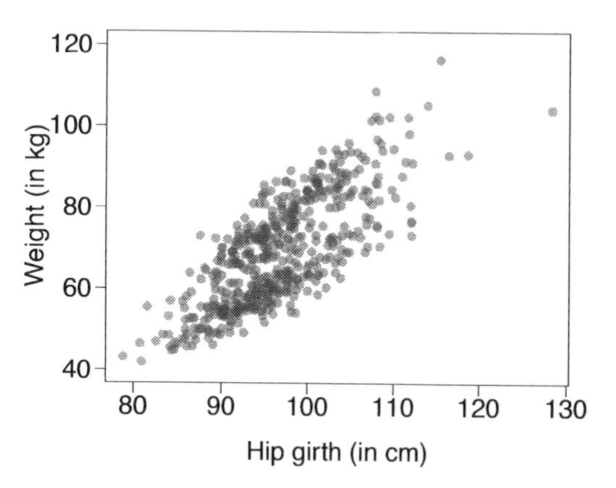

7.15 What would be the correlation between the ages of husbands and wives if men always married woman who were

(a) 3 years younger than themselves?

(b) 2 years older than themselves?

(c) half as old as themselves?

7.16 What would be the correlation between the annual salaries of males and females at a company if for a certain type of position men always made

(a) $5,000 more than women?

(b) 25% more than women?

(c) 15% less than women?

7.5.2 Fitting a line by least squares regression

7.17 The Association of Turkish Travel Agencies reports the number of foreign tourists visiting Turkey and tourist spending by year [47]. The scatterplot below shows the relationship between these two variables along with the least squares fit.

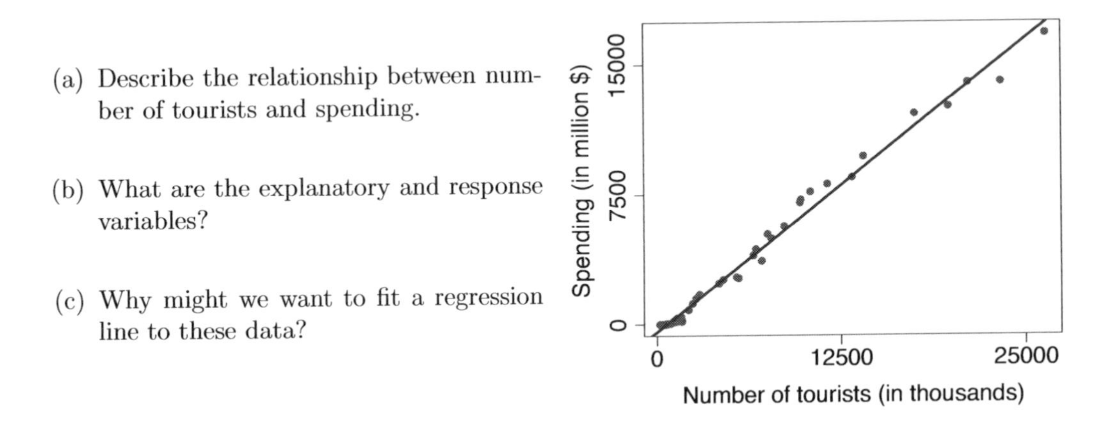

(a) Describe the relationship between number of tourists and spending.

(b) What are the explanatory and response variables?

(c) Why might we want to fit a regression line to these data?

7.18 The scatterplot below shows the relationship between number of calories and amount of carbohydrates (in grams) Starbucks food menu items contain [48]. Since Starbucks only lists the number of calories on the display items, we are interested in predicting the amount of carbs a menu item has based on its calorie content.

(a) Describe the relationship between number of calories and amount of carbohydrates (in grams) Starbucks food menu items contain.

(b) In this scenario, what are the explanatory and response variables?

(c) Why might we want to fit a regression line to these data?

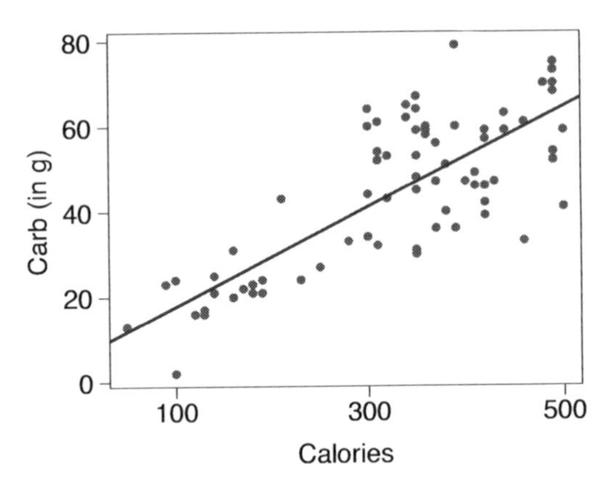

7.19 Does the tourism data plotted in Exercise 7.17 meet the conditions required for fitting a least squares line? In addition to the scatterplot provided in Exercise 7.17, use the residuals plot and the histogram below to answer this question.

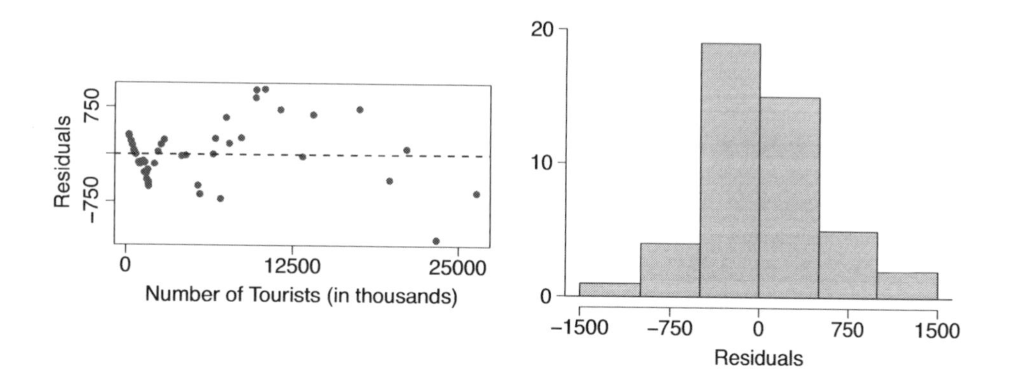

7.20 Does the Starbucks nutrition data plotted in Exercise 7.18 meet the conditions required for fitting a least squares line? In addition to the scatterplot provided in Exercise 7.18, use the residuals plot and the histogram below to answer this question.

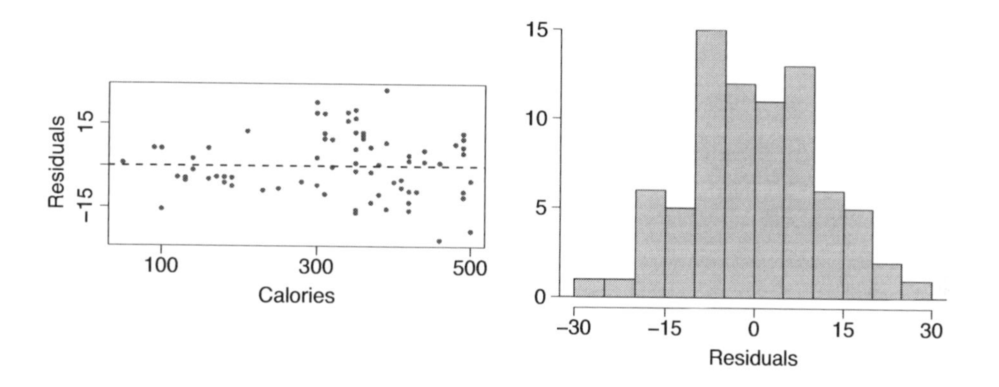

7.21 Exercise 7.11 introduces data on the Coast Starlight Amtrak train that runs from Seattle to Los Angeles. The mean travel time from one stop to the next on the Coast Starlight is 129 mins, with a standard deviation of 113 minutes. The mean distance traveled from one stop to the next is 107 miles with a standard deviation of 99 miles. The correlation between travel time and distance is 0.636.

(a) Write the equation of the regression line for predicting travel time.

(b) Interpret the slope and the intercept in context.

(c) The distance between Santa Barbara and Los Angeles is 103 miles. Use the model to estimate the time it takes for the Starlight to travel between these two cities.

(d) It actually takes the the Coast Starlight 168 mins to travel from Santa Barbara to Los Angeles. Calculate the residual and explain the meaning of this residual value.

(e) Suppose Amtrak is considering adding a stop to the Coast Starlight 500 miles away from Los Angeles. Would it be appropriate to use this linear model to predict the travel time from Los Angeles to this point?

7.22 Exercise 7.13 introduces data on shoulder girth and height of a group of individuals. The mean shoulder girth is 108.20 cm with a standard deviation of 10.37 cm. The mean height is 171.14 cm with a standard deviation of 9.41 cm. The correlation between height and shoulder girth is 0.666.

(a) Write the equation of the regression line for predicting height.

(b) Interpret the slope and the intercept in context.

(c) A randomly selected student from your class has a shoulder girth of 100 cm. Predict the height of this student.

(d) This student is actually 160 cm tall. Calculate the residual and explain what this residual means.

(e) A one year old has a shoulder girth of 56 cm. Would it be appropriate to use this linear model to predict the height of this child?

7.23 Based on the information given in Exercise 7.21, calculate R^2 of the regression line for predicting travel time from distance traveled for the Coast Starlight and interpret it in context.

7.24 Based on the information given in Exercise 7.22, calculate R^2 of the regression line for predicting height from shoulder girth and interpret it in context.

7.25 Data were collected on the number of hours per week students watch TV and the grade they earned in a statistics class (out of 100). Based on the scatterplot and the residuals plot provided, describe the relationship between the two variables and determine if a simple linear model is appropriate to predict grade from the number of hours per week the student watches TV.

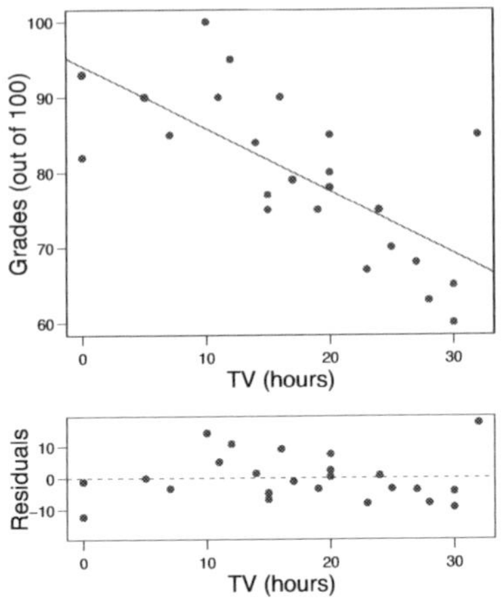

7.26 Exercise 7.18 introduces a data set on nutrition information on Starbucks food menu items. Based on the scatterplot and the residuals plot provided, describe the relationship between the protein content and calories of these menu items and determine if a simple linear model is appropriate to predict protein amount from the number of calories.

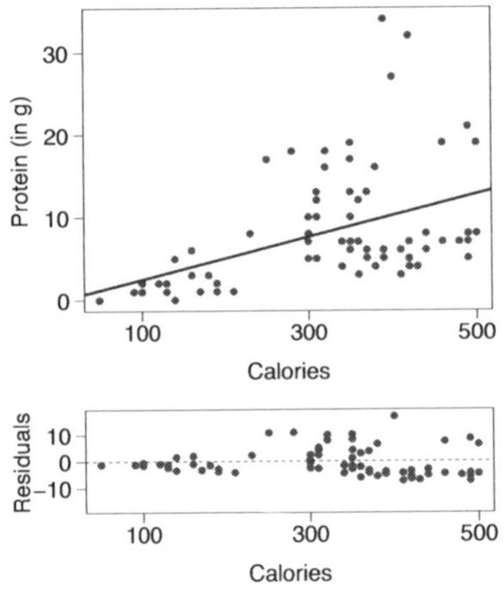

7.5.3 Types of outliers in linear regression

7.27 Identify the outliers in the scatterplots shown below and determine what type of outliers they are. Explain your reasoning.

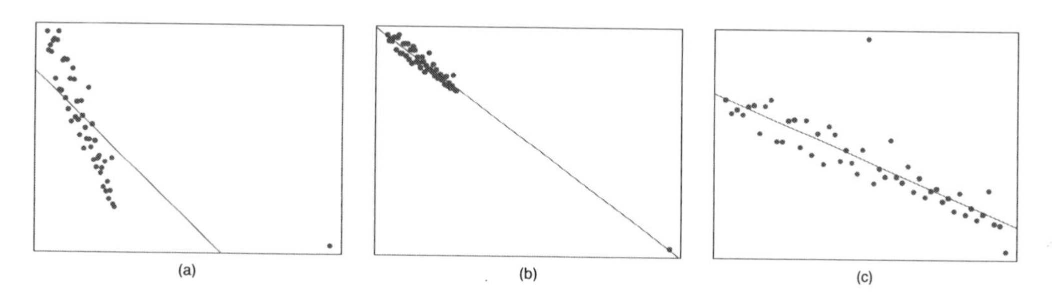

7.28 Identify the outliers in the scatterplots shown below and determine what type of outliers they are. Explain your reasoning.

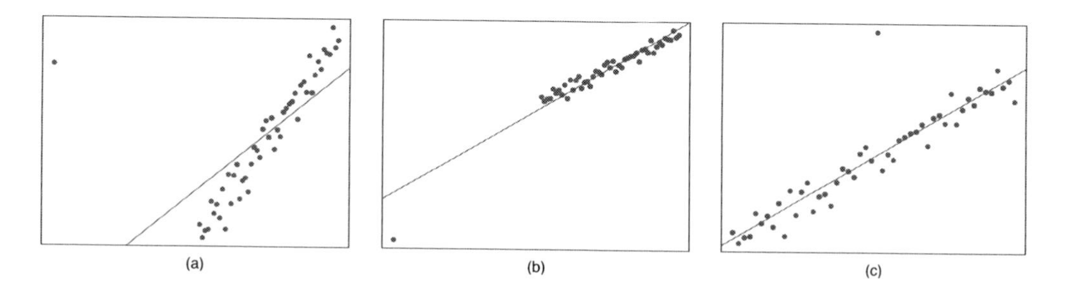

7.29 Exercise 7.12 introduces data on the average monthly temperature during the month babies first try to crawl (about 6 months after birth) and the average first crawling age for babies born in a given month. A scatterplot of these two variables reveals an outlying month when the average temperature is about 53 degrees Fahrenheit and average crawling age is about 28.5 weeks. What type of an outlier is this month? Explain your reasoning.

7.30 The scatterplot below shows the percent of families who own their home vs. the percent of the population living in urban areas in 2000 [49]. There are 52 observations, each corresponding to a state in the US (including Puerto Rico and District of Columbia).

(a) Describe the relationship between the percent of families who own their home and the percent of the population living in urban areas in 2000.

(b) The outlier at the bottom right corner is District of Columbia, where 100% of the population is considered urban. What type of an outlier is this observation?

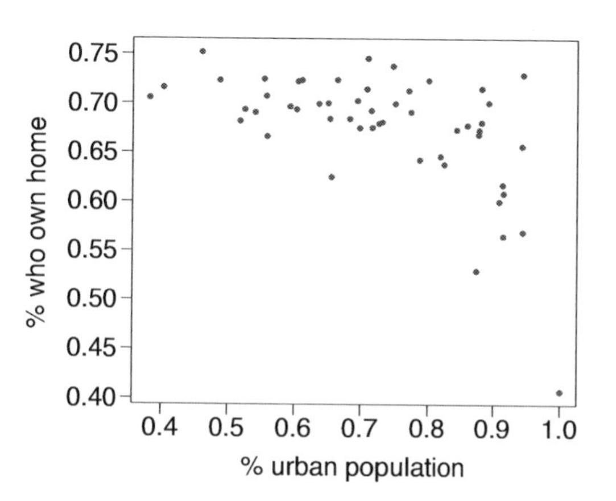

7.5.4 Inference for linear regression

7.31 How well does the number of beers a student drinks predict his or her blood alcohol content? Sixteen student volunteers at Ohio State University drank a randomly assigned number of cans of beer. Thirty minutes later, a police officer measured their blood alcohol content (BAC) in grams of alcohol per deciliter of blood [50]. Note that in all states of the U.S., the legal BAC limit is 0.08. In this experiment, the students were equally divided between men and women and differed in weight and usual drinking habits. Because of this variation, many students don't believe that the number of drinks predicts BAC well. In this problem, we examine how well the number of drinks predicts BAC when there are no other predictors available (Note: each persons tolerance is different, and you should not rely on these data to predict your BAC with high accuracy!) Given below is a scatterplot displaying the relationship between BAC and number of cans of beer as well as a summary output of the least squares fit for these data.

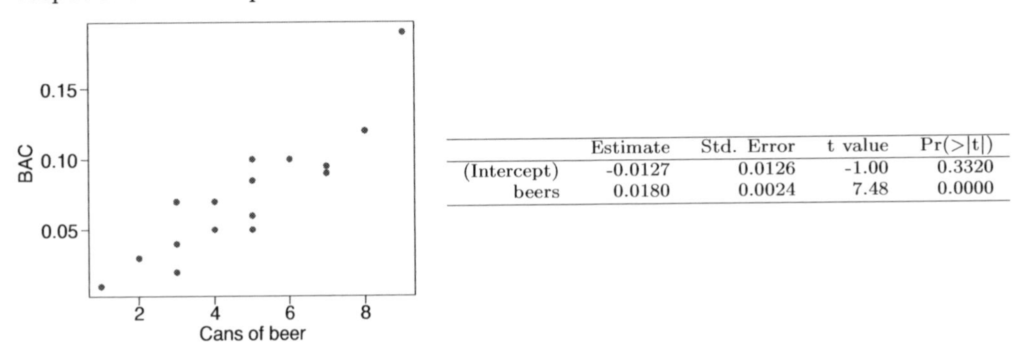

	Estimate	Std. Error	t value	Pr(>|t|)
(Intercept)	-0.0127	0.0126	-1.00	0.3320
beers	0.0180	0.0024	7.48	0.0000

(a) Describe the relationship between number of cans of beer and BAC.

(b) Write the equation of the regression line. Interpret the slope and intercept in context.

(c) Do the data provide strong evidence that drinking more beers is associated with an increase in blood alcohol? State the null and alternative hypotheses, report the p-value, and state your conclusion.

(d) The correlation coefficient for number of cans of beer and BAC is 0.89. Calculate R^2 and interpret it in context.

7.32 The scatterplot and least squares summary below show the relationship between weight measured in kilograms and height measured in centimeters based on data discussed in Exercise 7.13.

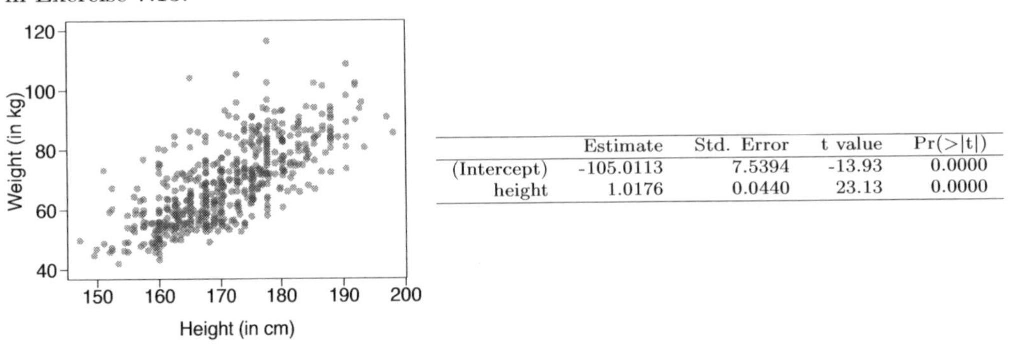

	Estimate	Std. Error	t value	Pr(>|t|)
(Intercept)	-105.0113	7.5394	-13.93	0.0000
height	1.0176	0.0440	23.13	0.0000

(a) Describe the relationship between height and weight.

(b) Write the equation of the regression line. Interpret the slope and intercept in context.

(c) Do the data provide strong evidence that an increase in height is associated with an increase in weight? State the null and alternative hypotheses, report the p-value, and state your conclusion.

(d) The correlation coefficient for height and weight is 0.72. Calculate R^2 and interpret it in context.

7.33 Exercise 7.6 presents a scatterplot displaying the relationship between husbands' and wives' ages in a random sample of 170 married couples in Britain. Given below is a summary output of the least squares fit for predicting wife's age from husband's age.

	Estimate	Std. Error	t value	Pr(>|t|)
(Intercept)	1.5740	1.1501	1.37	0.1730
ageHusband	0.9112	0.0259	35.25	0.0000

(a) Is there a statistically significant linear relationship between husbands' and wives' heights? State the hypotheses and include any information used to conduct the test.

(b) Write the equation of the regression line for predicting wife's age from husband's age.

(c) Interpret the slope and intercept in context.

7.34 Exercise 7.6 presents a scatterplot displaying the relationship between husbands' and wives' heights in a random sample of 170 married couples in Britain. Given below is a summary output of the least squares fit for predicting wife's height from husband's height.

	Estimate	Std. Error	t value	Pr(>|t|)
(Intercept)	43.5755	4.6842	9.30	0.0000
htHusband	0.2863	0.0686	4.17	0.0000

(a) Is there strong evidence that taller men marry taller women? State the hypotheses and include any information used to conduct the test.

(b) Write the equation of the regression line for predicting wife's age from husband's age.

(c) Interpret the slope and intercept in context.

7.35 Exercise 7.33 provides a summary output of the least squares fit for predicting wife's age from husband's age.

(a) Given that $R^2 = 0.88$, what is the correlation of husband and wife height in this data set?

(b) You meet a married man from Britain not included in the sample of 170 couples but who comes from the population that was sampled for this study and who is 55 years old. What would you predict his wife's height to be? How reliable is this prediction?

(c) You meet another married man from Britain not included in the sample of 170 couples but who comes from the population that was sampled for this study and who is 85 years old. Would it be wise to use the same linear model to predict his wife's age? Why or why not.

7.36 Exercises 7.6 and 7.34 provide a summary plot and a regression summary table for predicting wife's height from husband's height.

(a) Given that $R^2 = 0.09$, what is the correlation of husband and wife height in this data set?

(b) You meet a married man from Britain not included in the sample of 170 couples but who comes from the population that was sampled for this study and who is 5'9" (69 inches). What would you predict his wife's height to be? How reliable is this prediction?

(c) You meet another married man from Britain not included in the sample of 170 couples but who comes from the population that was sampled for this study and who is 6'7" (79 inches). Would it be wise to use the same linear model to predict his wife's height? Why or why not?

7.37 Exercise 7.30 gives a scatterplot displaying the relationship between the percent of families that own their home. Below is a summary of a least squares line for these data, excluding District of Columbia. There were 51 cases.

	Estimate	Std. Error	t value	Pr(>|t|)
(Intercept)	0.7976	0.0275	28.96	0.0000
percUrb	-0.1616	0.0374	-4.32	0.0001

(a) For these data, $R^2 = 0.28$. What is the correlation? How can you tell if it is positive or negative?

(b) Test the hypothesis that there is no linear association between percent home ownership and percent of the population living in an urban setting. State the hypotheses and include any information used to conduct the test.

(c) Calculate the predicted values percent home ownership for populations where 40% and 80% live in an urban setting. Use these two values to sketch the regression line on the scatterplot. State whether you believe a simple least squares line adequately fits these data.

(d) The residual plot for this regression is given below. How would you describe the trend visible in this plot? Based on this plot, should a simple least squares line be fit to these data?

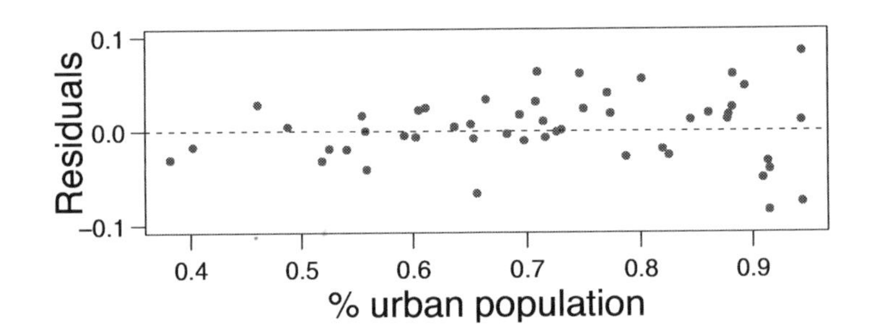

Chapter 8

Multiple regression and ANOVA

The principles of simple linear regression with one numerical predictor and one numerical response lay the foundation for more sophisticated regression methods used in a wide range of challenging settings. In Chapter 8, we explore multiple regression, which introduces the possibility of more than one predictor. We will also consider methods for analysis of variance (ANOVA), a tool useful both in practice and when learning about the mechanics of regression.

8.1 Introduction to multiple regression

Multiple regression extends the simple bivariate regression (two variables: x and y) to the case that still has one response but may have many predictors (denoted x_1, x_2, x_3, ...). The method is motivated by scenarios where many variables may be simultaneously connected to an output.

We will consider Ebay auctions of a video game called *Mario Kart* for the Nintendo Wii. The outcome variable of interest is the total price of an auction - the highest bid plus the shipping cost. But how is the total price related to characteristics of an auction? For instance, are longer auctions associated with a higher or lower prices? And, on average, how much more do buyers tend to pay for additional Wii wheels (plastic steering wheels that attach to the Wii controller) in auctions? Multiple regression will help us answer these and other questions.

The data set `marioKart` includes results from 143 auctions[1]. Four observations from this data set are shown in Table 8.1, and descriptions for each variable are shown in Table 8.2.

8.1.1 Using categorical variables with two levels as predictors

There are two predictor variables in the `marioKart` data set that are inherently categorical: a variable describing the condition of the game and the variable describing whether a stock photo was used for the auction. Two-level categorical variables are often coded using `0`'s

[1] Diez DM, Barr CD, and Çetinkaya M. 2011. *openintro*: OpenIntro data sets and supplemental functions. R package Version 1.2.

	totalPr	condNew	stockPhoto	duration	wheels
1	51.55	1	1	3	1
2	37.04	0	1	7	1
⋮	⋮	⋮	⋮	⋮	⋮
142	38.76	0	0	7	0
143	54.51	1	1	1	2

Table 8.1: Four observations from the `marioKart` data set.

variable	description
`totalPr`	the total of the final auction price and the shipping cost, in US dollars
`condNew`	a coded two-level categorical variable, which takes value `1` when the game is new and `0` if the game is used
`stockPhoto`	a coded two-level categorical variable, which takes value `1` if the primary photo used in the auction was a stock photo and `0` if the photo was unique to that auction
`duration`	the length of the auction, in days
`wheels`	the number of Wii wheels included with the auction (a *Wii wheel* is a plastic racing wheel that holds the Wii controller and is an optional but helpful accessory for playing Mario Kart Wii)

Table 8.2: Variables and their descriptions for the `marioKart` data set.

and `1`'s, which allows them to be incorporated into a regression model in the same way as a numerical predictor:

$$\widehat{\texttt{totalPr}} = \beta_0 + \beta_1 * \texttt{condNew}$$

If we fit this model for total price and game condition using simple linear regression, we obtain the following regression line estimate:

$$\widehat{\texttt{totalPr}} = 42.87 + 10.90 * \texttt{condNew} \qquad (8.1)$$

The `0-1` coding of the two-level categorical variable allows for a simple interpretation of the coefficient of `condNew`. When the game is in `used` condition, the `condNew` variable takes a value of zero, and the total auction price predicted from the model would be $\$42.87 + \$10.90 * (0) = \$42.87$. If the game is in `new` condition, then the `condNew` variable takes value one and the total price is predicted to be $\$42.87 + \$10.90 * (1) = \$53.77$. We now see clearly that the coefficient of `condNew` estimates the difference (\$10.90) in the total auction price when the game is new (\$53.77) versus used (\$42.87).

> **TIP: The coefficient of a two-level categorical variable**
> The coefficient of a binary variable corresponds to the estimated difference in the outcome between the two levels of the variable.

⊙ **Exercise 8.2** The best fitting linear model for the outcome `totalPr` and predictor `stockPhoto` is

$$\widehat{\texttt{totalPr}} = 44.33 + 4.17 * \texttt{stockPhoto} \qquad (8.3)$$

where the variable `stockPhoto` takes value `1` when a stock photo is being used and `0` when the photo is unique to that auction. Interpret the coefficient of `stockPhoto`.

Example 8.4 In Exercise 8.2, you found that auctions whose primary photo was a stock photo tended to sell for about \$4.17 more than auctions that feature a unique photo. Suppose a seller learns this and decides to change her Mario Kart Wii auction to have its primary photo be a stock photo. Will modifying her auction in this way earn her, on average, an additional \$4.17?

No, we cannot infer a causal relationship. It might be that there are inherent differences in auctions that use stock photos and those that do not. For instance, if we sorted through the data, we would actually notice that many of the auctions with stock photos tended to also include more Wii wheels. In this case, Wii wheels is a potential lurking variable.

8.1.2 Including and assessing many variables in a model

Sometimes there is underlying structure or relationship between the predictor variables. For instance, new games sold on Ebay tend to come with more Wii wheels, which may have led to higher prices for those auctions. We would like to fit a model that included all potentially important variables simultaneously, which would help us evaluate the relationship between a predictor variable and the outcome while controlling for the potential influence of other variables. This is the strategy used in **multiple regression**. While we remain cautious about making any causal interpretations using multiple regression, such models are a common first step in providing evidence of a causal connection.

Earlier we had constructed a simple linear model using `condNew` as a predictor and `totalPr` as the outcome. We also constructed a separate model using only `stockPhoto` as a predictor. Next, we want a model that uses both of these variables simultaneously and, while we're at it, we'll include the `duration` and `wheels` variables described Table 8.2:

$$\begin{aligned}\widehat{\texttt{totalPr}} &= \beta_0 + \beta_1 * \texttt{condNew} + \beta_2 * \texttt{stockPhoto} \\ &\quad + \beta_3 * \texttt{duration} + \beta_4 * \texttt{wheels} \\ \hat{y} &= \beta_0 + \beta_1 x_1 + \beta_2 x_2 + \beta_3 x_3 + \beta_4 x_4 \end{aligned} \tag{8.5}$$

where y represents the total price, x_1 is the game's condition, x_2 is whether a stock photo was used, x_3 is the duration of the auction, and x_4 is the number of Wii wheels included with the game. Just as with the single predictor case, this model may be missing important components or it might not properly represent the relationship between the total price and the available explanatory variables. However, while no model is perfect, we wish to explore the possibility that this one may fit the data reasonably well.

We estimate the parameters β_0, β_1, ..., β_4 in the same way as we did in the case of a single predictor. We select b_0, b_1, ..., b_4 that minimize the sum of the squared residuals:

$$SSE = \sum_{i=1}^{n} e_i^2 = \sum_{i=1}^{n} (y_i - \hat{y}_i)^2 \tag{8.6}$$

We typically use a computer to minimize this sum and compute point estimates, as shown in the sample output in Table 8.3. Using this output, we identify the point estimates b_i of each β_i, just as we did in the one-predictor case.

	Estimate	Std. Error	t value	Pr(>|t|)
(Intercept)	36.2110	1.5140	23.92	0.0000
condNew	5.1306	1.0511	4.88	0.0000
stockPhoto	1.0803	1.0568	1.02	0.3085
duration	-0.0268	0.1904	-0.14	0.8882
wheels	7.2852	0.5547	13.13	0.0000
				$df = 136$

Table 8.3: Output for the regression model where `totalPr` is the outcome and `condNew`, `stockPhoto`, `duration`, and `wheels` are the predictors.

Multiple regression model
A multiple regression model is a linear model with many predictors. In general, we write the model as

$$\hat{y} = \beta_0 + \beta_1 x_1 + \beta_2 x_2 + \cdots + \beta_p x_p$$

when there are p predictors. We often estimate the β_i parameters using a computer.

⊙ **Exercise 8.7** Write out the model in Equation (8.5) using the point estimates from Table 8.3. How many predictors are there in this model? Answers in the footnote[2].

⊙ **Exercise 8.8** What does β_4, the coefficient of variable x_4 (Wii wheels), represent? What is the point estimate of β_4? Answers in the footnote[3].

⊙ **Exercise 8.9** Compute the residual of the first observation in Table 8.1 on page 310. Hint: use the equation from Exercise 8.7. Answer in the footnote[4].

● **Example 8.10** The coefficients for x_1 (`condNew`) and x_2 (`stockPhoto`) are different than in the two simple linear models shown in Equations (8.1) and (8.3). Why might there be a difference?

If we examined the data carefully, we would see that some predictors are correlated. For instance, when we estimated the connection of the outcome `totalPr` and predictor `stockPhoto` using simple linear regression, we were unable to control for other variables like `condNew`. That model was biased by the lurking variable `condNew`. When we use both variables, this particular underlying and unintentional bias is reduced or eliminated (though bias from other lurking variables may still remain).

Example 8.10 describes a common issue in multiple regression: correlation among predictor variables. We say the two predictor variables are **collinear** when they are correlated, and this collinearity complicates model estimation. While it is impossible to prevent collinearity from arising in observational data, experiments are usually designed to prevent predictors from being correlated.

[2] $\hat{y} = 36.21 + 5.13x_1 + 1.08x_2 - 0.03x_3 + 7.29x_4$, and there are $p = 4$ predictor variables.

[3] It is the average difference in auction price for each additional Wii wheel included when holding the other variables constant. The point estimate is $b_4 = 7.29$.

[4] $e_i = y_i - \hat{y}_i = 51.55 - 49.62 = 1.93$, where 49.62 was computed using the predictor values for the observation and the equation identified in Exercise 8.7.

8.1.3 Adjusted R^2 as a better estimate of explained variance

We first used R^2 in Section 7.2 to determine the amount of variability in the response that was explained by the model:

$$R^2 = 1 - \frac{\text{variability in residuals}}{\text{variability in the outcome}} = 1 - \frac{Var(e_i)}{Var(y_i)}$$

where e_i represents the residuals of the model and y_i the outcomes. This equation remains valid in the multiple regression framework, but a small enhancement can often be even more informative.

⊙ **Exercise 8.11** The variance of the residuals for the model given in Exercise 8.9 is 23.34, and the variance of the total price in all the auctions is 83.06. Verify the R^2 for this model is 0.719.

This strategy for estimating R^2 is okay when there is just a single variable. However, it becomes less helpful when there are many variables. The regular R^2 is actually a biased estimate of the amount of variability explained by the model. To get a better estimate, we use the adjusted R^2.

Adjusted $\mathbf{R^2}$ as a tool for model assessment
The **adjusted $\mathbf{R^2}$** is computed as

$$R^2_{adj} = 1 - \frac{Var(e_i)/(n-p-1)}{Var(y_i)/(n-1)} = 1 - \frac{Var(e_i)}{Var(y_i)} \times \frac{n-1}{n-p-1}$$

where n is the number of cases used to fit the model and p is the number of predictor variables in the model.

Because p is never negative, the adjusted R^2 will be smaller – often times just a little smaller – than the unadjusted R^2. The reasoning behind the adjusted R^2 lies with the **degrees of freedom** associated with each variance[5].

⊙ **Exercise 8.12** There were $n = 141$ auctions in the `marioKart` data set and $p = 4$ predictor variables in the model. Use n, p, and the variances from Exercise 8.11 to verify $R^2_{adj} = 0.711$ for the Mario Kart model.

⊙ **Exercise 8.13** Suppose you added another predictor to the model, but the variance of the errors $Var(e_i)$ didn't go down. What would happen to the R^2? What would happen to the adjusted R^2? Answers in the footnote[6].

The idea that a predictor that doesn't explain any extra variance would actually "hurt" the adjusted R^2 highlights a common sentiment in statistics: avoid making a model more complicated than it needs to be.

---

[5] In multiple regression, the degrees of freedom associated with the variance of the estimate of the residuals is $n-p-1$, not $n-1$. For instance, if we were to make predictions for new data using our current model, we would find that the unadjusted R^2 is an overly optimistic estimate of the reduction in variance in the response, and using the degrees of freedom in the adjusted R^2 formula helps correct this bias.

[6] The unadjusted R^2 would stay the same and the adjusted R^2 would go down.

8.2 Model selection

The best model is not always the most complicated. Sometimes including variables that are not evidently important can actually reduce the accuracy of predictions. In this section we discuss model selection strategies, which will help us eliminate variables that are less important from the model.

In this section, and in practice, the model that includes all available explanatory variables is often referred to as the **full model**. Our goal is assess whether the full model is the best model. If it isn't, we want to identify a smaller model that is preferable.

8.2.1 Identifying variables that may not be helpful in the model

Table 8.4 provides a summary of the regression output for the full model. The last column of the table lists p-values that can be used to assess hypotheses of the following form:

H_0: $\beta_i = 0$ when the other explanatory variables are included in the model.

H_A: $\beta_i \neq 0$ when the other explanatory variables are included in the model.

	Estimate	Std. Error	t value	Pr(>|t|)
(Intercept)	36.2110	1.5140	23.92	0.0000
condNew	5.1306	1.0511	4.88	0.0000
stockPhoto	1.0803	1.0568	1.02	0.3085
duration	-0.0268	0.1904	-0.14	0.8882
wheels	7.2852	0.5547	13.13	0.0000
				$df = 136$

Table 8.4: The fit for the full regression model. This table is identical to Table 8.3.

● **Example 8.14** The coefficient of `condNew` has a t test statistic of $T = 4.88$ and a p-value for its corresponding hypotheses ($H_0 : \beta_1 = 0,\ H_A : \beta_1 \neq 0$) of about zero. How can this be interpretted?

If we keep all the other variables in the model and add no others, then there is strong evidence that a game's condition (new or used) has a real relationship with the total auction price.

● **Example 8.15** Is there strong evidence that using a stock photo is related to the total auction price?

The t test statistic for `stockPhoto` is $T = 1.02$ and the p-value is about 0.31. There is not strong evidence that using a stock photo in an auction is related to the total price of the auction. We might consider removing the `stockPhoto` variable from the model.

⊙ **Exercise 8.16** Identify the p-values for both the `duration` and `wheels` variables in the model. Is there strong evidence supporting the connection of these variables with the total price in the model?

There is not statistically significant evidence that either `stockPhoto` or `duration` are meaningfully contributing to the model. If the coefficients of these variables are not zero, their association with the outcome variable is probably weak. Next we consider common strategies for pruning such variables from a model.

> **TIP: Using adjusted R^2 instead of p-values for model selection**
> The adjusted R^2 may be used as an alternative to p-values for model selection, where a higher adjusted R^2 represents a better model fit. For instance, we could compare two models using their adjusted R^2, and the model with the higher adjusted R^2 would be preferred. This approach tends to include more variables in the final model when compared to the p-value approach.

8.2.2 Two model selection strategies

Two common strategies for adding or removing variables in a multiple regression model are called *backward-selection* and *forward-selection*. These techniques are often referred to as **stepwise** model selection strategies, because they add or delete one variable at a time as they "step" through the candidate predictors. We will discuss these strategies in the context of the p-value approach, however, the adjusted R^2 approach may be employed as an alternative.

The **backward-elimination** strategy starts with the model that includes all potential predictor variables. One-by-one variables are eliminated from the model until only variables with statistically significant p-values remain. The strategy within each elimination step is to drop the variable with the largest p-value, refit the model, and reassess the inclusion of all variables.

● **Example 8.17** Results corresponding to the *full model* for the `marioKart` data are shown in Table 8.4. How should we proceed under the backward-elimination strategy?

There are two variables with coefficients that are not statistically different from zero: `stockPhoto` and `duration`. We first drop the `duration` variable since it has a larger corresponding p-value, *then we refit the model.* A regression summary for the new model is shown in Table 8.5.

In the new model, there is not strong evidence that the coefficient for `stockPhoto` is different from zero (even though the p-value dropped a little) and the other p-values remain very small. So again we eliminate the variable with the largest non-significant p-value, `stockPhoto`, and refit the model. The updated regression summary is shown in Table 8.6.

In the latest model, we see that the two remaining predictors have statistically significant coefficients with p-values of about zero. Since there are no variables remaining that could be eliminated from the model, we stop. The final model includes only the `condNew` and `wheels` variables in predicting the total auction price:

$$\hat{y} = b_0 + b_1 x_1 + b_4 x_4 = 36.78 + 5.58 x_1 + 7.23 x_4$$

As an alternative description of how we could have performed this model selection strategy using adjusted R^2, please see the footnote[7].

[7] At each elimination step, we refit the model without each of the variables up for potential elimination (e.g. in the first step, we would fit four models, where each would be missing a different predictor). If one of these smaller models has a higher adjusted R^2 than our current model, we pick the smaller model with the largest adjusted R^2. Had we used the adjusted R^2 criteria, we would have kept the `stockPhoto` variable in this backwards-elimination example.

	Estimate	Std. Error	t value	Pr(>|t|)
(Intercept)	36.0483	0.9745	36.99	0.0000
condNew	5.1763	0.9961	5.20	0.0000
stockPhoto	1.1177	1.0192	1.10	0.2747
wheels	7.2984	0.5448	13.40	0.0000
				$df = 137$

Table 8.5: The output for the regression model where `totalPr` is the outcome and the duration variable has been eliminated from the model.

	Estimate	Std. Error	t value	Pr(>|t|)
(Intercept)	36.7849	0.7066	52.06	0.0000
condNew	5.5848	0.9245	6.04	0.0000
wheels	7.2328	0.5419	13.35	0.0000
				$df = 138$

Table 8.6: The output for the regression model where `totalPr` is the outcome and the duration and stock photo variables have been eliminated from the model.

Notice that the p-value for `stockPhoto` changed a little from the full model (0.309) to the model that did not include the `duration` variable (0.275). It is common for p-values of one variable to change, due to collinearity, after eliminating a different variable. This fluctuation emphasizes the importance of refitting a model after each variable elimination step. The p-values tend to change dramatically when the eliminated variable is highly correlated with another variable in the model.

The **forward-selection** strategy is the reverse of the backward-elimination technique. Instead of eliminating variables one-at-a-time, we add variables one-at-a-time until we cannot find any variables that present strong evidence of their importance in the model.

● **Example 8.18** Construct a model for the `marioKart` data set using the forward-selection strategy.

We start with the model that includes no variables. Then we fit each of the possible models with just one variable. That is, we fit the model including just the `condNew` predictor, then the model just including the `stockPhoto` variable, then a model with just `duration`, and a model with just `wheels`. Each of the four models (yes, we fit four models!) provides a p-value for the coefficient of the predictor variable. Out of these four variables, the `wheels` variable had the smallest p-value. Since its p-value is less than 0.05 (the p-value was smaller than 2e-16), we add the Wii wheels variable to the model. Once a variable is added in forward-selection, it will be included in all models considered and in the final model.

Since we successfully found a first variable to add, we consider adding another. We fit three new models: (1) the model including just the `condNew` and `wheels` variables (output in Table 8.6), (2) the model including just the `stockPhoto` and `wheels` variables, and (3) the model including only the `duration` and `wheels` variables. Of these models, the first had the lowest p-value for its new variable (the p-value corresponding to `condNew` was 1.4e-08). Because this p-value is below 0.05, we add the `condNew` variable to the model. Now the final model is guaranteed to include both the condition and Wii wheels variables.

We repeat the process a third time, fitting two new models: (1) the model including the `stockPhoto`, `condNew`, and `wheels` variables (output in Table 8.5) and (2) the model including the `duration`, `condNew`, and `wheels` variables. The p-value corresponding to `stockPhoto` in the first model (0.275) was smaller than the p-value corresponding to `duration` in the second model (0.682). However, since this smaller p-value was not below 0.05, there was not strong evidence that it should be included in the model. Therefore, neither variable is added and we are finished.

The final model is the same as that arrived at using the backward-selection strategy: we include the `condNew` and `wheels` variables into the final model. See the footnote for how we would have proceeded had we used the R^2_{adj} criteria instead of examining p-values[8].

> **Model selection strategies**
> The backward-elimination strategy begins with the largest model and eliminates variables one-by-one until we are satisfied that all remaining variables are important to the model. The forward-selection strategy starts with no variables included in the model, then it adds in variables according to their importance until no other important variables are found.

There is no guarantee that the backward-elimination and forward-selection strategies will arrive at the same final model regardless of whether we are using the p-value or R^2_{adj} criteria. If the backwards-elimination and forward-selection strategies are both tried and they arrive at different models, one option is to choose between the models using the R^2_{adj} criteria (other options exist but are beyond the scope of this book).

It is generally acceptable to use just one strategy, usually backward-elimination, and report the final model after verifying the conditions for fitting a linear model are reasonable.

8.3 Checking model assumptions using graphs

Multiple regression methods using the model

$$\hat{y} = \beta_0 + \beta_1 x_1 + \beta_2 x_2 + \cdots + \beta_p x_p$$

generally depend on the following four assumptions:

1. the residuals of the model are nearly normal,
2. the variability of the residuals is nearly constant,
3. the residuals are independent, and
4. each variable is linearly related to the outcome.

Simple and effective plots can be used to check each of these assumptions.

[8] Rather than look for variables with the smallest p-value, we look for the model with the largest R^2_{adj}. Using the forward-selection strategy, we start with the model with no predictors. Next we look at each model with a single predictor. If one of these models has a larger R^2_{adj} than the model with no variables, we use this new model. We repeat this procedure, adding one variable at a time, until we cannot find a model with a smaller R^2_{adj}. If we had done the forward-selection strategy using R^2_{adj}, we would have arrived at the model including `condNew`, `stockPhoto`, and `wheels`, which is a slightly larger model than we arrived at using the p-value approach.

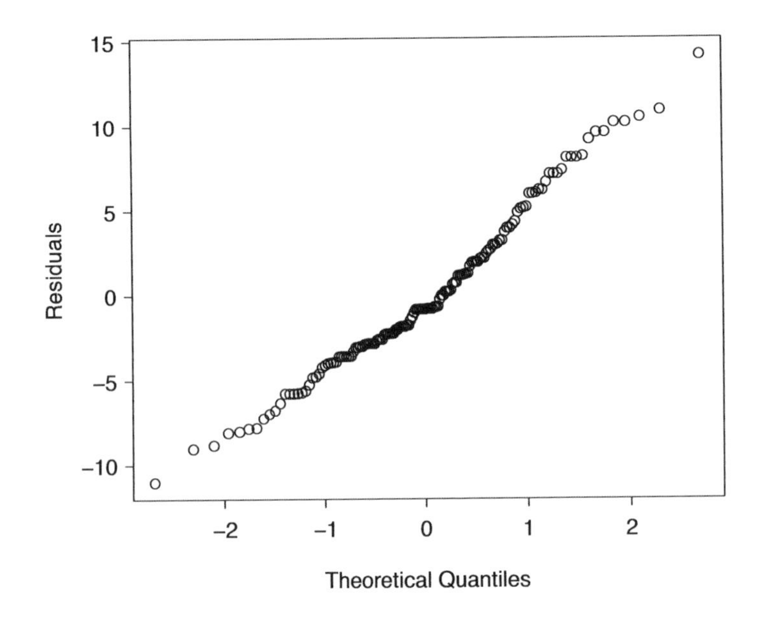

Figure 8.7: A normal probability plot of the residuals is helpful in identifying observations that might be outliers.

Normal probability plot. A normal probability plot of the residuals is shown in Figure 8.7. While the plot exhibits some minor irregularities, there are no outliers that might be cause for concern. In a normal probability plot for residuals, we tend to be most worried about residuals that appear to be outliers, since these indicate long tails in the distribution of residuals.

Absolute values of residuals against fitted values. A plot of the absolute value of the residuals against their corresponding fitted values ($\hat{y}_i$) is shown in Figure 8.8. This plot is helpful to check the condition that the variance of the residuals is approximately constant. We don't see any obvious deviations from constant variance in this example.

Residuals in order of their data collection. A plot of the residuals in the order their corresponding auctions were observed is shown in Figure 8.9. Such a plot is helpful in identifying any connection between cases that are close to one another, e.g. we could look for declining prices over time or if there was a time of the day when auctions tended to fetch a higher price. Here we see no structure that indicates a problem[9].

Residuals against each predictor variable. We consider a plot of the residuals against the `condNew` variable and the residuals against the `wheels` variable. These plots are shown in Figure 8.10. For the two-level condition variable, we are guaranteed not to see a trend, and instead we are verifying that the variability doesn't fluctuate across groups. In this example, when we consider the residuals against the `wheels` variable, we see structure. There appears to be curvature in the residuals, indicating the relationship is probably not linear.

[9] An especially rigorous check would use **time series** methods. For instance, we could check whether consecutive residuals are correlated. Doing so with these residuals yields no statistically significant correlations.

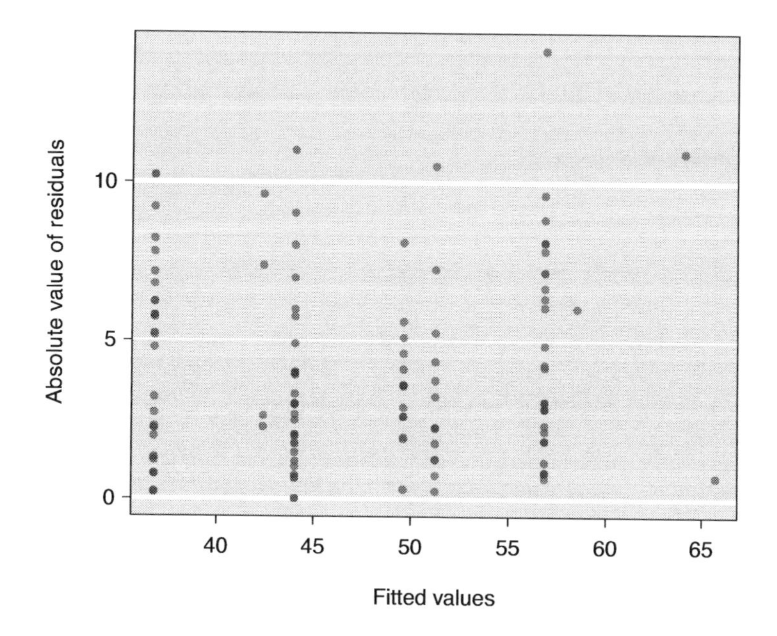

Figure 8.8: Comparing the absolute value of the residuals against the fitted values ($\hat{y}_i$) is helpful in identifying deviations from the constant variance assumption.

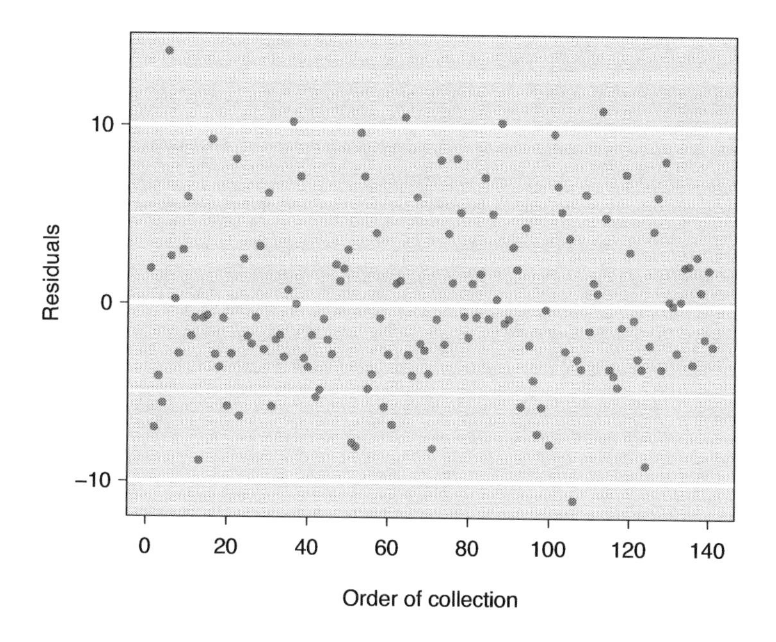

Figure 8.9: Plotting residuals in the order that their corresponding observations were collected helps identify connections between successive observations. If it seems that consecutive observations tend to be close to each other, this indicates the independence assumption of the observations would fail.

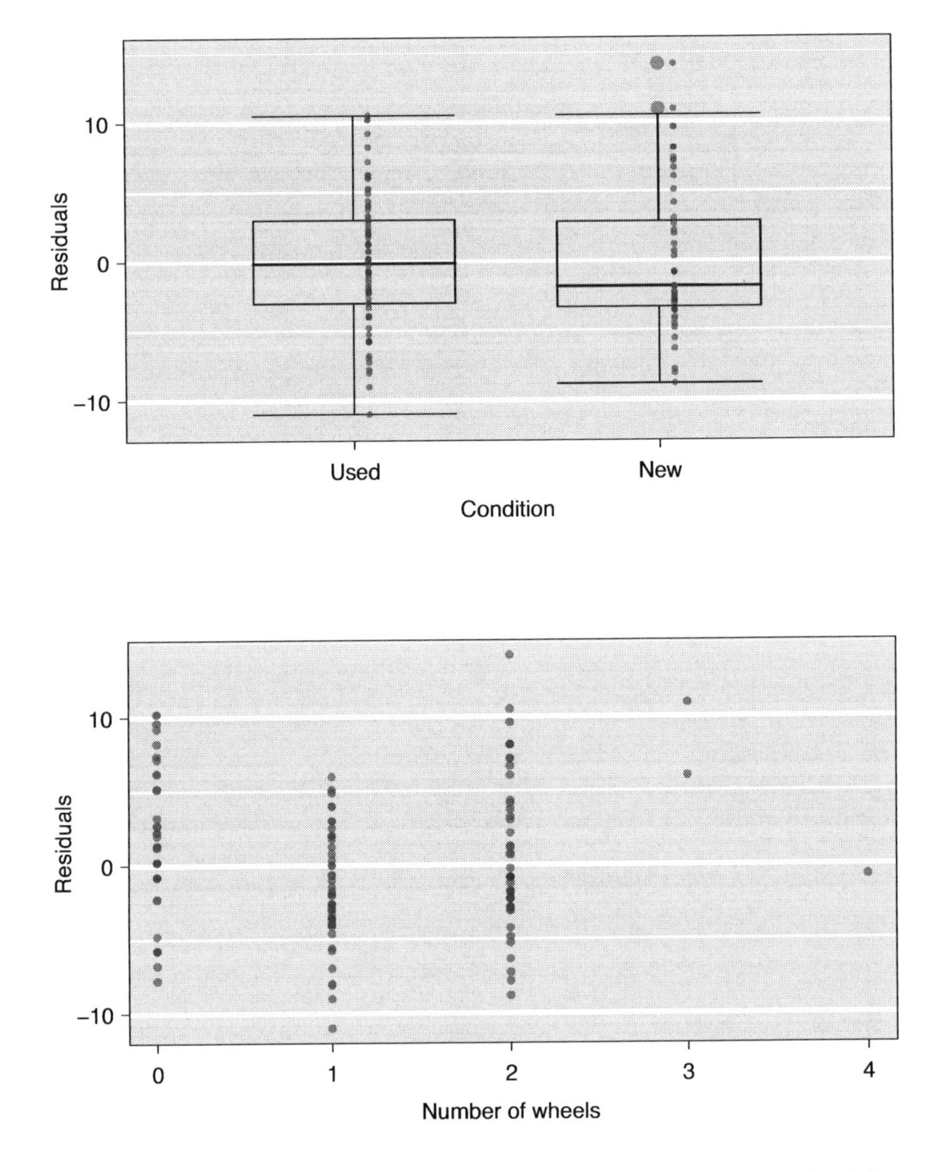

Figure 8.10: In the two-level variable for the game's condition, we check for differences in distribution shape or variability. For numerical predictors, we also check for trends or other structure. We see some slight bowing in the residuals against the `wheels` variable.

It is necessary to summarize diagnostics for any model fit. If the diagnostics support the model assumptions, this would improve credibility in the findings. If the diagnostic assessment shows remaining underlying structure in the residuals, we may still report the model but must also note its shortcomings. In the case of the auction data, we report that there may be a nonlinear relationship between the total price and the number of wheels included for an auction. This information would be important to buyers and sellers; omitting this information could be a setback to the very people who the model might assist.

"All models are wrong, but some are useful" -George E.P. Box
The truth is that no model is perfect. However, even imperfect models can be useful. Reporting a flawed model can be reasonable so long as we are clear and report the model's shortcomings.

Caution: Don't report results when assumptions are heavily violated
While there is a little leeway in model assumptions, don't go too far. If model assumptions are grossly violated, consider a new model, even if it means learning more statistical methods or hiring someone who can help.

TIP: Confidence intervals in multiple regression
Confidence intervals for coefficients in multiple regression can be computed using the same formula as in the single predictor model:

$$b_i \pm t_{df}^{\star} SE_{b_i}$$

where $t_{df}^{\star}$ is the appropriate t value corresponding to the confidence level and model degrees of freedom, $df = n - p - 1$.

8.4 ANOVA and regression with categorical variables

Fitting and interpreting models using categorical variables as predictors is similar to what we have encountered in simple and multiple regression. However, there is a twist: a single categorical variable will have multiple corresponding parameter estimates. To be precise, if the variable has C categories, then there will be $C - 1$ parameter estimates. Furthermore, it is not appropriate to use a Z or T score to determine the significance of the categorical variable as a predictor unless it only has $C = 2$ levels.

In this section, we will learn a new method called **analysis of variance (ANOVA)** and a new test statistic called F. ANOVA is used to assess whether the mean of the outcome variable is different for different levels of a categorical variable:

H_0: The mean outcome is the same across all categories. In statistical notation, $\mu_1 = \mu_2 = \cdots = \mu_k$ where μ_i represents the mean of the outcome for observations in category i.

H_A: The mean of the outcome variable is different for some (or all) groups.

These hypotheses are used to evaluate a model of the form

$$y_{i,j} = \mu_i + \epsilon_j \qquad (8.19)$$

where an observation $y_{i,j}$ belongs to group i and has error ϵ_j. Generally we make three assumptions in applying this model:

- the errors are independent,
- the errors are nearly normal, and
- the errors have nearly constant variance.

These conditions probably look familiar: they are the same conditions we used for multiple regression. When these three assumptions are reasonable, we may perform an ANOVA to determine whether the data provide strong evidence against the null hypothesis that all the μ_i are equal.

> **TIP: Level, category, and group are synonyms**
> We sometimes call the levels of a categorical variable its categories or its groups.

● **Example 8.20** College departments commonly run multiple lectures of the same introductory course each semester because of high demand. Consider a statistics department that runs three lectures of an introductory statistics course. We might like to determine whether there are statistically significant differences in first exam scores in these three classes (A, B, and C). Describe how the model and hypotheses above could be used to determine whether there are any differences between the three classes.

The hypotheses may be written in the following form:

H_0: The average score is identical in all lectures. Any observed difference is due to chance. Notationally, we write $\mu_A = \mu_B = \mu_C$.

H_A: The average score varies by class. We would reject the null hypothesis in favor of this hypothesis if there were larger differences among the class averages than what we might expect from chance alone.

We could label students in the first class as $y_{A,1}$, $y_{A,2}$, $y_{A,3}$, and so on. Students in the second class would be labeled $y_{B,1}$, $y_{B,2}$, etc. And students in the third class: $y_{C,1}$, $y_{C,2}$, etc. Then we could estimate the true averages (μ_A, μ_B, and μ_C) using the group averages: $\bar{y}_A$, $\bar{y}_B$, and $\bar{y}_C$.

Strong evidence favoring the alternative hypothesis in ANOVA is described by unusually large differences among the group means. We will soon learn that assessing the variability of the group means relative to the variability among individual observations within each group is key to ANOVA's success.

● **Example 8.21** Examine Figure 8.11. Compare groups I, II, and III. Can you visually determine if the differences in the group centers is due to chance or not? Now compare groups IV, V, and VI. Do these differences appear to be due to chance?

Any real difference in the means of groups I, II, and III is difficult to discern, because the data within each group are very volatile relative to any differences in the average outcome. On the other hand, it appears there are differences in the centers of groups IV, V, and VI. For instance, group IV appears to have a lower mean than that of the other two groups. Investigating groups IV, V, and VI, we see the differences in the groups' centers are noticeable because those differences are large *relative to the variability in the individual observations within each group*.

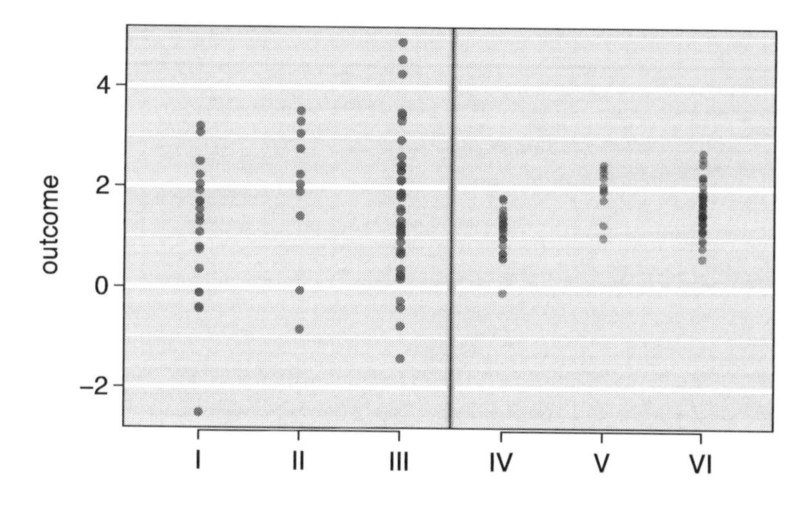

Figure 8.11: Side-by-side dot plot for the outcomes for six groups.

8.4.1 Is batting performance related to player position in MLB?

We would like to discern whether there are real differences between the batting performance of baseball players according to their position: outfielder (`OF`), infielder (`IF`), designated hitter (`DH`), and catcher (`C`). We will use a data set called `mlbBat10`, which includes batting records of 327 Major League Baseball (MLB) players from the 2010 season. Six of the 327 cases represented in `mlbBat10` are shown in Table 8.12, and descriptions for each variable are provided in Table 8.13. The measure we will use for the player batting performance (the outcome variable) is on-base percentage (`OBP`). The on-base percentage roughly represents the fraction of the time a player successfully gets on base or hits a home run.

	name	team	position	AB	H	HR	RBI	AVG	OBP
1	I Suzuki	SEA	OF	680	214	6	43	0.315	0.359
2	D Jeter	NYY	IF	663	179	10	67	0.270	0.340
3	M Young	TEX	IF	656	186	21	91	0.284	0.330
⋮	⋮	⋮	⋮	⋮	⋮	⋮	⋮		
325	B Molina	SF	C	202	52	3	17	0.257	0.312
326	J Thole	NYM	C	202	56	3	17	0.277	0.357
327	C Heisey	CIN	OF	201	51	8	21	0.254	0.324

Table 8.12: Six cases from the `mlbBat10` data matrix.

⊙ **Exercise 8.22** The null hypothesis under consideration is the following: $\mu_{\texttt{OF}} = \mu_{\texttt{IF}} = \mu_{\texttt{DH}} = \mu_{\texttt{C}}$. Write the null and corresponding alternative hypotheses in plain language. Answers in the footnote[10].

[10] H_0: The average on-base percentage is equal across the four positions. H_A: The average on-base percentage varies across some (or all) groups.

variable	description
name	Player name
team	The player's team, where the team names are abbreviated
position	The player's primary field position (OF, IF, DH, C)
AB	Number of opportunities at bat
H	Number of hits
HR	Number of home runs
RBI	Number of runs batted in
batAverage	Batting average, which is equal to H/AB

Table 8.13: Variables and their descriptions for the mlbBat10 data set.

● **Example 8.23** The player positions have been divided into four groups: outfield (OF), infield (IF), designated hitter (DH), and catcher (C). What would be an appropriate point estimate of the batting average by outfielders, $\mu_{\texttt{OF}}$?

A good estimate of the batting average by outfielders would be the sample average of batAverage for just those players whose position is outfield: $\bar{y}_{OF} = 0.334$.

Table 8.14 provides summary statistics for each group. A side-by-side box plot for the batting average is shown in Figure 8.15. Notice that the variability appears to be approximately constant across groups; nearly constant variance across groups is an important assumption that must be satisfied before we consider the ANOVA approach.

	OF	IF	DH	C
Sample size (n_i)	120	154	14	39
Sample mean ($\bar{y}_i$)	0.334	0.332	0.348	0.323
Sample SD (s_i)	0.029	0.037	0.036	0.045

Table 8.14: Summary statistics of on-base percentage, split by player position.

● **Example 8.24** The largest difference between the sample means is between the designated hitter and the catcher positions. Consider again the original hypotheses:

H_0: $\mu_{\texttt{OF}} = \mu_{\texttt{IF}} = \mu_{\texttt{DH}} = \mu_{\texttt{C}}$
H_A: The average on-base percentage (μ_i) varies across some (or all) groups.

Why might it be inappropriate to run the test by simply estimating whether the difference of $\mu_{\texttt{DH}}$ and $\mu_{\texttt{C}}$ is statistically significant at a 0.05 significance level?

The primary issue here is that we are inspecting the data before picking the groups that will be compared. It is inappropriate to examine all data by eye (informal testing) and only afterwards decide which parts to formally test. This is called **data snooping** or **data fishing**. Naturally we would pick the groups with the large differences for the formal test, leading to an unintentional inflation in the Type 1 Error rate. To understand this better, let's consider a slightly different problem.

Suppose we are to measure the aptitude for students in 20 classes in a large elementary school at the beginning of the year. In this school, all students are randomly assigned to classrooms, so any differences we observe between the classes at the start of the year are completely due to chance. However, with so many groups, we will probably

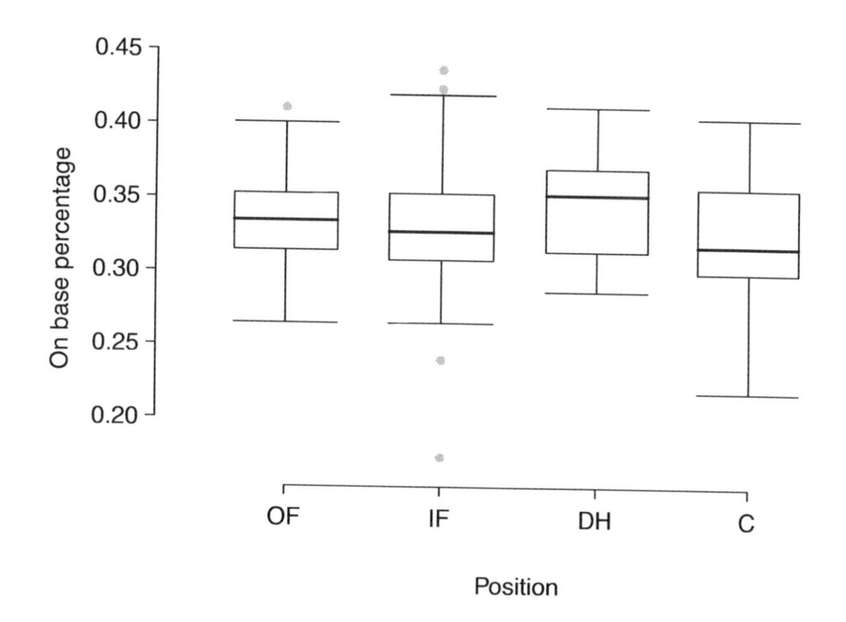

Figure 8.15: Side-by-side box plot of the on-base percentage for 327 players across four groups.

observe a few groups that look rather different from each other. If we select only these classes that look so different, we will probably make the wrong conclusion that the assignment wasn't random. While we might only formally test differences for a few pairs of classes, we informally evaluated the other classes by eye before choosing the most extreme cases for a comparison.

For additional reading on the ideas expressed in Example 8.24, we recommend reading about the **prosecutor's fallacy**[11].

In the next section we will learn how to use the F statistic and ANOVA to test whether differences in means could have happened just by chance.

8.4.2 Analysis of variance (ANOVA) and the F test

The method of analysis of variance focuses on answering one question: is the variability in the sample means so large that it seems unlikely to be from chance alone? This question is different from earlier testing procedures since we will *simultaneously* consider many groups, and evaluate whether their sample means differ more than we would expect from natural variation. We call this variability the **mean square between groups** (MSG), and it has an associated degrees of freedom, $df_G = k - 1$ when there are k groups. The MSG is sort of a scaled variance formula for means. If the null hypothesis is true, any variation in the sample means is due to chance and shouldn't be too large. Details of MSG calculations

[11]See, for example, http://www.stat.columbia.edu/~cook/movabletype/archives/2007/05/the_prosecutors.html.

are provided in the footnote[12], however, we typically use software for these computations.

The mean square between the groups is, on its own, quite useless in a hypothesis test. We need a benchmark value for how much variability should be expected among the sample means if the null hypothesis is true. To this end, we compute the mean of the squared errors, often abbreviated as the **mean square error** (MSE), which has an associated degrees of freedom value $df_E = n-k$. It is helpful to think of MSE as a measure of the variability of the residuals. Details of the computations of the MSE are provided in the footnote[13] for the interested reader.

When the null hypothesis is true, any differences among the sample means are only due to chance, and the MSG and MSE should be about equal. As a test statistic for ANOVA, we examine the fraction of MSG and MSE:

$$F = \frac{MSG}{MSE} \tag{8.25}$$

The MSG represents a measure of the between-group variability, and MSE the variability within each of the groups.

⊙ **Exercise 8.26** For the baseball data, $MSG = 0.00252$ and $MSE = 0.00127$. Identify the degrees of freedom associated with each mean square and verify the F statistic is 1.994.

We use the F statistic to evaluate the hypotheses in what is called an **F test**. We compute a p-value from the F statistic using an F distribution, which has two associated parameters: df_1 and df_2. For the F statistic in ANOVA, $df_1 = df_G$ and $df_2 = df_E$. An F distribution with 3 and 323 degrees of freedom, corresponding to the F statistic for the baseball hypothesis test, is shown in Figure 8.16.

The larger the observed variability in the sample means (MSG) relative to the residuals (MSE), the larger F will be and the stronger the evidence against the null hypothesis. Because larger values of F represent stronger evidence against the null hypothesis, we use the upper tail of the distribution to compute a p-value.

---

[12]Let $\bar{y}$ represent the mean of outcomes across all groups. Then the mean square between groups is computed as

$$MSG = \frac{1}{df_G} SSG = \frac{1}{k-1} \sum_{i=1}^{k} n_i \left(\bar{y}_i - \bar{y}\right)^2$$

where SSG is called the **sum of squares between groups** and n_i is the sample size of group i.

[13]Let $\bar{y}$ represent the mean of outcomes across all groups. Then the **sum of squares total** (SST) is computed as

$$SST = \sum_{i=1}^{n} \left(y_i - \bar{y}\right)^2$$

where the sum is over all observations in the data set. Then we compute the **sum of squared errors** (SSE) in one of three equivalent ways:

$$\begin{aligned} SSE &= SST - SSG \\ &= (n_1 - 1) * s_1^2 + (n_2 - 1) * s_2^2 + \cdots + (n_k - 1) * s_k^2 \\ &= \sum_{j=1}^{n} e_i^2 \end{aligned}$$

where s_i^2 is the sample variance (square of the standard deviation) of the residuals in group i, and the last expression represents the sum of the squared residuals across all groups. Then the MSE is the standardized form of SSE: $MSE = \frac{1}{df_E} SSE$.

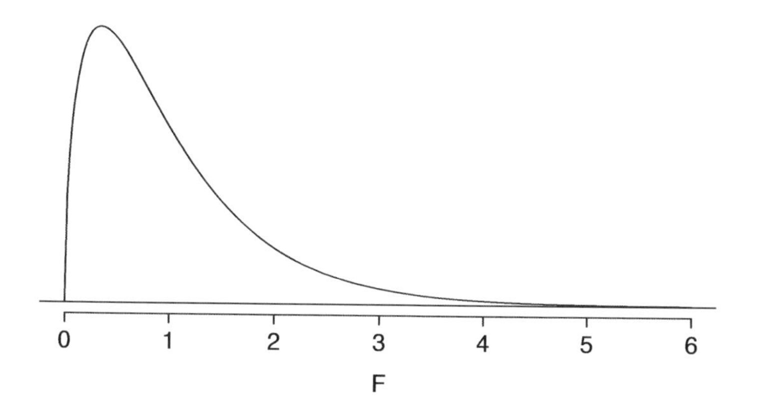

Figure 8.16: An F distribution with $df_1 = 3$ and $df_2 = 323$.

The F statistic and the F test
Analysis of variance (ANOVA) is used to test whether the mean outcome differs across 2 or more groups. ANOVA uses a test statistic F, which represents a standardized ratio of variability in the sample means relative to the variability of the residuals. If H_0 is true and the model assumptions are satisfied, the statistic F follows an F distribution with parameters $df_1 = k - 1$ and $df_2 = n - k$. The upper tail of the F distribution is used to represent the p-value.

⊙ **Exercise 8.27** The test statistic for the baseball example is $F = 1.994$. Shade the area corresponding to the p-value in Figure 8.16.

 **Example 8.28** The p-value corresponding to the solution for Exercise 8.27 is equal to about 0.115. Does this provide strong evidence against the null hypothesis?

The p-value is larger than 0.05, indicating the evidence is not sufficiently strong to reject the null hypothesis at a significance level of 0.05. That is, the data do not provide strong evidence that the average on-base percentage varies by player's primary field position.

8.4.3 Reading regression and ANOVA output from software

The calculations required to perform an ANOVA by hand are tedious and prone to human error. For these reasons it is common to use a statistical software to calculate the F statistic and p-value.

An ANOVA can be summarized in a table very similar to that of a regression summary. Table 8.17 shows an ANOVA summary to test whether the mean of on-base percentage varies by player positions in the MLB.

⊙ **Exercise 8.29** Earlier you verified that the F statistic for this analysis was 1.994, and the p-value of 0.115 was provided. Circle these values in Table 8.17 and notice the corresponding column name. Notice that both of these values are in the row labeled *position*, which corresponds to the categorical variable representing the player position variable.

	Df	Sum Sq	Mean Sq	F value	Pr(>F)
position	3	0.0076	0.0025	1.9943	0.1147
Residuals	323	0.4080	0.0013		
			$s_{pooled} = 0.036$ on $df = 323$		

Table 8.17: ANOVA summary for testing whether the average on-base percentage differs across player positions.

⊙ **Exercise 8.30** The $s_{pooled} = 0.036$ on $df = 323$ describes the estimated standard deviation associated with the residuals. Verify that s_{pooled} equals the square root of the MSE for the *Residuals* row.

8.4.4 Graphical diagnostics for an ANOVA analysis

There are three primary conditions we must check for an ANOVA analysis, all related to the residuals (errors) associated with the model. Recall that we assume the errors are independent, nearly normal, and have nearly constant variance across the groups.

Independence. If observations are collected in a particular order, we should plot the residuals in the order the corresponding observations were collected (e.g. see Figure 8.9 on page 319). For the baseball data, the data were collected from a sorted table, making such a review impossible. However, we can consider the nature of the data: Do we have reason to believe players are not independent? There are not obvious reasons why independence should not hold, so we will assume independence is reasonable in lieu of being able to examine this condition using data.

Approximately normal. The normality assumption for the residuals is especially important when the sample size is quite small. Figure 8.18 shows a normal probability plot for the residuals from the baseball data. We do see some deviation from normality at the low end, where there is a longer tail than what we would expect if the residuals were truly normal. While we should report this finding with the results of the hypothesis test, this slight deviation probably has little impact on the test results since there are so many players included in the sample and they are not spread thinly across many groups.

Constant variance. The last assumption is that the variance associated with the residuals is nearly constant from one group to the next. This assumption can be checked by examining a side-by-side box plot of the outcomes, as in Figure 8.15. In this case, the variability is similar in the four groups but not identical. We see in Table 8.14 on page 324 that the standard deviation varies a bit from one group to the next. Whether these differences are from natural variation is unclear, so we should report this uncertainty with the final results.

Caution: Diagnostics for an ANOVA analysis
Independence is always important to an ANOVA analysis. The normality condition is very important when the sample sizes for each group are relatively small. The constant variance condition is especially important when the sample sizes differ between groups.

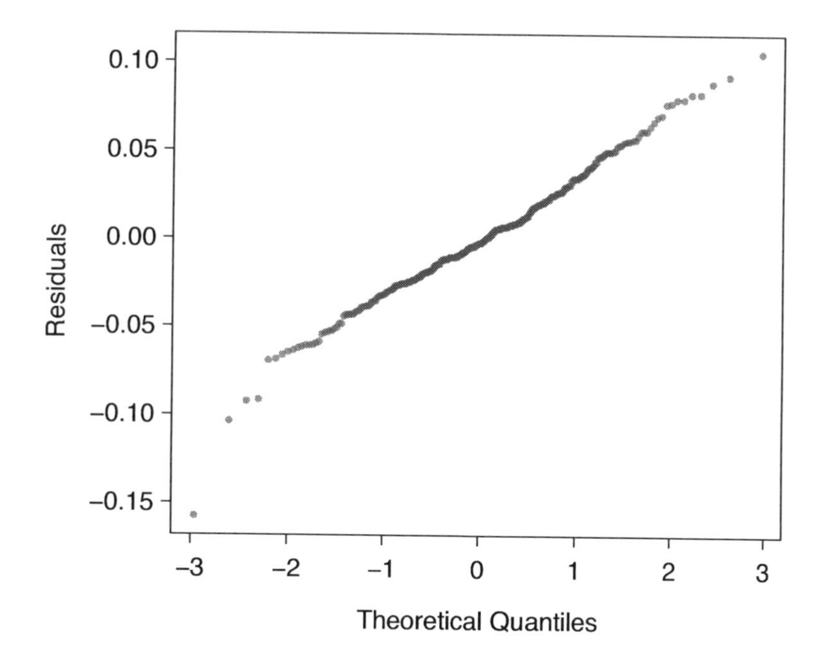

Figure 8.18: Normal probability plot of the residuals.

8.4.5 Multiple comparisons and controlling Type 1 Error rate

When we reject the null hypothesis in an ANOVA analysis, we might wonder, which of these groups have different means? To answer this question, we compare the means of each possible pair of groups. For instance, if there are three groups and there is strong evidence that there are some differences in the group means, there are three comparisons to make: group 1 to group 2, group 1 to group 3, and group 2 to group 3. These comparisons can be accomplished using a two-sample t test, but we must use a modified significance level and a pooled estimate of the standard deviation across groups.

● **Example 8.31** Example 8.20 on page 322 discussed three statistics lectures, all taught during the same semester. Table 8.19 shows summary statistics for these three courses, and a side-by-side box plot of the data is shown in Figure 8.20. We would like to conduct an ANOVA for these data. Do you see any deviations from the three conditions for ANOVA?

In this case (like many others) it is difficult to check independence in a rigorous way. Instead, the best we can do is use common sense to consider reasons the assumption of independence may not hold. For instance, the independence assumption may not be reasonable if there is a star teaching assistant that only half of the students may access; such a scenario would divide a class into two subgroups. After carefully considering the data, we believe that assuming independence may be acceptable.

The distributions in the side-by-side box plot appear to be roughly symmetric and show no noticeable outliers.

The box plots show approximately equal variability, which can be verified in Table 8.19, supporting the constant variance assumption.

Class i	A	B	C
n_i	58	55	51
$\bar{y}_i$	75.1	72.0	78.9
s_i	13.9	13.8	13.1

Table 8.19: Summary statistics for the first midterm scores in three different lectures of the same course.

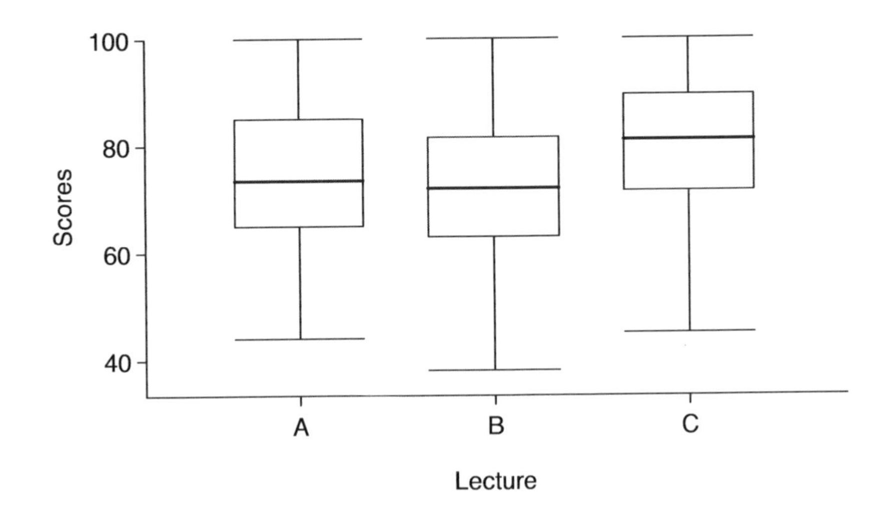

Figure 8.20: Side-by-side box plot for the first midterm scores in three different lectures of the same course.

⊙ **Exercise 8.32** An ANOVA was conducted for the midterm data, and a summary is shown in Table 8.21. What should we conclude?

	Df	Sum Sq	Mean Sq	F value	Pr(>F)
lecture	2	1290.11	645.06	3.48	0.0330
Residuals	161	29810.13	185.16		
			$s_{pooled} = 13.61$ on $df = 161$		

Table 8.21: ANOVA summary table for the midterm data.

There is strong evidence that the different means in each of the three classes is not simply due to chance. We might wonder, which of the classes are actually different? As discussed in earlier chapters, a two-sample t test could be used to test for differences in each possible pair of groups. However, one pitfall was discussed in Example 8.24 on page 324: when we run so many tests, the Type 1 Error rate increases. This issue is resolved by using a modified significance level.

Multiple comparisons and the Bonferroni correction for α
The scenario of testing many pairs of groups is called **multiple comparisons**. The **Bonferroni correction** suggests that a more stringent significance level is more appropriate for these tests:

$$\alpha^* = \alpha/K$$

where K is the number of comparisons being considered (formally or informally). If there are k groups, then usually all possible pairs are compared and $K = \frac{k(k-1)}{2}$.

Example 8.33 In Exercise 8.32, you found that the data showed strong evidence of differences in the average midterm grades between the three lectures. Complete the three possible pairwise comparisons using the Bonferroni correction and report any differences.

We use a modified significance level of $\alpha^* = 0.05/3 = 0.0167$. Additionally, we use the pooled estimate of the standard deviation: $s_{pooled} = 13.61$ on $df = 161$.

Lecture A versus Lecture B: The estimated difference and standard error are, respectively,

$$\bar{y}_A - \bar{y}_B = 75.1 - 72 = 3.1 \qquad SE = \sqrt{\frac{13.61^2}{58} + \frac{13.61^2}{55}} = 2.56$$

(See Section 6.2.4 on page 6.2.4 for additional details.) This results in a T score of 1.21 on $df = 161$ (we use the df associated with s_{pooled}) and a two-tailed p-value of 0.228. This p-value is larger than $\alpha^* = 0.0167$, so there is not strong evidence of a difference in the means of lectures A and B.

Lecture A versus Lecture C: The estimated difference and standard error are 3.8 and 2.61, respectively. This results in a T score of 1.46 on $df = 161$ and a two-tailed p-value of 0.1462. This p-value is larger than α^*, so there is not strong evidence of a difference in the means of lectures A and C.

Lecture B versus Lecture C: The estimated difference and standard error are 6.9 and 2.65, respectively. This results in a T score of 2.60 on $df = 161$ and a two-tailed p-value of 0.0102. This p-value is smaller than α^*. Here we find strong evidence of a difference in the means of lectures B and C.

We might summarize the findings of the analysis from Example 8.33 using the following notation:

$$\mu_A \stackrel{?}{=} \mu_B \qquad \mu_A \stackrel{?}{=} \mu_C \qquad \mu_B \neq \mu_C$$

The midterm mean in lecture A is not statistically distinguishable from those of lectures B or C. However, there is strong evidence that lectures B and C are different. In the first two pairwise comparisons, we did not have sufficient evidence to reject the null hypothesis. Recall that failing to reject H_0 does not imply H_0 is true.

Caution: Sometimes an ANOVA will reject the null but no groups will have statistically significant differences
It is possible to reject the null hypothesis using ANOVA and then to not subsequently identify differences in the pairwise comparisons. However, *this does not invalidate the ANOVA conclusion.* It only means we have not been able to successfully identify which groups differ in their means.

The ANOVA procedure examines the big picture: it considers all groups simultaneously to decipher whether there is evidence that some difference exists. Even if the test indicates that there is strong evidence of differences in group means, identifying with high confidence a specific difference as statistically significant is more difficult.

Consider the following analogy: we observe a Wall Street firm that makes large quantities of money based on predicting mergers. Mergers are generally difficult to predict, and if the prediction success rate is extremely high, that may be considered sufficiently strong evidence to warrant investigation by the Securities and Exchange Commission (SEC). While the SEC may be quite certain that there is insider trading taking place at the firm, the evidence against any single trader may not be very strong. It is only when the SEC considers all the data that they identify the pattern. This is effectively the strategy of ANOVA: stand back and consider all the groups simultaneously.

8.4.6 Using ANOVA for multiple regression

The ANOVA methodology can be extended to multiple regression, where we simultaneously incorporate categorical and numerical predictors into a model. The methods discussed so far – an outcome for a single categorical variable – is called **one-way ANOVA**. There are two extensions that we briefly discuss here: evaluating all variables in a model simultaneously, and using ANOVA in model selection where some variables are numerical and others categorical.

Some software will supply additional information about a multiple regression model fit beyond the regression summaries described in this textbook. This additional information can be used in an assessment of the utility of the full model. For instance, below is the full regression summary for the Mario Kart Wii game analysis from Section 8.2 (implemented with R statistical software[14]) using all four predictors:

```
Residuals:
     Min       1Q   Median       3Q      Max
-11.3788  -2.9854  -0.9654   2.6915  14.0346

Coefficients:
            Estimate Std. Error t value Pr(>|t|)
(Intercept) 36.21097    1.51401  23.917  < 2e-16 ***
condNew      5.13056    1.05112   4.881 2.91e-06 ***
stockPhoto   1.08031    1.05682   1.022    0.308
duration    -0.02681    0.19041  -0.141    0.888
wheels       7.28518    0.55469  13.134  < 2e-16 ***
---
Signif. codes:  0 *** 0.001 ** 0.01 * 0.05 . 0.1   1
```

[14] R is free and can be downloaded at www.r-project.org.

```
Residual standard error: 4.901 on 136 degrees of freedom
Multiple R-squared: 0.719,Adjusted R-squared: 0.7108
F-statistic: 87.01 on 4 and 136 DF,  p-value: < 2.2e-16
```

The main output labeled `Coefficients` should be familiar as the multiple regression summary. The last three lines are new and provide details about

- the standard deviation associated with the residuals (4.901),
- degrees of freedom (136),
- R^2 (0.719) and adjusted R^2 (0.7108), and
- also an F statistic (174.4 with $df_1 = 4$ and $df_2 = 136$) with an associated p-value (<2.2e-16, i.e. about zero).

The F statistic and p-value in the last line can be used for a test of the entire model. The p-value can be used to the answer the following question: Is there strong evidence that the model as a whole is significantly better than using no variables? In this case, with a p-value of less than 2.2×10^{-16}, there is extremely strong evidence that the variables included are helpful in prediction. Notice that the p-value does not verify that all variables are actually important in the model; it only considers the importance of all of of the variables simultaneously. This is similar to how ANOVA was earlier used to assess differences across all means without saying anything about the difference between a particular pair of means.

The second setting for ANOVA in the general multiple regression framework is one that is more delicate: model selection. We could compare the variability in the residuals of two models that differ by just one predictor using ANOVA as a tool to evaluate whether the data support the inclusion of that variable in the model. We postpone further details of this method to a later course.

8.5 Exercises

8.5.1 Introduction to multiple regression

8.1 In Chapter 6 you were introduced to a data set from an experiment to measure and compare the effectiveness of various feed supplements on the growth rate of chickens. Newly hatched chicks were randomly allocated into six groups, and each group was given a different feed supplement. Their weights in grams after six weeks are given along with feed types in the data set called `chickwts`. We are specifically interested in the effect of casein feed on the weights of these chicks, so we have created a variable called `casein` and coded chicks who were on casein feed as 1 and those who were on other diets as 0. The summary table below shows the results of a simple linear regression model for predicting `weight` from `casein`. [38]

	Estimate	Std. Error	t value	Pr(>|t|)
(Intercept)	248.64	9.54	26.06	0.0000
casein	74.94	23.21	3.23	0.0019

(a) Write the equation of the regression line.

(b) Interpret the slope in context, and calculate the predicted weight of chicks who are and who not are on another feed.

(c) Is there a statistically significant relationship between feed type (casein or other) and the average weight of chicks? State the hypotheses and include any information used to conduct the test. Note that if we look back at Exercise 6.19 on page 268, we would see that the variability within the casein group and the variability across the other groups are about equal and the distributions symmetric. With these conditions satisfied, it is reasonable to proceed with the test. (Note also that we don't need to check linearity since the predictor has only two levels.)

8.2 Vitamin C is believed to help promote dental health. One common way to get Vitamin C is by drinking orange juice. Another option is to take ascorbic acid tablets. An experiment was conducted to test if one source is more effective than the other. 60 guinea pigs were randomly assigned to these two delivery methods for Vitamin C, 30 in each group. The length of teeth in millimeters are given along with delivery methods in the data set called `ToothGrowth`. We created a variable called `OJ` and coded guinea pigs who were given orange juice as 1 and those who were given ascorbic acid as 0. The summary table below shows the results of a simple linear regression model for predicting the average tooth length, `len`, from `OJ`. [51]

	Estimate	Std. Error	t value	Pr(>|t|)
(Intercept)	16.96	1.37	12.42	0.0000
oj	3.70	1.93	1.92	0.0604

(a) Write the equation of the regression line.

(b) Interpret the slope in context, and calculate the predicted tooth length for guinea pigs who were given orange juice and those who were given ascorbic acid.

(c) Is there a statistically significant relationship between the average tooth length and delivery method of Vitamin C in guinea pigs? State the hypotheses and include any information used to conduct the test. Note that the variability within the orange juice and the ascorbic acid groups are about equal and the distributions symmetric. With these conditions satisfied, it is reasonable to proceed with the test.

8.3 The Child Health and Development Studies (CHDS) is a collection of studies, one of which considers all pregnancies between 1960 and 1967 among women in the Kaiser Foundation Health Plan in the San Francisco East Bay area. A random sample of these data are given in a data set called `babies`. We consider the relationship between smoking and weight of the baby. The variable `smoke` is coded 1 if the mother is a smoker, and 0 if not. The summary table below shows the results of a simple linear regression model for predicting the average birth weight of babies, measured in ounces (`bwt`), from `smoke`. [52]

	Estimate	Std. Error	t value	Pr(>|t|)
(Intercept)	123.05	0.65	189.60	0.0000
smoke	-8.94	1.03	-8.65	0.0000

The variability within the smokers and non-smokers are about equal and the distributions symmetric. With these conditions satisfied, it is reasonable to proceed with the test. (Note that we don't need to check linearity since the predictor has only two levels.)

(a) Write the equation of the regression line.

(b) Interpret the slope in context, and calculate the predicted birth weight of babies born to smoker and non-smoker mothers.

(c) Is there a statistically significant relationship between the average birth weight and smoking? State the hypotheses and include any information used to conduct the test.

8.4 Exercise 8.3 introduces a data set on birth weight of babies. Another variable we consider is `parity`, where 0 is first born, and 1 is otherwise. The summary table below shows the results of a simple linear regression model for predicting the average birth weight of babies, measured in ounces, from `parity`.

	Estimate	Std. Error	t value	Pr(>|t|)
(Intercept)	120.07	0.60	199.94	0.0000
parity	-1.93	1.19	-1.62	0.1052

(a) Write the equation of the regression line.

(b) Interpret the slope in context, and calculate the predicted birth weight of first borns and others.

(c) Is there a statistically significant relationship between the average birth weight and parity? State the hypotheses and include any information used to conduct the test.

8.5 The `babies` dataset used in Exercises 8.3 and 8.4 includes information on length of pregnancy in days (`gestation`), mother's age in years (`age`), mother's height in inches (`height`), and mother's pregnancy weight in pounds (`weight`), in addition to the `smoking` and `parity` variables considered earlier. Below are three observations from this data set.

	bwt	gestation	parity	age	height	weight	smoke
1	120	284	0	27	62	100	0
2	113	282	0	33	64	135	0
⋮	⋮	⋮	⋮	⋮	⋮	⋮	⋮
1236	117	297	0	38	65	129	0

The summary table below shows the results of a linear regression model for predicting the average birth weight of babies based on all of the variables included in the data set.

	Estimate	Std. Error	t value	Pr(>|t|)
(Intercept)	-80.41	14.35	-5.60	0.0000
gestation	0.44	0.03	15.26	0.0000
parity	-3.33	1.13	-2.95	0.0033
age	-0.01	0.09	-0.10	0.9170
height	1.15	0.21	5.63	0.0000
weight	0.05	0.03	1.99	0.0471
smoke	-8.40	0.95	-8.81	0.0000

(a) Write the equation of the regression line that includes all of the variables.

(b) Interpret the slopes of `gestation` and `age` in context.

(c) The coefficient for `parity` is different than in the simple linear model shown in Exercise 8.4. Why might there be a difference?

(d) Calculate the residual for the first observation in the data set.

8.6 Researchers interested in the relationship between absenteeism from school and certain demographic characteristics of children collected data from 146 randomly sampled students in rural New South Wales in a particular school year. These data are given in a data set called `quine`. Below are three observations from this data set.

	eth	sex	lrn	days
1	0	1	1	2
2	0	1	1	11
⋮	⋮	⋮	⋮	⋮
146	1	0	0	37

The summary table below shows the results of a linear regression model for predicting the average number of days absent based on ethnic background (`eth`: 0 - aboriginal, 1 - not aboriginal), sex (`sex`: 0 - female, 1 - male), and learner status (`lrn`: 0 - average learner, 1 - slow learner). [53]

	Estimate	Std. Error	t value	Pr(>|t|)
(Intercept)	18.93	2.57	7.37	0.0000
eth	-9.11	2.60	-3.51	0.0000
sex	3.10	2.64	1.18	0.2411
lrn	2.15	2.65	0.81	0.4177

(a) Write the equation of the regression line.

(b) Interpret each one of the slopes in context.

(c) Calculate the residual for the first observation in the data set.

8.7 The variance of the residuals for the model given in Exercise 8.5 is 249.28, and the variance of the birth weights of all babies in the data set is 332.57. Calculate the R^2 and the adjusted R^2. Note that there are 1236 observations in the data set.

8.8 The variance of the residuals for the model given in Exercise 8.6 is 240.57, and the variance of the number of absent days for all students in the data set is 264.17. Calculate the R^2 and the adjusted R^2. Note that there are 146 observations in the data set.

8.5.2 Model selection

8.9 Exercise 8.5 presents summary output for a regression model for predicting the average birth weight of babies based on six explanatory variables.

(a) Determine which variable(s) do not have a significant relationship with the outcome and should be candidates for removal from the model. If there is more than one such model, indicate which one should be removed first.

(b) The summary table below shows the results of the regression we refit after removing age from the model. Determine if any other variable(s) should be removed from the model.

	Estimate	Std. Error	t value	Pr(>|t|)
(Intercept)	-80.64	14.04	-5.74	0.0000
gestation	0.44	0.03	15.28	0.0000
parity	-3.29	1.06	-3.10	0.0020
height	1.15	0.20	5.64	0.0000
weight	0.05	0.03	2.00	0.0459
smoke	-8.38	0.95	-8.82	0.0000

8.10 Exercise 8.6 presents summary output for a regression model for predicting the average number of days absent based on three explanatory variables.

(a) Determine which variable(s) do not have a significant relationship with the outcome and should be candidates for removal from the model. If there is more than one such model, indicate which one should be removed first.

(b) The summary table below shows the results of the regression we refit after removing learner status from the model. Determine if any other variable(s) should be removed from the model.

	Estimate	Std. Error	t value	Pr(>|t|)
(Intercept)	19.98	2.22	9.01	0.0000
eth	-9.06	2.60	-3.49	0.0006
sex	2.78	2.60	1.07	0.2878

8.11 Exercise 8.5 provides regression output for the full model (including all explanatory variables available in the data set) for predicting birth weight of babies. In this exercise, we consider a forward-selection algorithm and add variables to the model one-at-a-time. The table below shows the p-value and adjusted R^2 of each model where we include only the corresponding predictor. Based on this table, which variable should be added to the model first?

variable	gestation	parity	age	height	weight	smoke
p-value	2.2×10^{-16}	0.1052	0.2375	2.97×10^{-12}	8.2×10^{-8}	2.2×10^{-16}
R^2_{adj}	0.1657	0.0013	0.0003	0.0386	0.0229	0.0569

8.12 Exercise 8.6 provides regression output for the full model (including all explanatory variables available in the data set) for predicting number of days absent from school. In this exercise, we consider a forward-selection algorithm and add variables to the model one-at-a-time. The table below shows the p-value and adjusted R^2 of each model where we include only the corresponding predictor. Based on this table, which variable should be added to the model first?

variable	ethnicity	sex	leaner status
p-value	0.0007	0.3142	0.5870
R^2_{adj}	0.0714	0.0001	0

8.5.3 Checking model assumptions using graphs

8.13 Exercise 8.9 presents a regression model for predicting the average birth weight of babies based on length of gestation, parity, height, weight, and smoke. Determine if the model assumptions are met using the plots below. If not, describe how to proceed with the analysis.

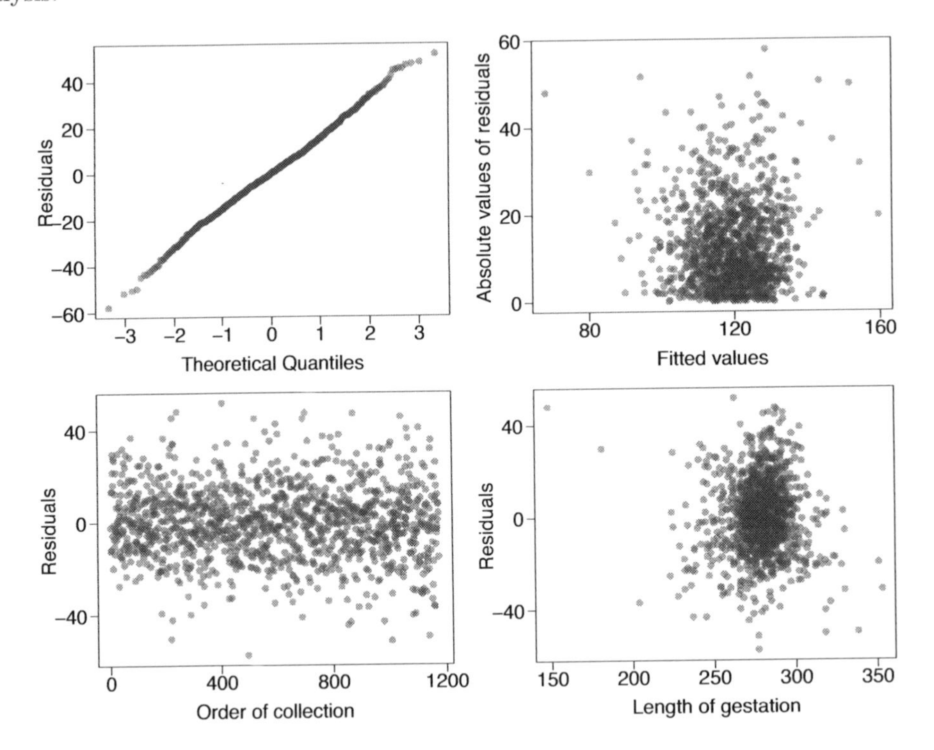

8.5.4 ANOVA and regression with categorical variables

8.15 In Exercise 8.1, we considered the effect of casein feed on chicks' weight. Instead of categorizing feed type as casein or other, we might also want to consider all feed types at once: casein, horsebean, linseed, meat meal, soybean, and sunflower. The ANOVA output below can be used to test for differences between the average weights of chicks on different diets.

	Df	Sum Sq	Mean Sq	F value	Pr(>F)
feed	5	231129.16	46225.83	15.36	0.0000
Residuals	65	195556.02	3008.55		

Conduct a hypothesis test to determine if these data provide strong evidence that the average weight of chicks varies across some (or all) groups. Refer to Exercise 6.19 on page 268 to assist in checking ANOVA conditions.

8.16 A professor who teaches a large introductory statistics class with eight discussion sections would like to test if student performance differs by discussion section. Each discussion section has a different teaching assistant. The summary table below shows the average final exam score for each discussion section as well as the standard deviation of scores and the number of students in each section.

	Sec 1	Sec 2	Sec 3	Sec 4	Sec 5	Sec 6	Sec 7	Sec 8
n_i	33	19	10	29	33	10	32	31
$\bar{x}_i$	92.94	91.11	91.80	92.45	89.30	88.30	90.12	93.35
s_i	4.21	5.58	3.43	5.92	9.32	7.27	6.93	4.57

The ANOVA output below can be used to test for differences between the average scores from the different discussion sections.

	Df	Sum Sq	Mean Sq	F value	Pr(>F)
section	7	525.01	75.00	1.87	0.0767
Residuals	189	7584.11	40.13		

Conduct a hypothesis test to determine if these data provide strong evidence that the average score varies across some (or all) groups. Check conditions and describe any assumptions you must make to conduct the test.

Appendix A

Bibliography

[1] Source: www.stats4schools.gov.uk, November 10, 2009.

[2] B. Ritz, F. Yu, G. Chapa, and S. Fruin, "Effect of air pollution on preterm birth among children born in Southern California between 1989 and 1993," *Epidemiology*, vol. 11, no. 5, pp. 502–511, 2000.

[3] J. McGowan, "Health Education: Does the Buteyko Institute Method make a difference?," *Thorax*, vol. 58, 2003.

[4] Gallagher, Visser, Sepulveda, Pierson, and H. Heymsfield, "How useful is body mass index for comparison of body fatness across age, sex, and ethnic groups?," *American Journal of Epidemiology*, vol. 143, no. 3, pp. 228–239, 1996.

[5] R. Fisher, "The use of multiple measurements in taxonomic problems," *Annals of Eugenics*, vol. 7, pp. 179–188, 1936.

[6] T. Allison and D. Cicchetti, "Sleep in mammals: ecological and constitutional correlates," *Arch. Hydrobiol*, vol. 75, p. 442, 1975.

[7] Source: Harvard Business Review, http://blogs.hbr.org/cs/2009/06/new_twitter_research_men_follo.html, April 1, 2011.

[8] Source: Yahoo News, http://news.yahoo.com/s/ac/20110315/tc_ac/8066912_happy_birthday_twitter, April 1, 2011.

[9] Source: CIA Factbook, https://www.cia.gov/library/publications/the-world-factbook/rankorder/2091rank.html, October 22, 2010.

[10] B. Turnbull, B. Brown, and M. Hu, "Survivorship of heart transplant data," *Journal of the American Statistical Association*, vol. 69, pp. 74–80, 1974.

[11] R. Rabin, "Risks: Smokers found more prone to dementia," October 29 2010. http://www.nytimes.com/2010/11/02/health/research/02risks.html.

[12] D. Graham, R. Ouellet-Hellstrom, T. MaCurdy, F. Ali, C. Sholley, C. Worrall, and J. Kelman, "Risk of acute myocardial infarction, stroke, heart failure, and death in elderly medicare patients treated with rosiglitazone or pioglitazone," *JAMA*, vol. 304, no. 4, p. 411, 2010.

[13] U.S. Census Bureau, 2005-2009 American Community Survey.

[14] Majority of Republicans No Longer See Evidence of Global Warming, October 27, 2010,
http://people-press.org/reports/questionnaires/669.pdf.

[15] USPSTF, "Screening for breast cancer: U.s. preventive services task force recommendation statement," *Annals of Internal Medicine*, vol. 151, pp. 716–726, 2009.

[16] J. A. Paulos, "Mammogram math," December 2009. New York Times, 13 December 2009.

[17] S. Johnson and D. Murray, "Empirical Analysis of Truck and Automobile Speeds on Rural Interstates: Impact of Posted Speed Limits," in *Transportation Research Board 89th Annual Meeting*, 2010.

[18] Source: SAMHSA, Office of Applied Studies, National Survey on Drug Use and Health, 2007 and 2008, http://www.oas.samhsa.gov/NSDUH/2k8NSDUH/tabs/Sect2peTabs1to42.htm#Tab2.5B.

[19] Source: Public Fact Sheet, Chickenpox (Varicella), http://www.mass.gov/Eeohhs2/docs/dph/cdc/factsheets/chickenpox.pdf.

[20] Source: What Frightens America's Youth?, http://www.gallup.com/poll/15439/What-Frightens-Americas-Youth.aspx.

[21] G. Heinz, L. Peterson, R. Johnson, and C. Kerk, "Exploring relationships in body dimensions," *Journal of Statistics Education*, vol. 11, no. 2, 2003.

[22] Source: http://www.ets.org/Media/Tests/GRE/pdf/gre_0809_interpretingscores.pdf.

[23] NAEP Data Explorer, April 16, 2011.

[24] A. Romero-Corral, V. Somers, J. Sierra-Johnson, R. Thomas, M. CollazoClavell, J. Korinek, T. Allison, J. Batsis, F. Sert-Kuniyoshi, and F. Lopez-Jimenez, "Accuracy of body mass index in diagnosing obesity in the adult general population," *International Journal of Obesity*, vol. 32, no. 6, pp. 959–966, 2008.

[25] Road Rules: Re-Testing Drivers at Age 65?, http://maristpoll.marist.edu/34-road-rules-re-testing-drivers-at-age-65.

[26] Civil War at 150: Still Relevant, Still Divisive, http://pewresearch.org/pubs/1958/civil-war-still-relevant-and-divisive-praise-confederate-leaders-flag.

[27] Is College Worth It?, http://pewresearch.org/pubs/1993/survey-is-college-degree-worth-cost-debt-college-presidents-higher-education-system.

[28] Public option gains support, October 20, 2009.

[29] Perceived Insufficient Rest or Sleep Among Adults United States, 2008, October 30, 2009.

[30] L. Ellis and C. Ficek, "Color preferences according to gender and sexual orientation," *Personality and Individual Differences*, vol. 31, no. 8, pp. 1375–1379, 2001.

[31] Drilling for oil and natural gas off the coast of California, http://www.surveyusa.com.

[32] Poll: 4 in 5 Support Full-Body Airport Scanners, November 15, 2010, http://www.cbsnews.com/8301-503544_162-20022876-503544.html.

[33] Four in 10 Americans Believe in Strict Creationism, December 17, 2010, http://www.gallup.com/poll/145286/Four-Americans-Believe-Strict-Creationism.aspx.

[34] R. Schmidt, R. Hansen, J. Hartiala, H. Allayee, L. Schmidt, D. Tancredi, F. Tassone, and I. Hertz-Picciotto, "Prenatal vitamins, one-carbon metabolism gene variants, and risk for autism," *Epidemiology*, vol. 22, no. 4, p. 476, 2011.

[35] R. Rabin, "Patterns: Prenatal vitamins may ward off autism," June 13 2011. http://www.nytimes.com/2011/06/14/health/research/14patterns.html?_r=1&ref=research.

[36] Facebook privacy, http://www.surveyusa.com.

[37] Source: Fuelecomy.gov, http://www.fueleconomy.gov/mpg/MPG.do?action=browseList2&make=Toyota&model=Prius.

[38] Source: R Dataset, http://stat.ethz.ch/R-manual/R-patched/library/datasets/html/chickwts.html.

[39] Source: U.S. Department of Energy, Fuel Economy Data, http://www.fueleconomy.gov/feg/download.shtml.

[40] R. Oldham-Cooper, C. Hardman, C. Nicoll, P. Rogers, and J. Brunstrom, "Playing a computer game during lunch affects fullness, memory for lunch, and later snack intake," *The American journal of clinical nutrition*, vol. 93, no. 2, p. 308, 2011.

[41] Gallup, "Americans' views of egypt sharply more negative," February 8, 2011. http://www.gallup.com/poll/File/146006/Egypt_Favorability_Feb_08_2011.pdf.

[42] CDC, "2008 assisted reproductive technology report." http://www.cdc.gov/art/ART2008/index.htm.

[43] Source: Mythbusters, Season 3, Episode 28, http://www.yourdiscovery.com/video/mythbusters-top-10-is-yawning-contagious.

[44] D. Hand, *A handbook of small data sets.* Chapman & Hall/CRC, 1994.

[45] Source: R Dataset, http://stat.ethz.ch/R-manual/R-patched/library/datasets/html/trees.html.

[46] J. Benson, "Season of birth and onset of locomotion: Theoretical and methodological implications," *Infant behavior and development*, vol. 16, no. 1, pp. 69–81, 1993.

[47] Source: Association of Turkish Travel Agencies, http://www.tursab.org.tr/en/statistics/foreign-visitors-figure-tourist-spendings-by-years_1083.html.

[48] Source: Starbucks.com, data collected on March 10, 2011, http://www.starbucks.com/menu/nutrition.

[49] Source: American Fact Finder, generated on December 27, 2010, http://www.factfinder.census.gov.

[50] J. Malkevitch and L. Lesser, *For All Practical Purposes: Mathematical Literacy in Today's World.* WH Freeman & Co, 2008.

[51] Source: R Dataset, http://stat.ethz.ch/R-manual/R-devel/library/datasets/html/ToothGrowth.html.

[52] Source: http://www.ma.hw.ac.uk/~stan/aod/library.

[53] Source: R Dataset, http://stat.ethz.ch/R-manual/R-patched/library/MASS/html/quine.html.

Appendix B

End of chapter exercise solutions

1 Introduction to data

1.1 (a) Control: 56%. Treatment: 70%. (b) There is a 14% difference between the pain reduction rates in the two groups. It appears that patients in the treatment group are more likely to show improvement and, at a first glance, acupuncture appears to be an effective treatment for migraines. (c) It's hard to say. The difference is somewhat large, but the sample is somewhat small.

1.3 (a) 143,196 eligible subjects who were born in Southern California between 1989 and 1993. (b) The variables are measurements of CO, NO_2, ozone, and particulate matter less than $10\mu m$ (PM_{10}) collected at air-quality-monitoring stations as well as the birth weights of the babies. All of these variables are continuous numerical variables. (c) Does air pollution exposure have an effect on preterm births?'

1.5 (a) 202 black and 504 white adults who resided in or near New York City, were ages 20-94 years, and had BMIs of 18-35 kg/m^2. (b) Age (numerical, continuous), sex (categorical), ethnicity (categorical), weight, height, waist and hip circumference, length of tibia, body density and volume, total body water (numerical, continuous). (c) How useful is BMI for predicting body fatness across age, sex and ethnic groups?

1.7 (a) A participant in the survey. (b) 1,691 participants. (c) gender (gender of the participant), age (age of the participant, in years), marital (marital status of the participant), grossIncome (gross income of the participant, in £), smoke (whether or not the participant smokes), amtWeekends (number of cigarettes smoked on weekend, # of cigarettes / day), amtWeekdays (number of cigarettes smoked on a week day, # of cigarettes / day).

1.9 gender (categorical), age (originally numerical, continuous, though it was recorded as a discrete numerical variable), maritalStatus (categorical), grossIncome (originally numerical, continuous, but recorded as categorical), smoke (categorical), amtWeekends (numerical, discrete), amtWeekdays (numerical, discrete).

1.11 We would expect productivity to increase as stress increases, but up to a point, after that productivity would decrease as stress continued to increase. The exact shape of your plot may be a little different.

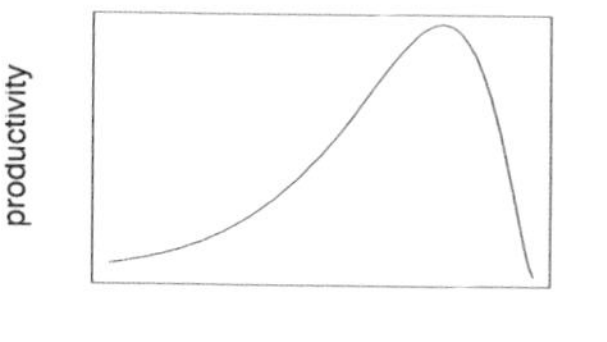

1.13 (a) Population mean, $\mu_x = 5.5$; sample mean, $\bar{x} = 6.25$. (b) Population mean, $\mu_x = 52$; sample mean, $\bar{x} = 58$.

1.15 (a) Decrease. (b) 73.6. (c) The new score, x_{25}, is more than 1 standard deviation away from the previous mean, and this will tend to increase the standard deviation of the data. While possible, it is mathematically rather tedious to calculate the new standard deviation.

1.17 The distribution of amount of cigarettes smoked on weekends and on weekdays are both right skewed. The median of both distributions is between 10 and 15 cigarettes, the first quartile is between 5 and 10 cigarettes, and the third quartile is between 15 and 20 cigarettes. Hence the IQR of both distributions is roughly about 10 cigarettes. There are potential outliers above 40 cigarettes per day, giving both distributions a long right tail. We can also see that there are more respondents who smoke only a few cigarettes (0 to 5) on the weekdays, about 80 people, than on weekends, about 60 people. Another feature that is visible from the histograms are peaks at 10 and 20 cigarettes. This may be because most people do not keep track of exactly how many cigarettes they smoke, but round their answers to half a pack (10 cigarettes) or a whole pack (20 cigarettes). Due to these peaks, the distributions could be classified as bimodal.

1.19 $s_{amtWeekends} = 0, s_{amtWeekdays} = 4.18$. Variability of the amount of cigarettes smoked is higher on weekdays than on the weekends for this sample.

1.21 (a) 6 (b) 6.5

1.23 Plot below.

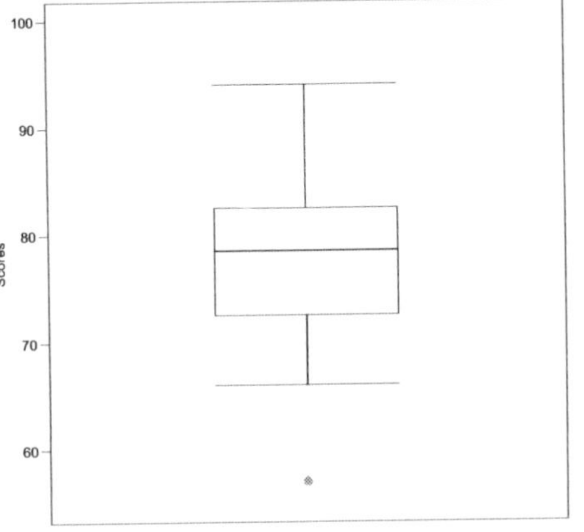

1.25 (a) The distribution is unimodal and symmetric with a mean around 60 and a standard deviation of roughly 3; matches the box plot (2). (b)The distribution is uniform and values range from 0 to 100; matches box plot (3) which shows a symmetric distribution in this range. Also, each 25% chunk of the box plot have about the same width and there are no suspected outliers. (c) The distribution is unimodal and right skewed with a median between 1 and 2. The IQR of the distribution is roughly 1; matches box plot (1).

1.27 (a) Since median is defined as the 50th percentile and about 50% of the data is in the first bar, we would expect median to be between 0 and 20. Q1 is also between 0 and 20 as the 25th percentile is in the first bar as well. Q3, defined as the 75th percentile, is located between 40 and 60. (b) The distribution is right-skewed, so the long tail will pull the mean above the median.

1.29 It appears that marathon times decreased greatly between 1970-1975 and remained somewhat steady thereafter. Males consistently had shorter marathon times than females throughout the years. From the box plots of males and females, we could tell that males ran faster "on average", however, we could not tell that the winning male time for each year was better than the winning female time. We also could not tell from the histogram or the box plot that marathon times have been decreasing for males and females throughout the years.

1.31 (a) The distribution is right skewed with potential outliers on the positive end, therefore the median and the IQR are appropriate measures of center and spread. (b) The distribution is somewhat symmetric and probably does not have outliers, therefore the mean and the standard deviation are appropriate measures of center and spread. (c) The distribution would be right skewed. There would be some students who do not consume any alcohol but this is the minimum (there cannot be students who consume fewer than 0 drinks). There would be a few students who consume many more drinks than their peers, giving the distribution a long right tail. Due to the skew, the median and the IQR would be appropriate measures of center and spread. (d) The distribution would be right skewed. Most employees would make something on the order of the median salary, but we would expect to have some high level executives making a lot more. The distribution would have a long right tail, and the median and the IQR would be more more appropriate measures of center or spread.

1.33 (a) As well as the order of the categories, we can also see the relative frequencies in the bar plot. These proportions are not readily available in the pie chart. (b) None. (c) Bar plot, so that we can also see the relative frequencies of the categories in this graph.

1.35 (a) Proportion of patients who are alive at the end of the study is higher in the treatment group than in the control group. Therefore survival is not independent of whether or not the patient got a transplant. (b) The shape of the distribution of survival times in both groups is right skewed with outliers on the high end. The median survival time for the control group is much lower than the median survival time for the treatment group; patients who got a transplant typically lived longer. The maximum survival time for the treatment group is much higher (about 5 years) than the maximum survival time for the control group. Even though the maximum survival time for the control group is about 4 years, this observation is an outlier. Overall, very few patients without transplants made it beyond a year while nearly half of the transplant patients survived at least one year. It should also be noted that while the first and third quartiles of the treatment group is higher than those for the control group, the IQR for the treatment group is much bigger, indicating that there is more variability in survival times in the treatment group.

1.37 (a) The population is all adults 20 and older living in the greater New York City area. The sample is the 202 black and 504 white men and women who resided in or near New York City and had BMIs of 18-35 kg/m^2. (b) The population is all Californians registered to vote in the 2010 midterm elections. The sample is the 1000 registered California voters who were surveyed for this study.

1.39 (a) This is an observational study. (b) Wealth is one lurking variable. Countries with individuals who can widely afford internet probably also can afford basic medical care. (Note: Answers may vary.)

1.41 (a) Simple random sample. Non-response bias, if only those people who have strong opinions about the survey responds his sample may not be representative of the population. (b) Convenience sample. Under coverage bias, his sample may not be representative of the population since it consists only of his friends.

1.43 (a) Non-responders are most likely parents who have busier schedules and have difficulty spending time with their kids after school. (b) The women who are not reached 3 years later are most likely renters (as opposed to homeowners) who may be in a lower socio-economic status. (c) There is no control group and there may be lurking variables. For example, it may be that these people who go running are generally healthier and/or do other exercises.

1.45 No, this was an observational study, and we cannot make such a causal statement based on an observational study.

1.47 Prepare two cups for each participant one containing regular Coke and the other containing Diet Coke. Make sure the cups are identical and contain equal amounts of soda. Label the cups A (regular) and B (diet). (Be sure to randomize A and B for each trial!) Give each participant the two cups, one cup at a time, in random order, and ask the participant to record a value that indicates how much she liked the beverage. Be sure that neither the participant nor the person handing out the cups knows the identity of the beverage to make this a double-blind experiment. (Answers may vary.).

1.49 (a) Experiment. (b) Treatment: exercise twice a week, control: no exercise. (c) Yes, the blocking variable is age. (d) No. (e) Since this is an experiment, we can make a causal statement. Since the sample is random, the causal statement can be generalized to the population at large. However, we should be cautious about making a causal statement because of a possible placebo effect. Note that this study could not actually be conducted since people cannot be required to participate in a clinical trial.

1.51 (a) False. Instead of comparing counts, we should compare percentages of people in each group who suffered a heart attack. (b) True. (c) False. Association does not imply causation. We cannot infer a causal relationship based on an observational study. (We cannot say changing the drug a person is on affects her risk, which is why part (b) is true.) (d) True.

2 Probability

2.1 False. The tosses are independent trials.

2.3 (a) 10 tosses. With a low number of flips the variability in the number of heads observed is much larger, so a result further from 50% is more likely. (b) 100 tosses. With more flips, the observed proportion of heads would probably be closer to 50% and therefore above 40%. (c) 100 tosses. The more flips, the less variability away from 50%. (d) 10 tosses. Fewer flips mean more volatility and a greater chance of getting far from 50% and below 30%.

2.5 (a) 1/1024. (b) 1/1024. (c) 1023/1024.

2.7 (a) Figure below. (b) 5% (c) 70% (d) 95% (e) 5% (f) No, there are bloggers who own both types of cameras.

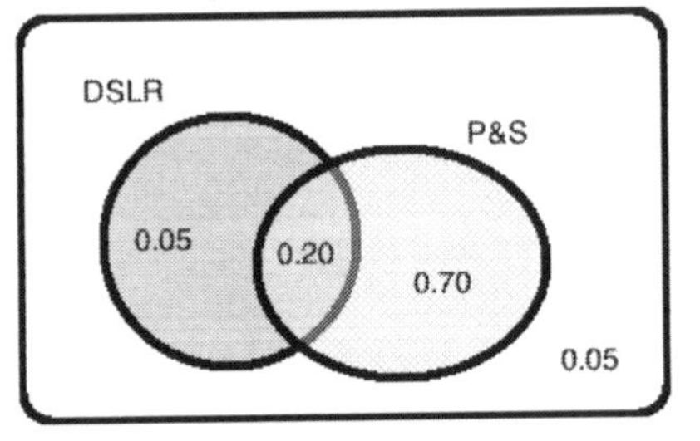

2.9 (a) Not mutually exclusive. If the class is not graded on a curve, then independent. If graded on a curve, dependent. (b) Not mutually exclusive, most likely dependent. (c) No. See the answer to (a) when the course is not graded on a curve.

2.11 (a) 0.26. (b) 0.23. (c) Assuming that the education level of the husband and wife are independent, 0.0598. (d) Independence, which may not be a reasonable assumption since people often marry others with a comparable level of education.

2.13 (a) Sum greater than 1. (b) OK mathematically. (c) Sum less than 1. (d) Negative probabilities make no sense. (e) OK. (f) Probabilities cannot be less than 0 or greater than 1.

2.15 Approximate answers are OK. Answers are only estimates based on the sample. (a) 0.42. (b) 0.15. (c) 0.37. (d) 0.06.

2.17 (a) The distribution is right skewed, with a median between $35,000 and $49,999. The IQR of the distribution is about $27,500. There are probably outliers on the high end due to the nature of the data. (b) 62.2%. (c) Assuming gender and income are independent: 25.5%. (d) P(less than $50,000 and female) = 29.4%. The independence assumption does not appear to be valid. If gender and income were independent, we would expect the 25.5% of the sample to be female and make less than \$50,000, but actually a higher proportion fall into this category.

2.19 No, P(DSLR | point&shoot) = 0.22, which is not equal to P(DSLR).

2.21 (a) 0.2825. (b) 0.1905. (c) 0.4167. (d) No, because P(black hair | brown eyes) $\neq$ P(black hair | blue eyes). (Other explanations are possible.)

2.23 (a) 0.65. (b) 0.72. (c) Under the assumption of independence of gender and hamburger preference: 0.468. While it is possible there is some mysterious connection between burger choice and finding a partner, independence is probably a reasonable assumption. (e) 0.514.

2.25 Female, most cats smaller than 2.5kg are female.

2.27 0.6049.

2.29 (a) Tree diagram below. (b) 0.68. (c) 0.32. (d) Your test results have come in. While the test came back positive, this is not conclusive. A positive test result can occur even when a patient has no disease; occasionally a test will be wrong. For this reason, we will need to run some additional tests.

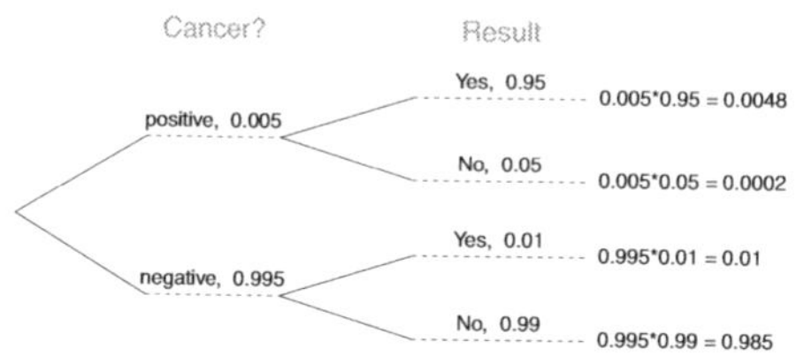

2.31 (a) 0.3. (b) 0.3. (c) 0.3. (d) 0.09. (e) Yes, each draw is from the same set of marbles.

2.33 (a) 0.0909. (b) 0.3182. (c) 0.4545. (d) 0. (e) 0.2879.

2.35 0.0519.

2.37 (a) 13. (b) No, this would be unreliable. The students are not a random sample.

2.39 (a) Table below. Expected winnings: \$3.59. SD: 3.37. (b) EV: -\$1.41, SD: \$3.37. (c) No. The expected net profit is negative, so on average you expect to lose money.

Event	3 hearts	3 blacks	Else
X	\$50	\$25	\$0
$P(X)$	0.0129	0.1176	0.8695
$X * P(X)$	0.65	2.94	0
$(X - E(X))^2 P(X)$	0.1115	0.0497	11.2062

2.41 (a) EV: -\$0.16, SD: \$2.99. (b) EV: -\$0.16, SD: \$1.73. (c) Expected values are the same but the standard deviations are different. The standard deviation from the game where winnings and losses are tripled is higher, making this game riskier.

2.43 (a) Table to the right. Expected winnings: -\$0.54 (b) No, he is expected to lose money on average.

Event	2,...,9	J, Q, K	Ace	A♣
X	-2	1	3	23
$P(X)$	0.6923	0.2308	0.0577	0.0192

2.45 \$4.26.

2.47 (a) Mean: \$3.90, SD: \$0.34. (b) Mean: \$27.30, SD: \$0.89.

3 Distributions of random variables

3.1 Plots below. (a) 0.0885. (b) 0.0694. (c) 0.5886. (d) 0.0456.

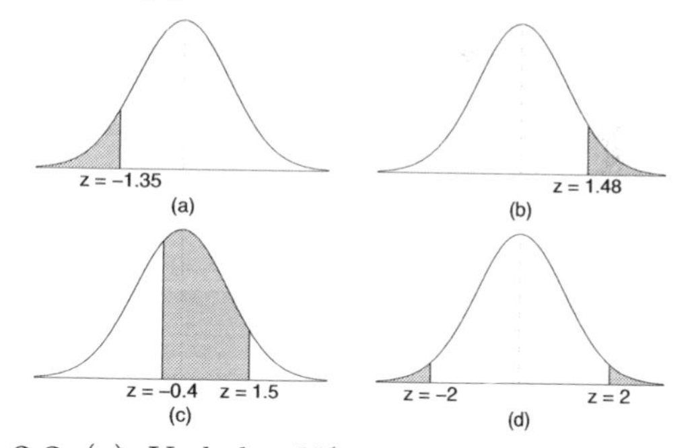

3.3 (a) Verbal: $N(\mu = 462, \sigma = 119)$, Quant: $N(\mu = 584, \sigma = 151)$. (b) $Z_{VR} = 1.33$, $Z_{QR} = 0.57$. Plots below. (c) She scored 1.33 standard deviations above the mean on the Verbal Reasoning section and 0.57 standard deviations above the mean on the Quantitative Reasoning section. (d) $Perc_{VR} = 91\%$, $Perc_{QR} = 72\%$. (e) Verbal Reasoning. (f) VR: 9%, QR: 28%. (g) We cannot compare the raw scores since they are on different scales. Her scores will be measured relative to the merits of other students on each exam, so it is helpful to consider the Z score. Comparing her percentiles is more appropriate for determining how well she did compared to others.

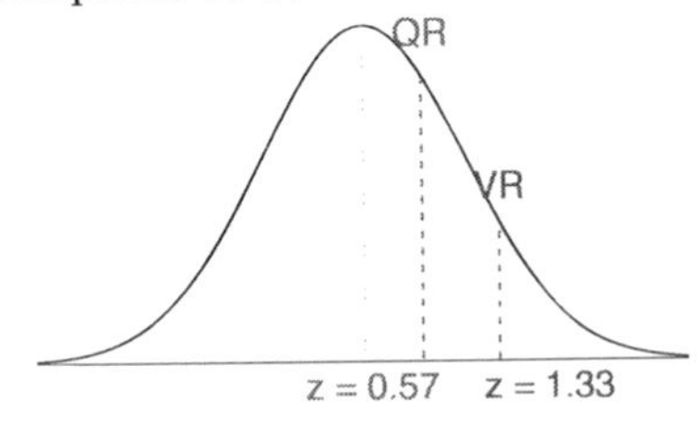

3.5 Answers to (b) and (c) would not change, though we would not draw a Normal curve on which to show these scores. We could not answer parts (d) and (e) since the normal probability table is only valid for the normal model.

3.7 (a) 711. (b) 400.

3.9 Figures below. (a) 0.1210. (b) 0.1558. (c) 62.68 inches. (d) 43.25%

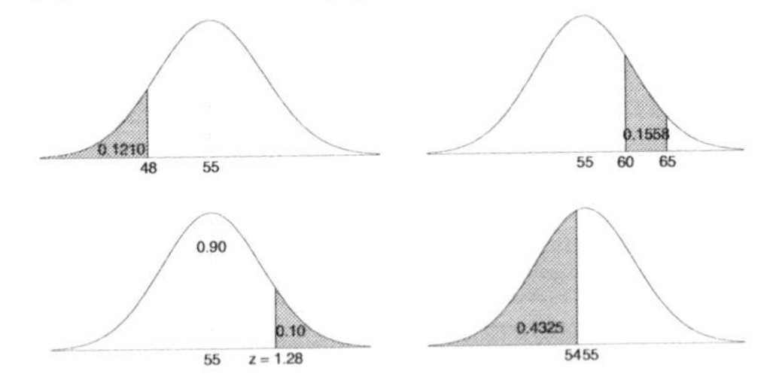

3.11 (a) 0.1401. (b) 70.6°F or colder.

3.13 (a) 0.67. Using 0.68 is also okay, but your answers for part (c) will differ a little from the listed solution. (b) $x = \$1800$, $\mu = \$1650$. (c) $\sigma = \$223.88$.

3.15 (a) 0.2327. Figure below. (b) If you are bidding on only one auction and set a maximum bid price that is too low, chances are someone will outbid you and you won't win the auction. If your maximum bid price is too high, you may win the auction but you may be paying more than is necessary. If you are bidding on more than one auction and your maximum bid price is too low, chances are you won't win any of the auctions. However, if your maximum bid price is too high, you may win more than one auction and end up with multiple copies of the book. (c) An answer roughly equal to the 10th percentile would be reasonable. Regrettably, no percentile cutoff point guarantees beyond any possible event that you win at least one auction. However, you may pick a higher percentile if you want to be more sure of winning an auction. (d) Using the 10th percentile: \$69.80. Answers may vary but should correspond to the answer given in part (c).

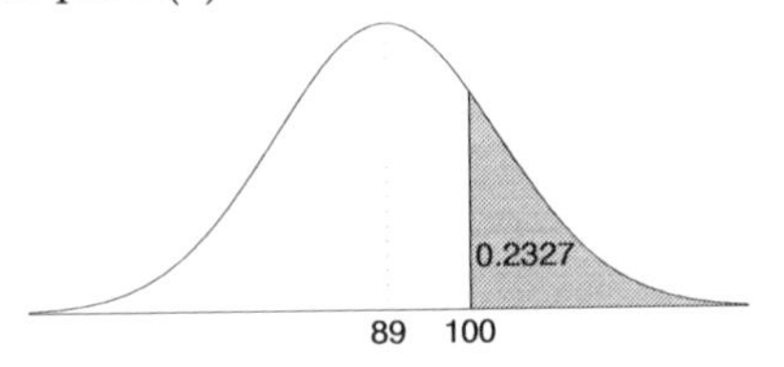

3.17 70% of the data are within 1 SD, 95% are within 2 SD, and 100% are within 3 SD of the mean. The data approximately follow the 68-95-99.7% Rule.

3.19 The distribution is unimodal, symmetric, and approximately follows the 68-95-99.7% Rule. The superimposed normal curve seems to approximate the distribution pretty well. The points on the Normal probability plot also seem to follow a straight line. There is one possible outlier on the lower end that is apparent in both graphs, but it is not too extreme. We can say that the distribution is nearly normal.

3.21 No, in poker cards are dealt without replacement, and they have more than two categories.

3.23 Approximate answers are OK. (a) 0.13. (b) 0.12. (c) $\mu = 2.04$, $\sigma = 1.46$. (d) $\mu = 3.33$, $\sigma = 2.79$. (e) When p was smaller, i.e. the event was rarer, the expected number of trials before a success and the standard deviation increased.

3.25 (a) 0.096. (b) $\mu = 8$, $\sigma = 7.48$.

3.27 (a) Yes, it meets the four required conditions. (b) 0.203. (c) 0.203. (d) 0.167. (e) 0.997.

3.29 (a) $\mu = 34.85$, $\sigma = 3.25$. (b) Yes, since 45 is more than 3 standard deviations from the mean. (c) 0.0015 (an answer of 0.0009 would be okay if using a normal approximation, since the conditions for the approximation are satisfied). In part (b), we had determined that it would be unusual to observe 45 or more 18-20 year olds who have consumed alcoholic beverages among a random sample of 50, and the we calculated a very low probability for this event.

3.31 (a) 0.5160. (b) 0.1234. (c) 0.8483. (d) No, otherwise there is a 15.17% chance that 2 or more will be afraid of spiders in any particular tent.

3.33 (a) 0.109. (b) 0.219. (c) 0.137. (d) 0.551. (e) 0.084. (f) Since 2 is $\frac{2-4}{1.06} = -1.89$ standard deviations below the expected number of brown eyed children, strictly speaking this would not be considered unusual. However, it should be noted that the z-score for this value is pretty close to 2, making this observation borderline unusual.

3.35 The probability model is below.

Y	-3	-1	1	3
$P(Y)$	0.1458	0.3936	0.3543	0.1063

3.37 (a) (1/5)*(1/4)*(1/3)*(1/2)*(1/1) = 1/(5!) = 1/120. (e) 120 = 5!. (c) 8! = 40,320.

3.39 (a) 0.0804 using the geometric distribution. (b) 0.0322 using the binomial distribution. (c) 0.0193 using the negative binomial distribution.

3.41 (a) Negative Binomial (n = 4, p = 0.55): Of the four trials considered here, the last trial must be a success and there were exactly 2 successes. (b) 0.1838. (c) $\binom{3}{1} = \frac{3!}{2!1!} = 3$. (d) In the binomial model we have no restrictions on the outcome of the last trial while in the negative binomial model the last trial is fixed. Therefore we are interested in the number of ways of orderings of the other $k-1$ successes in the first $n-1$ trials.

3.43 (a) Poisson with $\lambda = 75$. (b) $\mu = \lambda = 75$, $\sigma = \sqrt{\lambda} = 8.66$. (c) No, since 60 is within 2 standard deviations of the mean.

3.45 $P(X = 70) = \frac{75^{70}e^{-75}}{70!} = 0.0402$

4 Foundations for inference

4.1 (a) Mean. (b) Mean. (c) Proportion. (d) Mean. (e) Proportion.

4.3 The point estimates are the corresponding sample values. (a) $\bar{x} = 13.65$, median= 14. (b) $s = 1.91$, $IQR_{estimate} = 2$. (c) Use the Z score to evaluate ($Z_{16} = 1.23$, $Z_{18} = 2.28$), so 18 credits is unusually high but 16 is not, where we use 2 standard deviations from the mean as a cutoff for deciding what is unusual.

4.5 No, sample point estimates only approximate the population parameter, and they vary from one sample to another.

4.7 Standard error, $SE_{\bar{x}} = \frac{1.91}{\sqrt{100}} = 0.191$.

4.9 (a) $SE_{\bar{x}} = 2.89$ (b) The Z score is 1.73 (absolute value is less than 2), so $80 is consistent.

4.11 (a) Independence is met by the random sampling assumption and that the sample is less than 10% of the population. The sample size is also sufficiently large. We cannot check the assumption that the distribution isn't extremely skewed. (b) (19.862, 20.058). (c) We are 90% confident that the true mean amount of coffee in Starbucks venti cups is between 19.862 ounces and 20.058 ounces. (d) 90% of random samples of size 50 will yield confidence intervals that capture the true mean amount of coffee in Starbucks venti cups. (e) Yes, 20 ounces is included in the interval. (f) A 95% confidence interval would be wider. All else kept constant, when confidence level increases so does the margin of error and hence the interval becomes wider. We cast a wider interval.

4.13 (a) Less. (b) We can infer from the sample statistics that the distribution is skewed, so no we cannot. (c) The only condition that may not be met for normality of the mean relates to skew: it is unclear if the distribution is extremely skewed or not. We'll suppose the skew is strong but not too extreme, something we may like to look into further. Solution: 0.0985. (d) Decreases the standard error by a factor $\sqrt{2}$.

4.15 When the confidence level increases, so does the margin of error and the width of the interval. A wide interval may be undesirable even if the confidence level is higher.

4.17 (a) False, since we need only check whether the skew is not too extreme. (b) False, we are 100% sure the average for *these* patients is in this interval. (c) True. (d) False, the confidence interval is not about sample means. (e) False, as the confidence level increases, so does the width of the interval. (f) True. (g) False, since in calculation of the standard error we divide the standard deviation by square root of the sample size, we would need to quadruple the sample size.

4.19 (a) 0.0004. (b) Since the sample is random and the 10% condition is met, we can assume the that how much one penny weighs is independent of another. Since the population distribution is normal, and hence not extremely skewed, sampling distribution of means will be nearly normal even though $n < 50$. $N(\mu = 2.5, \sigma_{\bar{x}} = 0.0095)$. (c) Approximately 0. (d) Plot below. (e) The sample or sampling distributions would not be approximately normal.

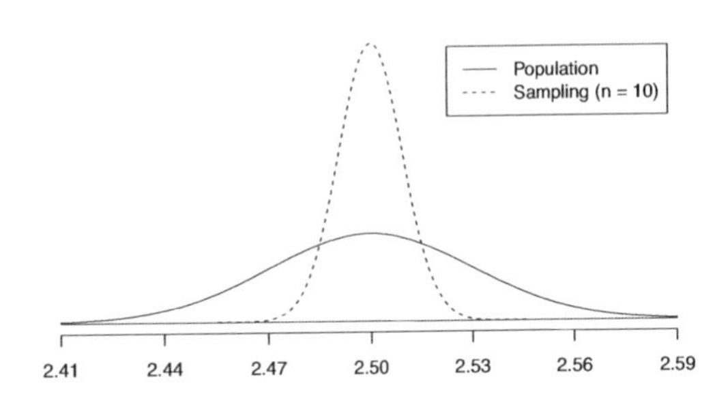

4.21 (a) From the histogram: $P(X > 5) = \frac{350+100+25+20+5}{3000} = \frac{500}{3000} = 0.17$. It's okay if your answer differs a little. (b) Two different answers are reasonable. 1) The conditions are reasonably met. We know the population standard deviation, so we can know the standard error (SD of $\bar{x}$) with certainty. The population distribution is also only slightly skewed, so a sample of 15 would probably have a sampling distribution for the mean that is nearly normal. Solution: 0.0956. 2) If you had said the normality condition for $\bar{x}$ was questionable because the population distribution was not normal, that is also an acceptable answer. (c) Assumptions/conditions are certainly met. Solution: 0.1788.

4.23 (a) H_0: $\mu = 8$ (On average New Yorkers sleep 8 hrs a night), H_A: $\mu < 8$ (On average New Yorkers sleep less than 8 hrs a night). (b) H_0: $\mu = 15$ (The average amount of company time spent not working is 15 minutes), H_A: $\mu > 15$ (The average amount of company time spent not working is greater than 15 minutes).

4.25 The hypotheses should be about the population mean (μ), not the sample mean. If he believes that \$1.3 million is an overestimation, the alternative hypothesis should be *less than* and not *greater than*. The correct way to set up these hypotheses is as follows: H_0: $\mu = \$1.3$ *million*, H_A: $\mu < \$1.3$ *million*.

4.27 (a) 180 minutes is not in the interval, so this is implausible. (b) 2.2 hours (132 minutes) is in the interval, so we conclude the estimated wait time of 2.2 hours is reasonable. (c) A 99% confidence interval will be wider than a 95% confidence interval. Hence even without calculating the interval we can tell that 132 minutes would be in it.

4.29 (a) H_0: Anti-depressants do not work for the treatment of Fibromyalgia. H_A: Anti-depressants work for the treatment of Fibromyalgia. (b) Concluding that anti-depressants work for the treatment of Fibromyalgia when they actually do not. (c) Concluding that anti-depressants do not work for the treatment of Fibromyalgia when they actually do. (d) If she makes a Type I error, she will continue taking medication that does not actually treat her disorder. If she makes a Type II error, she will stop taking medication that could treat her disorder.

4.31 (a) Yes, if we assume there isn't too much skew, which is certainly reasonable once we realize the possible percentages, which must be between 0% and 100%, are bounded within 3 standard deviations from the mean. (b) H_0: $\mu = 0.25$, H_A: $\mu \neq 0.25$. $Z = -7.71 \rightarrow$ two-sided p-value ≈ 0. Reject H_0: the evidence indicates that the percentage of time college students spend on the internet for coursework has changed over the last decade. (c) If the percentage of time college students spend on the Internet for course work has actually remained at 25%, the probability of getting a random sample of 238 college students where the average percentage of time they spend on the Internet for course work is 10% or less or 40% or more is approximately 0. (d) Type I, since we may have incorrectly rejected H_0.

4.33 H_0: $\mu = 7$, H_A: $\mu \neq 7$. $Z = -1.04 \rightarrow$single tail$= 0.1492 \rightarrow$ p-value $= 2 * 0.1492 = 0.2984$. There isn't sufficient evidence that the average lifespan of all ball bearings produced by this machine is not 7 hours. The manufacturer's claim is not implausible.

4.35 $\bar{x} = 36.05$.

4.37 (a) The distribution is unimodal and right skewed, median is between 5 and 10 years old, and the IQR is roughly 10. There are potential outliers on the higher end. (b) When the sample size is small, the sampling distribution is right skewed, just like the population distribution. As the sample size increases, the sampling distribution gets more and more unimodal and symmetric, just like the CLT suggests.

4.39 (a) If the skew is not too strong, the assumptions are met. (b) H_0: $\mu = 432$, H_A: $\mu < 432$. $Z = -3.28 \rightarrow$ p-value (single tail) $= 0.0005$. Since the p-value $< \alpha$, we reject H_0. There is evidence that the average amount savings of all customers who switch their insurance is less than \$432. (c) Yes, the insurance company's claim may be an overestimate since the hypothesis test result indicated there was strong evidence that the average savings is less than the advertised amount. (d) (\$376.47, \$413.53). (e) Yes, the hypothesis test was statistically significant and \$432 was not in the confidence interval.

4.41 (a) The only condition we cannot check is for extreme skew. Here, we will assume this is not an issue; in practice, this is something we should verify. (b) H_0: $\mu = 500$, H_A: $\mu \neq 500$. $Z = -3.86 \rightarrow$single tail$\approx 0 \rightarrow$ p-value $\approx 2 * 0 = 0$. Since the p-value $< \alpha$ (0.05), we reject H_0. The data provide strong evidence that the average increase in reading speed is not 500% (it is below 500% based on the data). (c) No, the company's claim of an average of 500% increase in reading speed does not appear to be accurate. (d) 371.88% to 458.12%. (e) Yes. The hypothesis test rejected that the average increase was 500%, and 500% was not in the confidence interval.

4.43 $n \geq 693$.

---

5 Large sample inference

5.1 (a) Hypothesis test for paired data. (b) H_0: $\mu_{diff} = 0$ (There is no difference in average daily high temperature between January 1, 1968 and January 1, 2008), H_A: $\mu_{diff} > 0$ (Average daily high temperature in January 1, 1968 was lower than average daily high temperature in January, 2008.) (c) Independence is satisfied since we have a random sample that is less than 10% of the possible locations we could collect such measurements in the continental U.S. There is also a one-to-one correspondence between the observations in the data set, making it appropriate for a paired analysis. The sample size is sufficiently large ($n = 51$). If we had the data in hand, we would also check for extreme skew. $Z = 1.60$, p-value $= 0.0548$. (d) Fail to reject H_0. The data do not provide strong evidence of temperature warming in the continental US. However, it should be noted that the p-value is very close to 0.05. (e) Type II. If we made such an error and concluded that there isn't strong evidence for temperature warming in the continental US, but in reality average temperature on January 1, 2008 is higher than average temperature on January 1, 1968. (f) Yes.

5.3 (a) H_0: $\mu_B = \mu_A$, another way to write this is $\mu_B - \mu_A = 0$ (The population mean of number of cigarettes smoked per day did not change after the Surgeon General's report), H_A: $\mu_1 > \mu_2$, another way to write this is $\mu_1 - \mu_2 > 0$ (The population mean of number of cigarettes smoked per day decreased after the Surgeon General's report) (b) Independence is satisfied since we have two random samples that are less than 10% of their respective populations. The sample sizes are sufficiently large. If we had the data in hand, we would also check for extreme skew. $Z = 1.89$, p-value $= 0.0294$. (c) Reject H_0. There is sufficient evidence that the number of cigarettes smoked per day decreased after the Office of the Surgeon General's report. (d) No, we cannot make a causal connection because this is observational data. (e) Type I, since we may have incorrectly rejected H_0.

5.5 Independence is satisfied since we have two independent random samples that are each less than 10% of the population. The sample sizes are sufficiently large. If we had the data in hand, we would also check for extreme skew. We are 90% confident that the average score in 2004 was 0.16 to 5.84 points lower than the average score in 2008.

5.7 H_0: $\mu_M = \mu_W$, H_A: $\mu_M < \mu_W$. $Z = -97.35$, p-value ≈ 0. Reject H_0. The data provide strong evidence that average body fat percentage for women is higher.

5.9 (a) False, for proportions need to check success/failure condition, not $n \geq 50$. (b) True. (c) False, only 1.65 standard errors away from the mean, and we use 2 as a cutoff for what is called unusual. (d) True. (e) False, standard error would decrease only by a factor of $\sqrt{2}$.

5.11 (a) True. (b) False, standard error would decrease only by a factor of $\sqrt{2}$. (c) True. (d) True. (e) False, success/failure condition is not satisfied.

5.13 (a) Paremeter: proportion of all graduates from this university who found a job within one year of graduating. Point estimate: 0.87. (b) Independence is satisfied since we have a random sample that is less than 10% of the population. Normality is satisfied since the success-failure condition is met. CI: (0.837 , 0.903). (c) We are 95% confident that the true proportion of graduates from this university who found a job within one year of completing their undergraduate degree is between 83.7% and 90.3%. (d) 95% of random samples of 400 would produce a confidence interval that includes the true proportion of students at this university who found a job within one year of graduating from college. (e) It would be wider. (f) It would be narrower.

5.15 (a) She needs a minimum of 3,394 subjects and therefore needs to set aside a minimum of $67,880. (b) It will be wider.

5.17 (a) $ME = 1.96 * \sqrt{0.66 * 0.34/1,1018} \approx 0.03$. (b) No, for two reasons. The point estimate is slightly below 67%, and 67% is contained in the interval.

5.19 (a) H_0: $p = 0.5$, H_A: $p > 0.5$. Assuming the sample is random. The sample is simple random and from <10% of the population, so independence is reasonable. The success/failure condition is also met. $Z = 4.66$, p-value ≈ 0. Since the p-value is small, we reject H_0. The data provide strong evidence that majority of the Americans think the Civil War is still relevant. (b) If in fact only 50% of Americans thought the Civil War is still relevant, the probability of obtaining a random sample of 1,507 Americans where 56% think it is still relevant would be approximately 0. (c) We are 90% confident that 54% to 58% of all Americans think that the Civil War is still relevant. This agrees with the conclusion of the earlier hypothesis test since the interval lies above 50%.

5.21 (a) H_0: $p = 0.5, H_A : p < 0.5$. The assumptions and conditions are satisfied. $Z = -0.73$, p-value $= 0.2327$. Since the p-value is large, we fail to reject H_0. The data do not provide strong evidence that less than half of American adults who decide to not go to college make this decision because they cannot afford college. (b) Yes, since we failed to reject H_0.

5.23 (a) The assumptions and conditions are satisfied. We are 80% confident that the 44.5% to 51.5% of all Americans who decide not to go to college do so because they cannot afford it. This agrees with the conclusion of the earlier hypothesis test since the interval includes 50%. (b) 1,818.

5.25 (a) H_0: $p = 0.3$, H_A: $p > 0.3$. The assumptions and conditions are satisfied. $Z = 1.89$, p-value $= 0.0294$. Since the p-value small, we reject H_0. The data provide strong evidence that the rate of sleep deprivation for New Yorkers is higher than the rate of sleep deprivation in the population at large. (b) If in fact 30% of New Yorkers were sleep deprived, the probability of getting a random sample of 300 New Yorkers where more than 105 are sleep deprived would be 0.0294.

5.27 (a) H_0: $p = 0.18$, H_A: $p \neq 0.18$. The assumptions and conditions are satisfied. $Z = 0.74$, p-value $= 0.4592$. Since the p-value is large, we fail to reject H_0. The data do not provide strong evidence that the percentage of students at this university who smoke has changed over the last five years. (b) Type II, since we may have incorrectly failed to reject H_0.

5.29 (a) H_0: $p = 0.65$, H_A: $p > 0.65$. Assuming that the 250 $<$ 10% of high school graduates at this school district, all conditions and assumptions are satisfied. $Z = 1.26$, p-value $= 0.1038$. Since the p-value is large, we fail to reject H_0. The data do not provide strong evidence that the percentage of students in this rural school district who go out of state for college has increased. (b) If in fact 65% of students in this school district went out of state for college, the probability of getting a random sample of 250 students where 172 or more of them go out of state for college would be 0.1038.

5.31 (a) The assumptions and conditions are satisfied. 95% CI: (0.138, 0.270). (b) We are 95% confident that the proportion of students from the rural school district who plan to go out of state for college is 13.8% to 27% higher than the proportion of students from the urban school district who do.

5.33 (a) H_0: $p_D = p_I$, H_A: $p_D > p_I$. The assumptions and conditions are satisfied. $Z = 11.29$, p-value ≈ 0. Since the p-value is very small, we reject H_0. The data provide strong evidence that the proportion of Democrats who support the plan is higher than the proportion of Independents who support the plan. (b) Type I, since we may have incorrectly rejected H_0. (c) No, rejecting the null hypothesis of $p_1 = p_2$ is equivalent to rejecting that $p_1 - p_2 = 0$. Therefore we would not expect a confidence interval for the difference between the two proportions to include 0. (d) We are 95% confident that the proportion of Democrats who support the plan is 23% to 33% higher than the proportion of Independents who do. (e) True.

5.35 The assumptions and conditions are satisfied. We are 95% confident that the proportion of Californians who are sleep deprived is 1.7% less to 0.1% more than the proportion of Oregonians who are sleep deprived. Since the confidence interval includes 0, we would not reject a null hypothesis that the two population proportions equal to each other.

5.37 (a) True. (b) False, the interval only estimates the difference in population parameters. (c) False, to get the 95% confidence interval for $(p_{placebo} - p_{medication})$, all we have to do is to swap the bounds of the original confidence interval and take their negatives. (d) True. (e) False, the confidence interval for the difference between the proportions of success includes 0, so we cannot reject the hypothesis of no difference.

5.39 (a) College grads: 35.2%. Non-grads: 33.9%. (b) H_0: $p_{CG} = p_{NCG}$, H_A: $p_{CG} \neq p_{NCG}$. The assumptions and conditions are satisfied. $Z = 0.37$, p-value $= 0.7114$. Since the p-value is large, we fail to reject H_0. The data do not provide strong evidence of a difference between the proportions of college graduates and non-college graduates who support off-shore drilling in California.

5.41 (a) We are 90% confident that the proportion of Republicans who support the use of full-body scans at airports is 3% lower to 7% higher than the proportion of Democrats who do. (b) No, this does not prove it; though the data does not provide strong evidence to the contrary.

5.43 (a) H_0: The distribution of the format of the book used by the students follows the professor's predictions. H_A: The distribution of the format of the book used by the students does not follow the professor's predictions. (b) $E_{hard\ copy} = 75.6$, $E_{print} = 31.5$, $E_{online} = 18.9$. (c) We are not told explicitly that the sample is random, however, we have no reason to believe that this class is not representative of all introductory statistics students. We can safely assume that $126 < 10\%$ of all introductory statistics students. We may think it is reasonable to suppose the students are independent. However, the professor probably should have included a question asking whether the student decisions relied on any other students' decisions when they purchased, printed, or read the book online. All expected counts are at least 10. Format of the book used is a categorical variable. (d) $\chi^2 = 2.32$, $df = 2$, p-value > 0.3. (e) Since the p-value is large, we reject H_0. The data do not provide strong evidence indicating the professor's predictions were statistically inaccurate.

5.45 (a) 47.5. (b) 296.6. (c) 21.0.

5.47 (a) H_0: There is no difference in the rates of autism of children of mothers who did and did not use prenatal vitamins during the first three months before pregnancy. H_A: There is some difference in the rates of autism of children of mothers who did and did not use prenatal vitamins during the first three months before pregnancy. (b) $E_{row\ 1,col\ 1} = 95.2, E_{row\ 1,col\ 2} = 85.8, E_{row\ 2,col\ 1} = 158.8, E_{row\ 2,col\ 2} = 143.2$. The assumptions and conditions are satisfied. $\chi^2 = 8.85$, $df = 1, 0.001 <$ p-value < 0.005. Since the p-value is small, we reject H_0. There is strong evidence a difference in the rates of autism of children of mothers who did and did not use prenatal vitamins during the first three months before pregnancy. (c) The title of this newspaper article makes it sound like using prenatal vitamins can prevent autism, which is a causal statement. Since this is an observational study, we cannot make causal statements based on the findings of the study. A more accurate title would be "Mothers who use prenatal vitamins before pregnancy are found to have children with a lower rate of autism".

5.49 H_0: The opinion of college grads and non-grads is not different on the topic of drilling for oil and natural gas off the coast of California. H_A: Opinions regarding the drilling for oil and natural gas off the coast of California has an association with college education. $E_{row\ 1,col\ 1} = 151.5, E_{row\ 1,col\ 2} = 134.5, E_{row\ 2,col\ 1} = 162.1, E_{row\ 2,col\ 2} = 143.9, E_{row\ 3,col\ 1} = 124.5, E_{row\ 3,col\ 2} = 110.5$. The assumptions and conditions are satisfied. $\chi^2 = 11.46$, $df = 2, 0.001 <$ p-value < 0.005. Since the p-value is small, we reject H_0. There is strong evidence that there is some difference in rate of support for drilling for oil and natural gas off the Coast of California based on whether or not the respondent graduated from college. Support for off-shore drilling and having graduated from college do not appear to be independent.

6 Small sample inference

6.1 (a) $t^*_{41} = 1.68$ (b) $t^*_{20} = 2.53$ (c) $t^*_{28} = 2.05$ (d) $t^*_{11} = 3.11$

6.3 With a larger critical value, the confidence interval ends up being wider.

6.5 (a) H_0: $\mu = 8$ (New Yorkers sleep 8 hrs per night on average.), H_A: $\mu < 8$ (New Yorkers sleep less than 8 hrs per night on average.) (b) Independence is satisfied since the sample is random and less than 10% of the population. The distribution doesn't appear to be strongly skewed. $T = -1.75$, $df = 24$. (c) If in fact the true population mean of the amount New Yorkers sleep per night was 8 hours, the probability of getting a random sample of 25 New Yorkers where the average amount of sleep is 7.73 hrs per night or less is between 0.025 and 0.05. (d) Reject H_0, the data provide strong evidence that New Yorkers sleep less than 8 hours per night on average. (e) No.

6.7 (a) We are 90% confident that New Yorkers on average sleep 7.47 to 7.99 hours per night. (b) Yes.

6.9 (a) H_0: $\mu = 1900$, H_A: $\mu \neq 1900$. Independence assumption is met since the sample is random and less than 10% of the population. We are told to assume normality. $T = -1.66$, $df = 29, 0.10 <$ p-value < 0.20. Since the p-value > 0.05, we fail to reject H_0. The data do not provide strong evidence of a change in the average calorie intake of diners at this restaurant. (b) We are 95% confident that diners at this restaurant consume an average of 1690 calories to 1922 calories per meal. (c) Yes.

6.11 $\bar{x} = 56.91$.

6.13 No, distributions are extremely skewed.

6.15 (a) p-value < 0.005, we reject H_0. (b) p-value is about 0.01, we reject H_0. (c) $0.025 <$p-value< 0.05, we reject H_0. (d) p-value > 0.20, we fail to reject H_0.

6.17 (a) We are 95% confident that those in the group that got the weight loss pill lost 0.92 lbs less to 4.92 lbs more than those in the placebo group. (c) No. (d) No.

6.19 (a) Chicken that were fed linseed on average weigh 218.75 grams while those that were given horsebean weigh on average 160.20 grams. Both distributions are relatively symmetric with no apparent outliers. There is more variability in the weights of chicken that were given linseed. (b) H_0: $\mu_L = \mu_H$, H_A: $\mu_L \neq \mu_H$. Independence is satisfied since both samples are random and less than 10% of their prospective populations. The distributions do not appear to be extremely skewed and the samples are independent of each other. $T = 3.02$, $df = 10, 0.01 <$ p-value < 0.02. Reject H_0, the data provide strong evidence of a difference between the average weights of chicken that were fed linseed and horsebean. (c) Type I, since we may have incorrectly rejected H_0. (d) Yes.

6.21 (a) H_0: $\mu_A = \mu_M$, H_A: $\mu_A \neq \mu_M$. $T = 5.46$, $df = 25$, p-value < 0.01. Reject H_0, the data provide strong evidence that there is a difference in the average city mileage between cars with automatic and manual transmissions.

6.23 We are 95% confident that on the highway cars with manual transmission get on average 5.53 to 10.33 MPG more than cars with automatic transmission.

6.25 H_0: $\mu_T = \mu_C$, H_A: $\mu_T \neq \mu_C$. $T = 2.69$, $df = 21, 0.01 <$ p-value < 0.02. Since the p-value < 0.05, we reject H_0. The data provide strong evidence that the amount of biscuits consumed by the patients in the treatment and control groups are different.

6.27 (a) H_0: $p = 0.69$, H_A: $p \neq 0.69$. (b) $\hat{p} = 0.57$. (c) The success-failure condition is not satisfied. (d) Each student can be represented with a card. Take 100 cards, 69 black cards representing those who follow the news about Egypt and 31 red cards representing those who do not. Shuffle the cards and draw with replacement (shuffling each time in between draws) 30 cards representing the 30 high school students. Calculate the proportion of black cards in this sample, $\hat{p}_{sim}$, i.e. the proportion of those who follow the news. Repeat 10,000 times and plot the resulting sample proportions. The p-value will be two times the proportion of simulations where $\hat{p}_{sim} \leq 0.57$. (Note: answers may vary, and in practice we would use a compute to simulate.) (e) p-value ≈ 0.27 (Note: answers may vary a little.) Fail to reject H_0. The data do not provide strong evidence that the proportion of high school students who followed the news about Egypt is different than the proportion of American adults who did.

6.29 (a) H_0: $p_P = p_C$, H_A: $p_P \neq p_C$. (b) -0.35. (c) Doubling the one tail, the p-value is about 0.03. Reject H_0. The data provide strong evidence that people react differently under the two scenarios.

---

7 Introduction to linear regression

7.1 (a) The relationship is linear therefore the residuals plot will show randomly distributed residuals around 0 with constant variance. (b) The scatterplot shows a fan shape, with higher variability in y for lower x. Therefore the residuals plot will also show a fan shape, wider around lower x, narrower around higher x. There may also be characteristics indicating nonlinearity for points on the left.

7.3 (2) and (5) show a strong correlation. Even though (1) and (4) show a strong association, the relationship is not linear therefore correlation would not be strong. (3) and (6) show very weak or no relationship. Answers may vary slightly, e.g. one persons *moderate* may be equivalent to another persons *strong*.

7.5 (a) Exam 2, since the points cluster closer to the line in the second scatterplot. (b) Exam 2 and the final are relatively close to each other chronologically, or Exam 2 may be cumulative so has greater similarities in material to the final exam.

7.7 (a) 4. (b) 3. (c) 1. (d) 2.

7.9 (a) The relationship is positive, weak, and possibly linear. There appears to be one outlier, a student who is about 63 inches tall whose fastest speed is 0 mph. This is probably a student who doesn't drive. (b) There is no obvious explanation why simply being tall should lead a person to drive faster. However, one possible outside factor may be gender. Males tend to be taller than females on average, and and personal experiences (anecdotal) may suggest they drive faster (confirmed in sociological studies). (c) It appears that males are taller on average than females and they also drive faster. The gender variable is a lurking variable for the positive association we observe between fastest driving speed and height.

7.11 (a) There is a somewhat weak, positive, possibly linear relationship between the distance traveled and travel time. (b) Changing the units will not change the form, direction or strength of the relationship between the two variables. (c) Since changing units doesn't affect correlation, $R = 0.636$.

7.13 (a) There is a moderately strong, positive, linear relationship between shoulder girth and height. (b) Changing the units, even if just for one of the variables, will not change the form, direction or strength of the relationship between the two variables.

7.15 (a) $R = 1$. (b) $R = 1$. (c) $R = 1$.

7.17 (a) There is a positive, very strong, linear association between number of tourists and spending. (b) Explanatory: number of tourists (in thousands), response: spending (in million \$). (c) We can predict spending for a given number of tourists using a regression line. This may be useful information for determining how much the country may want to spend in advertising abroad, or to forecast expected revenues from tourism.

7.19 Even though the relationship appears linear in the scatterplot, the residuals plot actually shows a non-linear relationship, therefore we should not fit a least squares line to these data.

7.21 (a) $\widehat{travel\ time} = 51 + 0.726 * distance$. (b) b_1: For each additional mile in distance, the model predicts an additional 0.726 minutes in travel time. b_0: When the distance traveled is 0 miles, the travel time is expected to be 51 minutes. It does not make sense to have a travel distance of 0 miles. Here, the y-intercept serves only to adjust the height of the line and is meaningless by itself. (c) 126 minutes. (d) 42 minutes, underestimate. (e) No, extrapolation.

7.23 Approximately 40% of the variability in travel time is accounted for by the model, i.e. explained by distance traveled.

7.25 No, there is an outlier that appears to have substantial pull on the line. We'll see more on this topic in the next section. The residuals does not show a random scatter around 0, which further suggests that a linear model may not be appropriate.

7.27 (a) Influential. (b) Leverage. (c) Neither influential nor leverage.

7.29 Neither influential nor high leverage.

7.31 (a) The relationship appears to be strong, positive and linear. There is one potential outlier, the student who had 9 cans of beer. (b) $\widehat{BAC} = -0.0127 + 0.0180 * beers$. b_1: For each additional can of beer consumed, the model predicts an additional 0.0180 grams per deciliter BAC. b_0: Students who don't have any beer are expected to have a blood alcohol content of -0.0127. It is not possible to have a negative blood alcohol content. Here, the y-intercept serves only to adjust the height of the line and is meaningless by itself. (c) H_0: $\beta_1 = 0$, H_A: $\beta_1 > 0$. p-value ≈ 0. Reject H_0. Number of cans of beer consumed and blood alcohol content are positively correlated and the true slope parameter is indeed greater than 0. (d) Approximately 79% of the variability in blood alcohol content can be explained by number of cans of beer consumed.

7.33 (a) H_0: $\beta_1 = 0$, H_A: $\beta_1 \neq 0$. $T = 35.25$, $df = 168$, p-value ≈ 0. Reject H_0. Wives' and husbands' ages are correlated and the true slope parameter is indeed greater than 0. (b) $\widehat{ageWife} = 1.5740 + 0.9112 * ageHusband$. (c) b_1: For each additional year in husband's age, the model predicts an additional 0.9112 years in wife's age. b_0: Men who are 0 years old are expected to have wives who are on average 1.5740 years old. The intercept here is meaningless and serves only to adjust the height of the line.

7.35 (a) $R = 0.94$. The slope is positive, so R must also be positive. (b) 51.69, since R^2 is high, the prediction based on this regression model is reliable. (c) No, extrapolation.

7.37 (a) $R = -0.53$. The slope is negative, so R must also be negative. (b) H_0: $\beta_1 = 0$, H_A: $\beta_1 \neq 0$. $T = 4.32$, $df = n - 2 = 49$, p-value ≈ 0.0001. Reject H_0. Percent homeownership and percent of the population living in an urban setting are correlated and the true slope parameter is indeed greater than 0. (c) The calculations and plotted line are not shown. The regression line does not adequately fit these data. (d) There is a fan shaped pattern apparent in this plot, which indicates non-constant variability in the residuals (little variability when x is small, more variability when x is large). Since the residuals have changing variability as we move across the plot, we should seek more appropriate statistical methods if we want to obtain a reliable estimate of the best fitting straight line.

---

8 Multiple regression and ANOVA

8.1 (a) $\widehat{weight} = 248.64 + 74.94 * casein$. (b) The estimated mean weight of chicks who are on casein feed is 74.94 grams higher than those who are given other feeds. Casein: 323.58 grams, No casein: 248.64 grams. (c) H_0: The true coefficient for `casein` is zero ($\beta_1 = 0$). H_A: The true coefficient for `casein` is not zero ($\beta_1 \neq 0$). $T = 3.23$, and the p-value is approximately 0.0019. With such a low p-value, we reject H_0. The data provide strong evidence that the true slope parameter is different than 0, and hence there appears to be a statistically significant relationship between feed type (casein or other) and the average weight of chicks.

8.3 (a) $\widehat{bwt} = 123.05 - 8.94 * smoke$. (b) The estimated body weight of babies born to smoking mothers is 8.94 ounces lower than those who are born to non-smoking mothers. Smoker: 114.11 ounces, Non-smoker: 123.05 ounces. (c) H_0: The true coefficient for `smoke` is zero ($\beta_1 = 0$). H_A: The true coefficient for `smoke` is not zero ($\beta_1 \neq 0$). $T = -8.65$, and the p-value is approximately 0. Since p-value is very small we reject H_0. The data provide strong evidence that the true slope parameter is different than 0. There is strong evidence that the linear relationship between birth weight and smoking is real.

8.5 (a) $\widehat{bwt} = -80.41 + 0.44 * gestation - 3.33 * parity - 0.01 * age + 1.15 * height + 0.05 * weight - 8.40 * smoke$. (b) β_1: The model predicts a 0.44 ounce increase in the birth weight of the baby for each additional day in length of pregnancy, all else held constant. β_3: The model predicts a 0.01 ounce decrease in the birth weight of the baby for each additional year in mother's age, all else held constant. (c) Parity might be correlated with one of the other variables in the model, which introduces collinearity and complicates model estimation. (d) -0.58.

8.7 (a) $R^2 = 1 - (249.28/332.57) = 0.2504$. $R^2_{adj} = 1 - (249.28/(1236 - 6 - 1))/(332.57/(1236 - 1)) = 0.2468$.

8.9 (a) There does not appear to be a significant relationship between the age of the mother and the birth weight of the baby since the p-value for the age variable is relatively high. We might consider removing this variable from the model. (b) No, all variables in the model now appear to have a significant relationship with the outcome therefore we would not need to removed any more variables.

8.11 Based on the p-value alone, either gestation or smoke should be added to the model first. However, since the adjusted R^2 for the model with gestation is higher, it would be preferable to add gestation in the first step of the forward-selection algorithm. (Other explanations are possible. For instance, it would be reasonable to only use the adjusted R^2.)

8.13 (1) Normality of residuals: The normal probability plot shows a nearly straight line of points, providing evidence that the nearly normal assumption is reasonable. (2) Constant variance of residuals: The scatterplot of the absolute values of residuals versus the fitted values suggests that there may be a few outliers, some with lower than average fitted values and some with higher than average fitted values. (3) The residuals should be independent: The scatterplot of residuals versus the order of data collection shows a random scatter, suggesting that this assumption is met. (4) Each variable should be linearly related to the outcome (i.e. we don't see any nonlinear trends): No nonlinear trends are evident. However, there are some outliers at the extremes of length of gestation and weight of the mother, so we should carefully examine these particular cases. There is some concern regarding constant variance across the parity groups.

We have two main concerns: outliers and constant variance. None of the outliers are exceptionally extreme, and there are a very large number of observations, so the influence of the outliers is probably mitigated (though we may want to study them more carefully, if possible). Additionally, while the constant variance assumption is violated across the parity groups, this violation is not very extreme. It is probably still reasonable to report the results while noting this model violation.

8.15 Based on the side-by-side boxplots shown in Exercise 6.19, the constant variance assumption appears to be reasonable. Because the chicks were randomly assigned to their groups (and presumably kept separate from one another), independence of observations is also reasonable. H_0: $\mu_1 = \mu_2 = \cdots = \mu_6$. H_A: The average weight (μ_i) varies across some (or all) groups. $F_{5,65} = 15.36$ and the p-value is approximately 0. With such a small p-value, we reject H_0. The data provide strong evidence that the average weight of chicks varies across some (or all) groups.

Appendix C

Distribution tables

C.1 Normal Probability Table

The area to the left of Z represents the percentile of the observation. The normal probability table always lists percentiles.

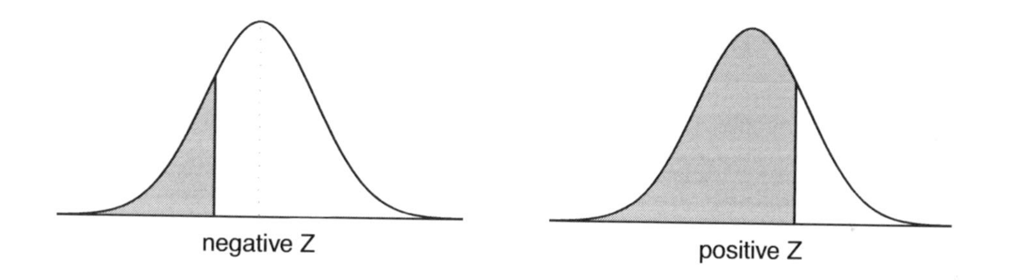

To find the area to the right, calculate 1 minus the area to the left.

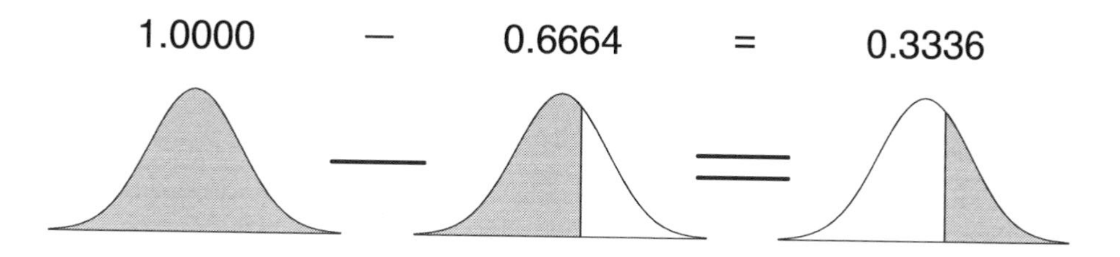

For additional details about working with the normal distribution and the normal probability table, see Section 3.1.

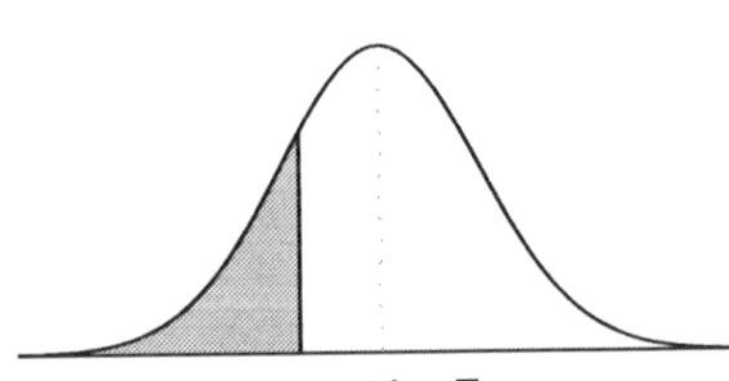

Second decimal place of Z										
0.09	0.08	0.07	0.06	0.05	0.04	0.03	0.02	0.01	0.00	Z
0.0002	0.0003	0.0003	0.0003	0.0003	0.0003	0.0003	0.0003	0.0003	0.0003	−3.4
0.0003	0.0004	0.0004	0.0004	0.0004	0.0004	0.0004	0.0005	0.0005	0.0005	−3.3
0.0005	0.0005	0.0005	0.0006	0.0006	0.0006	0.0006	0.0006	0.0007	0.0007	−3.2
0.0007	0.0007	0.0008	0.0008	0.0008	0.0008	0.0009	0.0009	0.0009	0.0010	−3.1
0.0010	0.0010	0.0011	0.0011	0.0011	0.0012	0.0012	0.0013	0.0013	0.0013	−3.0
0.0014	0.0014	0.0015	0.0015	0.0016	0.0016	0.0017	0.0018	0.0018	0.0019	−2.9
0.0019	0.0020	0.0021	0.0021	0.0022	0.0023	0.0023	0.0024	0.0025	0.0026	−2.8
0.0026	0.0027	0.0028	0.0029	0.0030	0.0031	0.0032	0.0033	0.0034	0.0035	−2.7
0.0036	0.0037	0.0038	0.0039	0.0040	0.0041	0.0043	0.0044	0.0045	0.0047	−2.6
0.0048	0.0049	0.0051	0.0052	0.0054	0.0055	0.0057	0.0059	0.0060	0.0062	−2.5
0.0064	0.0066	0.0068	0.0069	0.0071	0.0073	0.0075	0.0078	0.0080	0.0082	−2.4
0.0084	0.0087	0.0089	0.0091	0.0094	0.0096	0.0099	0.0102	0.0104	0.0107	−2.3
0.0110	0.0113	0.0116	0.0119	0.0122	0.0125	0.0129	0.0132	0.0136	0.0139	−2.2
0.0143	0.0146	0.0150	0.0154	0.0158	0.0162	0.0166	0.0170	0.0174	0.0179	−2.1
0.0183	0.0188	0.0192	0.0197	0.0202	0.0207	0.0212	0.0217	0.0222	0.0228	−2.0
0.0233	0.0239	0.0244	0.0250	0.0256	0.0262	0.0268	0.0274	0.0281	0.0287	−1.9
0.0294	0.0301	0.0307	0.0314	0.0322	0.0329	0.0336	0.0344	0.0351	0.0359	−1.8
0.0367	0.0375	0.0384	0.0392	0.0401	0.0409	0.0418	0.0427	0.0436	0.0446	−1.7
0.0455	0.0465	0.0475	0.0485	0.0495	0.0505	0.0516	0.0526	0.0537	0.0548	−1.6
0.0559	0.0571	0.0582	0.0594	0.0606	0.0618	0.0630	0.0643	0.0655	0.0668	−1.5
0.0681	0.0694	0.0708	0.0721	0.0735	0.0749	0.0764	0.0778	0.0793	0.0808	−1.4
0.0823	0.0838	0.0853	0.0869	0.0885	0.0901	0.0918	0.0934	0.0951	0.0968	−1.3
0.0985	0.1003	0.1020	0.1038	0.1056	0.1075	0.1093	0.1112	0.1131	0.1151	−1.2
0.1170	0.1190	0.1210	0.1230	0.1251	0.1271	0.1292	0.1314	0.1335	0.1357	−1.1
0.1379	0.1401	0.1423	0.1446	0.1469	0.1492	0.1515	0.1539	0.1562	0.1587	−1.0
0.1611	0.1635	0.1660	0.1685	0.1711	0.1736	0.1762	0.1788	0.1814	0.1841	−0.9
0.1867	0.1894	0.1922	0.1949	0.1977	0.2005	0.2033	0.2061	0.2090	0.2119	−0.8
0.2148	0.2177	0.2206	0.2236	0.2266	0.2296	0.2327	0.2358	0.2389	0.2420	−0.7
0.2451	0.2483	0.2514	0.2546	0.2578	0.2611	0.2643	0.2676	0.2709	0.2743	−0.6
0.2776	0.2810	0.2843	0.2877	0.2912	0.2946	0.2981	0.3015	0.3050	0.3085	−0.5
0.3121	0.3156	0.3192	0.3228	0.3264	0.3300	0.3336	0.3372	0.3409	0.3446	−0.4
0.3483	0.3520	0.3557	0.3594	0.3632	0.3669	0.3707	0.3745	0.3783	0.3821	−0.3
0.3859	0.3897	0.3936	0.3974	0.4013	0.4052	0.4090	0.4129	0.4168	0.4207	−0.2
0.4247	0.4286	0.4325	0.4364	0.4404	0.4443	0.4483	0.4522	0.4562	0.4602	−0.1
0.4641	0.4681	0.4721	0.4761	0.4801	0.4840	0.4880	0.4920	0.4960	0.5000	−0.0

*For $Z \leq -3.50$, the probability is less than or equal to 0.0002.

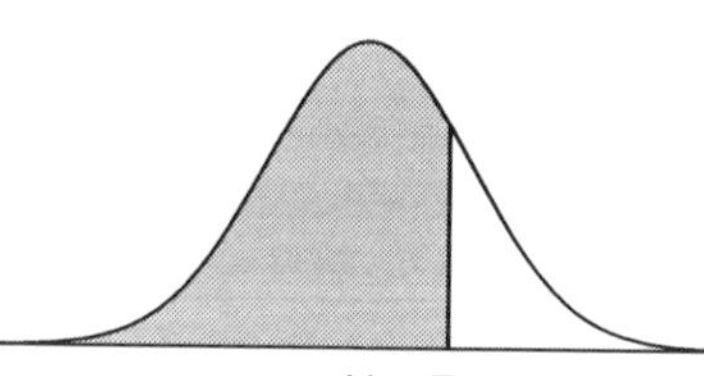

Z	Second decimal place of Z									
	0.00	0.01	0.02	0.03	0.04	0.05	0.06	0.07	0.08	0.09
0.0	0.5000	0.5040	0.5080	0.5120	0.5160	0.5199	0.5239	0.5279	0.5319	0.5359
0.1	0.5398	0.5438	0.5478	0.5517	0.5557	0.5596	0.5636	0.5675	0.5714	0.5753
0.2	0.5793	0.5832	0.5871	0.5910	0.5948	0.5987	0.6026	0.6064	0.6103	0.6141
0.3	0.6179	0.6217	0.6255	0.6293	0.6331	0.6368	0.6406	0.6443	0.6480	0.6517
0.4	0.6554	0.6591	0.6628	0.6664	0.6700	0.6736	0.6772	0.6808	0.6844	0.6879
0.5	0.6915	0.6950	0.6985	0.7019	0.7054	0.7088	0.7123	0.7157	0.7190	0.7224
0.6	0.7257	0.7291	0.7324	0.7357	0.7389	0.7422	0.7454	0.7486	0.7517	0.7549
0.7	0.7580	0.7611	0.7642	0.7673	0.7704	0.7734	0.7764	0.7794	0.7823	0.7852
0.8	0.7881	0.7910	0.7939	0.7967	0.7995	0.8023	0.8051	0.8078	0.8106	0.8133
0.9	0.8159	0.8186	0.8212	0.8238	0.8264	0.8289	0.8315	0.8340	0.8365	0.8389
1.0	0.8413	0.8438	0.8461	0.8485	0.8508	0.8531	0.8554	0.8577	0.8599	0.8621
1.1	0.8643	0.8665	0.8686	0.8708	0.8729	0.8749	0.8770	0.8790	0.8810	0.8830
1.2	0.8849	0.8869	0.8888	0.8907	0.8925	0.8944	0.8962	0.8980	0.8997	0.9015
1.3	0.9032	0.9049	0.9066	0.9082	0.9099	0.9115	0.9131	0.9147	0.9162	0.9177
1.4	0.9192	0.9207	0.9222	0.9236	0.9251	0.9265	0.9279	0.9292	0.9306	0.9319
1.5	0.9332	0.9345	0.9357	0.9370	0.9382	0.9394	0.9406	0.9418	0.9429	0.9441
1.6	0.9452	0.9463	0.9474	0.9484	0.9495	0.9505	0.9515	0.9525	0.9535	0.9545
1.7	0.9554	0.9564	0.9573	0.9582	0.9591	0.9599	0.9608	0.9616	0.9625	0.9633
1.8	0.9641	0.9649	0.9656	0.9664	0.9671	0.9678	0.9686	0.9693	0.9699	0.9706
1.9	0.9713	0.9719	0.9726	0.9732	0.9738	0.9744	0.9750	0.9756	0.9761	0.9767
2.0	0.9772	0.9778	0.9783	0.9788	0.9793	0.9798	0.9803	0.9808	0.9812	0.9817
2.1	0.9821	0.9826	0.9830	0.9834	0.9838	0.9842	0.9846	0.9850	0.9854	0.9857
2.2	0.9861	0.9864	0.9868	0.9871	0.9875	0.9878	0.9881	0.9884	0.9887	0.9890
2.3	0.9893	0.9896	0.9898	0.9901	0.9904	0.9906	0.9909	0.9911	0.9913	0.9916
2.4	0.9918	0.9920	0.9922	0.9925	0.9927	0.9929	0.9931	0.9932	0.9934	0.9936
2.5	0.9938	0.9940	0.9941	0.9943	0.9945	0.9946	0.9948	0.9949	0.9951	0.9952
2.6	0.9953	0.9955	0.9956	0.9957	0.9959	0.9960	0.9961	0.9962	0.9963	0.9964
2.7	0.9965	0.9966	0.9967	0.9968	0.9969	0.9970	0.9971	0.9972	0.9973	0.9974
2.8	0.9974	0.9975	0.9976	0.9977	0.9977	0.9978	0.9979	0.9979	0.9980	0.9981
2.9	0.9981	0.9982	0.9982	0.9983	0.9984	0.9984	0.9985	0.9985	0.9986	0.9986
3.0	0.9987	0.9987	0.9987	0.9988	0.9988	0.9989	0.9989	0.9989	0.9990	0.9990
3.1	0.9990	0.9991	0.9991	0.9991	0.9992	0.9992	0.9992	0.9992	0.9993	0.9993
3.2	0.9993	0.9993	0.9994	0.9994	0.9994	0.9994	0.9994	0.9995	0.9995	0.9995
3.3	0.9995	0.9995	0.9995	0.9996	0.9996	0.9996	0.9996	0.9996	0.9996	0.9997
3.4	0.9997	0.9997	0.9997	0.9997	0.9997	0.9997	0.9997	0.9997	0.9997	0.9998

*For $Z \geq 3.50$, the probability is greater than or equal to 0.9998.

C.2 t Distribution Table

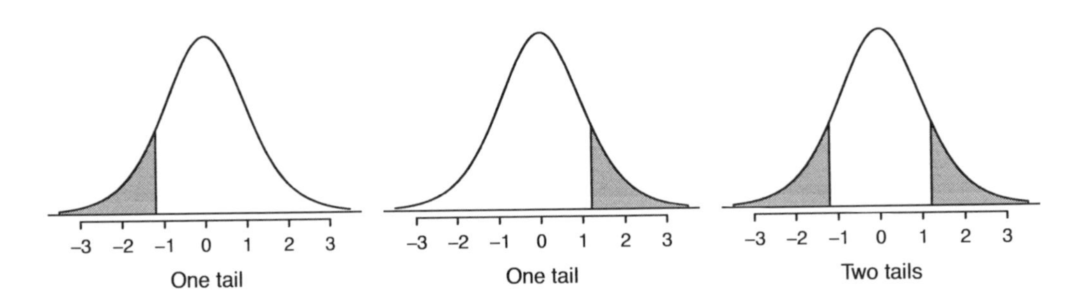

Figure C.1: Three t distributions.

one tail	0.100	0.050	0.025	0.010	0.005
two tails	0.200	0.100	0.050	0.020	0.010
df 1	3.08	6.31	12.71	31.82	63.66
2	1.89	2.92	4.30	6.96	9.92
3	1.64	2.35	3.18	4.54	5.84
4	1.53	2.13	2.78	3.75	4.60
5	1.48	2.02	2.57	3.36	4.03
6	1.44	1.94	2.45	3.14	3.71
7	1.41	1.89	2.36	3.00	3.50
8	1.40	1.86	2.31	2.90	3.36
9	1.38	1.83	2.26	2.82	3.25
10	1.37	1.81	2.23	2.76	3.17
11	1.36	1.80	2.20	2.72	3.11
12	1.36	1.78	2.18	2.68	3.05
13	1.35	1.77	2.16	2.65	3.01
14	1.35	1.76	2.14	2.62	2.98
15	1.34	1.75	2.13	2.60	2.95
16	1.34	1.75	2.12	2.58	2.92
17	1.33	1.74	2.11	2.57	2.90
18	1.33	1.73	2.10	2.55	2.88
19	1.33	1.73	2.09	2.54	2.86
20	1.33	1.72	2.09	2.53	2.85
21	1.32	1.72	2.08	2.52	2.83
22	1.32	1.72	2.07	2.51	2.82
23	1.32	1.71	2.07	2.50	2.81
24	1.32	1.71	2.06	2.49	2.80
25	1.32	1.71	2.06	2.49	2.79
26	1.31	1.71	2.06	2.48	2.78
27	1.31	1.70	2.05	2.47	2.77
28	1.31	1.70	2.05	2.47	2.76
29	1.31	1.70	2.05	2.46	2.76
30	1.31	1.70	2.04	2.46	2.75

one tail	0.100	0.050	0.025	0.010	0.005
two tails	0.200	0.100	0.050	0.020	0.010
df 31	1.31	1.70	2.04	2.45	2.74
32	1.31	1.69	2.04	2.45	2.74
33	1.31	1.69	2.03	2.44	2.73
34	1.31	1.69	2.03	2.44	2.73
35	1.31	1.69	2.03	2.44	2.72
36	1.31	1.69	2.03	2.43	2.72
37	1.30	1.69	2.03	2.43	2.72
38	1.30	1.69	2.02	2.43	2.71
39	1.30	1.68	2.02	2.43	2.71
40	1.30	1.68	2.02	2.42	2.70
41	1.30	1.68	2.02	2.42	2.70
42	1.30	1.68	2.02	2.42	2.70
43	1.30	1.68	2.02	2.42	2.70
44	1.30	1.68	2.02	2.41	2.69
45	1.30	1.68	2.01	2.41	2.69
46	1.30	1.68	2.01	2.41	2.69
47	1.30	1.68	2.01	2.41	2.68
48	1.30	1.68	2.01	2.41	2.68
49	1.30	1.68	2.01	2.40	2.68
50	1.30	1.68	2.01	2.40	2.68
60	1.30	1.67	2.00	2.39	2.66
70	1.29	1.67	1.99	2.38	2.65
80	1.29	1.66	1.99	2.37	2.64
90	1.29	1.66	1.99	2.37	2.63
100	1.29	1.66	1.98	2.36	2.63
150	1.29	1.66	1.98	2.35	2.61
200	1.29	1.65	1.97	2.35	2.60
300	1.28	1.65	1.97	2.34	2.59
400	1.28	1.65	1.97	2.34	2.59
500	1.28	1.65	1.96	2.33	2.59
∞	1.28	1.65	1.96	2.33	2.58

C.3 Chi-Square Probability Table

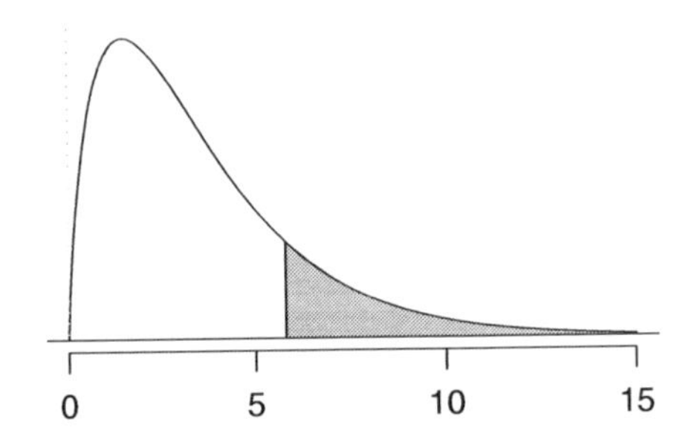

Figure C.2: Areas in the chi-square table always refer to the right tail.

Upper tail		0.3	0.2	0.1	0.05	0.02	0.01	0.005	0.001
df	1	1.07	1.64	2.71	3.84	5.41	6.63	7.88	10.83
	2	2.41	3.22	4.61	5.99	7.82	9.21	10.60	13.82
	3	3.66	4.64	6.25	7.81	9.84	11.34	12.84	16.27
	4	4.88	5.99	7.78	9.49	11.67	13.28	14.86	18.47
	5	6.06	7.29	9.24	11.07	13.39	15.09	16.75	20.52
	6	7.23	8.56	10.64	12.59	15.03	16.81	18.55	22.46
	7	8.38	9.80	12.02	14.07	16.62	18.48	20.28	24.32
	8	9.52	11.03	13.36	15.51	18.17	20.09	21.95	26.12
	9	10.66	12.24	14.68	16.92	19.68	21.67	23.59	27.88
	10	11.78	13.44	15.99	18.31	21.16	23.21	25.19	29.59
	11	12.90	14.63	17.28	19.68	22.62	24.72	26.76	31.26
	12	14.01	15.81	18.55	21.03	24.05	26.22	28.30	32.91
	13	15.12	16.98	19.81	22.36	25.47	27.69	29.82	34.53
	14	16.22	18.15	21.06	23.68	26.87	29.14	31.32	36.12
	15	17.32	19.31	22.31	25.00	28.26	30.58	32.80	37.70
	16	18.42	20.47	23.54	26.30	29.63	32.00	34.27	39.25
	17	19.51	21.61	24.77	27.59	31.00	33.41	35.72	40.79
	18	20.60	22.76	25.99	28.87	32.35	34.81	37.16	42.31
	19	21.69	23.90	27.20	30.14	33.69	36.19	38.58	43.82
	20	22.77	25.04	28.41	31.41	35.02	37.57	40.00	45.31
	25	28.17	30.68	34.38	37.65	41.57	44.31	46.93	52.62
	30	33.53	36.25	40.26	43.77	47.96	50.89	53.67	59.70
	40	44.16	47.27	51.81	55.76	60.44	63.69	66.77	73.40
	50	54.72	58.16	63.17	67.50	72.61	76.15	79.49	86.66

Index

16116231R10202

Made in the USA
Lexington, KY
05 July 2012